Alexander V. Bobylev

Kinetic Equations

De Gruyter Series in Applied and Numerical Mathematics

Volume 5/1

Alexander V. Bobylev

Kinetic Equations

Volume 1: Boltzmann Equation, Maxwell Models, and Hydrodynamics beyond Navier–Stokes

DE GRUYTER

Mathematics Subject Classification 2010
35-02, 65-02, 65C30, 65C05, 65N35, 65N75, 65N80

Author
Prof. Dr. Alexander V. Bobylev
Russian Academy of Sciences
Keldysh Institute of Applied Mathematics
Miusskaya Sq. 4
125047 Moscow
Russian Federation

ISBN 978-3-11-055012-2
e-ISBN (PDF) 978-3-11-055098-6
e-ISBN (EPUB) 978-3-11-055017-7
ISSN 2512-1820

Library of Congress Control Number: 2020942736

Bibliographic information published by the Deutsche Nationalbibliothek
The Deutsche Nationalbibliothek lists this publication in the Deutsche Nationalbibliografie; detailed bibliographic data are available on the Internet at http://dnb.dnb.de.

Typesetting: VTeX UAB, Lithuania
Printing and binding: CPI books GmbH, Leck

www.degruyter.com

To IRINA & IRINA

Preface

Classical kinetic equations by Boltzmann, Vlasov, and Landau are used for a mathematical description of physical processes in rarefied gases and plasmas. Generally speaking, kinetic equations describe the behavior of a large number N of particles, which obey some given laws of motion, in the limit when N tends to infinity and another parameter, which characterizes the interaction between particles (e. g., the size of particles), tends to zero. Classical kinetic equations correspond to laws of classical mechanics. On the other hand, there have been many attempts in the last two decades to apply this kind of equations to N-particle systems, where particles have a much more complex structure (in applications to chemistry, biology, and even socio-economic systems). From a mathematical point of view, the kinetic equations are usually non-linear integro-differential equations for the so-called distribution function (a probability density for distribution of particles in the phase space). The most famous example is the classical Boltzmann equation, firstly published in 1872. Since then the extensive development of new methods and ideas of kinetic theory has transformed this field of science into an interesting and important part of contemporary mathematical physics.

It is difficult to find another scientific area that unites such different parts of mathematics as non-linear ordinary differential equations, partial differential equations (PDEs), integral equations, functional analysis, probability theory, etc. At the same time, detailed information about the mathematical properties of kinetic equations is usually not given in standard university courses in mathematical physics. This creates difficulties for beginners. Another difficulty is that mathematics students are sometimes not very familiar with physics. In principle, the first part of this book can be used as a brief introduction to the theory of kinetic equations for beginners. It is based on graduate courses that I gave in some universities in Germany, Italy, and Sweden. On the other hand, it was not my purpose to cover in this book all important areas of the theory. As explained below, only a few topics are presented in detail. Naturally they are close to my personal research. In my view, significant progress has been made in these topics in the last decades. A criterion for selection was also to present results that have not yet appeared in monograph form. No preliminary knowledge in physics, except, perhaps, some elementary properties of Hamiltonian equations in classical mechanics, is assumed. I also preferred to use more elementary mathematical methods when possible.

The book can be roughly divided into three parts. The first part (Chapters 1 and 2) is a brief introduction to the Boltzmann equation and its formal connection with classical mechanics and probability theory. The second part (Chapters 3–6) is devoted to the general theory of homogeneous Maxwell kinetic models. This part plays a central role in the book. The third part (Chapter 7) is about hydrodynamic equations beyond the Navier–Stokes level.

https://doi.org/10.1515/9783110550986-201

I hope that the book will be interesting for both physicists and mathematicians whose work is related to kinetic theory and applications.

I am pleased to express my gratitude to my wife Irina Potapenko for her constant support and invaluable assistance in preparing the manuscript. The second volume of this book, devoted to other kinetic equations and kinetic models, will hopefully appear in the not very distant future.

Moscow, September 2020 A. V. Bobylev

Acknowledgment

I am grateful to all colleagues from the Keldysh Institute of Applied Mathematics and the international kinetic community for collaboration and very important discussions at formal and informal meetings.

In particular, I would like to thank Leif Arkeryd, Carlo Cercignani, Irene Gamba, Giampiero Spiga and Giuseppe Toscani for their long-time collaboration with me. Some of important results presented in this book were obtained jointly with them.

The work on the "hydrodynamic" Chapter 7 of the book was supported by the Russian Science Foundation grant No. 18-11-00238 "Hydrodynamics-type equations: symmetries, conservation laws, invariant difference schemes".

https://doi.org/10.1515/9783110550986-202

Contents

Introduction

The methods of kinetic theory of gases are used now in many different fields of science and technology. They are closely connected with particle methods of numerical modeling of various processes on modern computers. It is worth to stress that:

1. The kinetic theory bridges the microscopic and macroscopic description of continuum media. It is also a bridge between reversible dynamics and irreversible thermodynamics.
2. Kinetic equations constitute one main type of models in a number of frontline research areas in technology and science (shuttle and satellite technology, plasmas, semiconductors, nuclear engineering, astrophysics, cosmology, and others).
3. Kinetic equations are of central importance from a purely mathematical perspective, containing the limiting cases of almost any dynamical system's behavior, including as their own limits all classical equations of fluid dynamics and taking up a middle ground between stochastic and deterministic points of view.

This book is intended to present some new information in this field to both beginners and specialists. We give a brief summary of the contents of the book below. It can be informally divided into three parts. The first part (Chapters 1 and 2) of the book contains more or less standard material with all necessary details for beginners. We begin with the N-particle system described by Newton equations in Hamiltonian form. Then we introduce the notion of N-particle distribution function and the Liouville equation. The two-body problem and pair collisions are discussed in detail because of their significance for rarefied gases. Following the Grad version of BBGKY-hierarchy, we finally present a formal derivation of the Boltzmann equation for hard spheres at the end of Chapter 1. The formal generalization of that equation to the case of any reasonable intermolecular potential is made at the beginning of Chapter 2. Then we study all general properties of the Boltzmann equation, including conservation laws and the H-theorem. The Boltzmann equation for mixtures is introduced and discussed at the end of Chapter 2.

The second part (Chapters 3–6) of the book is devoted to the general theory of kinetic Maxwell models. In particular, Chapters 3–5 present the theory of the spatially homogeneous Boltzmann equation for Maxwell molecules based on the Fourier transform in the velocity space. This approach is a bit unusual for kinetic equations. Note that the linear integral transforms are usually not very useful for non-linear equations. Therefore, it is not surprising that a significant simplification of the Boltzmann collision integral by Fourier transform remained unknown for roughly 100 years. After the author's first publication [17] in 1975, this approach was spread not very rapidly. For example, the well-known book by Truesdell and Muncaster [144] was published in 1980. A big part of the book is devoted to Maxwell molecules, but the authors did not use the Fourier transform because they were not aware of its advantages. Making parallels

https://doi.org/10.1515/9783110550986-001

with PDEs one can say that the role of Maxwell molecules (in the wide sense) for kinetic equations of Boltzmann's type is, to some extent, similar to the role of PDEs with constant coefficients in the general theory of PDEs. Today's generation of researchers is well familiar with this approach and there are many new results in this area. At the same time, it is impossible to find a book which systematically presents the existing theory of Maxwell models based on Fourier transform.

Chapters 3–6 can be considered as an attempt to fill this gap. The main attention is paid to the concept of self-similar solutions and their role in the large time asymptotic behavior of solutions to the Cauchy problem. In particular, Chapter 6 contains results on inelastic and generalized Maxwell models of kinetic equations. The results were obtained in the last two decades by the author in collaboration with Carlo Cercignani and Irene Gamba.

It is important to stress the traditional difference in the approach of physicists and mathematicians to non-linear kinetic equations of Boltzmann type. For example, the first problem for mathematicians is to prove the existence and uniqueness theorem. Physicists would simply postulate these properties and try to get more specific information about solutions and its asymptotic behavior, sometimes ignoring necessary demands of mathematical rigor. Fortunately, in the case of Maxwell models it is possible to unify these two different approaches. In all parts of the book we try to distinguish clearly between mathematically rigorous and formal results.

The third part (Chapter 7) of the book can be read independently of Chapters 3–6, though it contains a few references to Chapter 3. This part is about the status of "higher" equations of hydrodynamics obtained by the famous Chapman–Enskog expansion. We try to understand how to use this expansion in order to improve the results obtained by solving the classical Euler and Navier–Stokes equations for rarefied gas. The Burnett equations, which appeared in 1930s [74] as the next step after the Navier–Stokes level, were always considered by the fluid mechanics community with certain skepticism. A clear mathematical confirmation of that point of view was obtained in 1982 [23]. It was shown by a simple linear stability analysis of Burnett equations that they are ill-posed. Since then there have been many attempts by different authors to overcome this obstacle. We show in Chapter 7 how to regularize the Chapman–Enskog expansion and to find a proper replacement for Burnett equations. The mathematically rigorous results for the Burnett-type approximation are obtained for the linearized Boltzmann equation. The optimal, in some sense, set of non-linear equations of hydrodynamics, which we call "generalized Burnett equations," is also derived and discussed in detail in Chapter 7.

As a rule, standard mathematical notation will be used without comment. Vectors and real or complex numbers are denoted by letters $x, v, t, w, H, \omega, \dots$. Sets, operators and matrices are always denoted by capital letters $Q, L, M, A, B, \dots$. The notation $\log x$ means natural logarithm of x. Theorems, lemmas, and formulas are referred to chapter, section and number. For example the fourth equation in Section 3.2 of Chapter 3 will be numbered as (3.2.4).

1 From particle dynamics to the Boltzmann equation

1.1 N-Particle dynamics and modeling of rarefied gases

We consider $N \geq 1$ identical particles with mass $m = 1$. This system is characterized by a $6N$-dimensional phase vector $Z_N = \{z_1, \dots, z_N\}$ with components $z_i = (x_i, v_i)$, where $x_i \in \mathbb{R}^3$ and $v_i \in \mathbb{R}^3$ denote respectively a position and a velocity of the ith particle, $i = 1, \dots, N$. Usually it is assumed below that the particles interact via a given pair potential $\Phi(r)$, where $r > 0$ stands for the distance between two interacting particles. We also assume that $\Phi(r) \to 0$ if $r \to \infty$. The equations of motion of the system have the following Hamiltonian form (see any textbook in classical mechanics, e. g., [110]):

$$\begin{aligned} \partial_t x_i &= \partial H_N / \partial v_i, \quad \partial_t v_i = -\partial H_N / \partial x_i, \\ H_N &= \frac{1}{2} \sum_{i=1}^{N} |v_i|^2 + \sum_{1 \leq i < j \leq N} \Phi(|x_i - x_j|), \\ x_i(0) &= x_i^{(0)}, \quad v_i(0) = v_i^{(0)}, \quad i, j = 1, \dots, N. \end{aligned} \tag{1.1.1}$$

Thus the temporal evolution of the system can be understood as the motion of the phase point $Z_N = \{z_1, \dots, z_N\}$ in the phase space $\mathbb{R}^{6N}$. This motion obeys conservation laws of energy $E_N \in \mathbb{R}$ and momentum $P_N \in \mathbb{R}^3$, i. e.,

$$E_N = H_N[Z_N(t)] = \text{const.}, \quad P_N = \sum_{i=1}^{N} v_i(t) = \text{const.} \tag{1.1.2}$$

The laws (1.1.2) follow directly from equations (1.1.1).

In principle, the above described N-particle system can be used for modeling of real gases or liquids. Then the main problem is that the number of particles N is of order of 10^{23} (Avogadro's number). It is intuitively clear that the case of rarefied gas is easier for description than the case of dense gas. Indeed, in the low density limit we get a free molecular flow, i. e., each particle moves independently with constant velocity. Consequently our main assumption will be the following: A typical distance between particles is much greater than the effective diameter d of the potential. It is known that the typical "size" d of a molecule of the air is roughly equal to $d \approx 3.7 \cdot 10^{-8}$ cm, whereas the number density $\bar{n}$ of the air is about $\bar{n} \approx 2.5 \cdot 10^{19}$ cm^{-3} under normal conditions. The inequality

$$\delta = \bar{n} d^3 \ll 1 \tag{1.1.3}$$

is the well-known criterion for ideal gas. Note that $\delta \approx 10^{-3}$ for air under normal conditions on the surface of the Earth. This parameter δ is decreasing with height. Therefore, kinetic equations (in particular, the classical Boltzmann equation for rarefied gases) are important for applications in space science and technology.

https://doi.org/10.1515/9783110550986-002

The simplest model intermolecular potential $\Phi(r)$ corresponds to particles interacting like hard spheres of diameter d. Then we formally obtain

$$\Phi_{\mathrm{HS}}(r) = \begin{cases} \infty, & \text{if } 0 < r \le d, \\ 0, & \text{otherwise.} \end{cases} \tag{1.1.4}$$

Another well-known model corresponds to power-like repulsive potentials,

$$\Phi(r) = \frac{\alpha}{r^n}, \quad \alpha > 0, \ n \ge 1, \tag{1.1.5}$$

including the Coulomb case $n = 1$ with any sign of α.

The scattering problem for equations (1.1.1) with $N = 2$ is considered in Section 1.4. In the next section we discuss some probabilistic aspects of kinetic theory of gases.

1.2 Distribution functions and the Liouville equation

We introduce an important notion of one-particle distribution function $f(x, v, t)$ (the words "one-particle" are usually omitted below for the sake of brevity). The physical meaning of this function is the following: The average number of particles in any measurable set $\Delta \in \mathbb{R}^3 \times \mathbb{R}^3$ is given by the equality

$$n_\Delta(t) = \int_\Delta dx dv f(x, v, t). \tag{1.2.1}$$

In other words, $f(x, v, t)$ is the density of the number of particles in the phase space. Usually we assume that the initial data

$$f(x, v, 0) = f^{(0)}(x, v) \tag{1.2.2}$$

are given. How can we find the distribution function $f(x, v, t)$ for $t > 0$? This is, in a sense, the main problem of kinetic theory. It can be shown that for some special physical systems, like rarefied gases, the temporal evolution of the distribution function $f(x, v, t)$ is described by the so-called "kinetic equation"

$$f_t = A(f), \tag{1.2.3}$$

where $A(f)$ is a non-linear operator acting on f. We usually assume that the initial value problem (1.2.2)–(1.2.3) has a unique solution $f(x, v, t)$ on some time interval $0 \le t \le T$.

Let us consider the simplest kinetic equation connected with the system (1.1.1). Omitting index $i = 1$, we obtain equations of free motion

$$x_t = 0, \quad v_t = 0. \tag{1.2.4}$$

The motion of one particle can also be described by the distribution function $f(x, v, t)$, having in that case the meaning of probability density if

$$\int\limits_{\mathbb{R}^3\times\mathbb{R}^3} dxdvf^{(0)}(x, v) = 1, \tag{1.2.5}$$

in the notation of (1.2.1). The solution of equations (1.2.4) is obvious:

$$x(t) = x(0) + v(0)t, \quad v(t) = v(0).$$

Therefore, $f(x, v, t)$, satisfying conditions (1.2.1), reads

$$f(x, v, t) = f^{(0)}(x - vt, v) = \exp(-tv \cdot \partial_x) f^{(0)}(x, v)\,. \tag{1.2.6}$$

Here and below, the dot denotes the scalar product in $\mathbb{R}^3$. We can check by differentiation that

$$f_t + v \cdot \partial_x f = 0. \tag{1.2.7}$$

This is the simplest kinetic equation. Note that kinetic equation (1.2.7) is exactly equivalent to dynamical equation (1.2.4). The probabilistic description is caused only by uncertainty in initial conditions.

Let us now extend these arguments to the case of N non-interacting particles. We consider equations (1.1.1) with $\Phi(r) \equiv 0$ and obtain

$$\begin{aligned} &\partial_t x_i = v_i, \quad \partial_t v_i = 0; \\ &x_i(0) = x_i^{(0)}, \quad v_i(0) = v_i^{(0)}, \quad i = 1, \ldots, N. \end{aligned} \tag{1.2.8}$$

Thus we have N independent vector equations for each particle. It is natural to introduce the N-particle distribution function $F_N(z_1, \ldots, z_N;\, t)$, $z_i = (x_i, v_i)$, $1 \le i \le N$, with a meaning of a probability density in the N-particle phase space $\mathbb{R}^{6N}$. The initial condition reads

$$\begin{aligned} &F_N|_{t=0} = F_N^{(0)}(z_1, \ldots, z_N), \\ &\int\limits_{\mathbb{R}^6\times\cdots\times\mathbb{R}^6} dz_1 \ldots dz_n\, F_N^{(0)}(z_1, \ldots, z_N) = 1. \end{aligned} \tag{1.2.9}$$

Remark. Here and below we use notations like $F_N(z_1, \ldots, z_N)$ (with capital F) for various multi-particle distribution functions, which have a meaning of probability density. These functions are always normalized by one in the whole phase space. The notations like $f_N(z_1, \ldots, z_N)$ will be used for a slightly different class of functions related to equality (1.2.1). The difference disappears for the trivial case $N = 1$.

Then it is easy to see that

$$F_N(z_1,\dots,z_N;t) = F_N^{(0)}[z_1(t),\dots,z_N(t)],$$
$$z_i(t) = (x_i - v_i t, v_i), \quad i = 1,\dots,N.$$

Note that

$$\left(\partial_t + \sum_{i=1}^{N} v_i \cdot \partial_{x_i}\right) F_N(x_1, v_1, \dots, x_N, v_N; t) = 0. \tag{1.2.10}$$

Thus we have the simplest version of the Liouville equation.

We assume that all particles are identical and independently distributed at $t = 0$, i. e.,

$$F_N|_{t=0} = \prod_{i=1}^{N} f^{(0)}(x_i, v_i). \tag{1.2.11}$$

Then the similar factorization holds for all $t > 0$,

$$F_N(z_1,\dots,z_N;t) = \prod_{i=1}^{N} f(z_i,t) = \prod_{i=1}^{N} f^{(0)}(x_i - v_i t, v_i). \tag{1.2.12}$$

This property is known as "the propagation of chaos" [101]. It is self-evident for non-interacting particles, but it also can be proved as an asymptotic property of more complex multi-particle systems.

What changes if we consider the Hamiltonian system (1.1.1) with the non-zero potential $\Phi(r) \neq 0$? Then we can still use the N-particle distribution function $F_N(z_1,\dots,z_N;t)$. We shall see below that the equation for F_N reads

$$\left[\partial_t + \sum_{i=1}^{N}\left(v_i \cdot \partial_{x_i} - \frac{\partial \Phi_N}{\partial x_i} \cdot \partial_{v_i}\right)\right] F_N(x_1, v_1, \dots, x_N, v_N;\, t) = 0\,,$$
$$\Phi_N = \sum_{1 \le i < j \le N} \Phi(|x_i - x_j|)\,; \quad i = 1,\dots,N. \tag{1.2.13}$$

This is the famous Liouville equation [110]. It can be derived very easily. For a moment we simplify our notations in the following way:

$$z = \{z_1,\dots,z_N\} \in \mathbb{R}^{6N}, \quad z_i = (x_i, v_i);$$
$$w(z) = \{w_1,\dots,w_N\} \in \mathbb{R}^{6N}, \quad w_i = \left(\frac{\partial H}{\partial v_i}, -\frac{\partial H}{\partial x_i}\right),$$
$$H = H_N = \frac{1}{2}\sum_{i=1}^{N} |v_i|^2 + \sum_{1 \le i < j \le N} \Phi(|x_i - x_j|);$$
$$F(z;t) = F_N(z_1,\dots,z_N;\, t).$$

Then we can treat $F(z,t)$ as a density of a fluid in $\mathbb{R}^{6N}$, which moves in accordance with a dynamical system

$$z_t = w(z), \tag{1.2.14}$$

where $w(z)$ is assumed to be a "nice" function.

Then the density $F(z;t)$ satisfies the continuity equation

$$F_t + \operatorname{div}_z F\,w = 0, \tag{1.2.15}$$

where div_z denotes the divergence with respect to z. Simple calculation yields

$$\operatorname{div}_z w(z) = \sum_{i=1}^{N}\left(\frac{\partial}{\partial x_i}\cdot\frac{\partial H}{\partial v_i} - \frac{\partial}{\partial v_i}\cdot\frac{\partial H}{\partial x_i}\right) = 0.$$

Hence, we obtain

$$F_t + \sum_{1}^{N}\left(\frac{\partial H}{\partial v_i}\cdot\frac{\partial F}{\partial x_i} - \frac{\partial H}{\partial x_i}\cdot\frac{\partial F}{\partial v_i}\right) = 0, \tag{1.2.16}$$

i. e., the Liouville equation (1.2.13) in slightly different notations. The sum in (1.2.16) is usually called the Poisson brackets $\{F,H\}$ (only for particles with unit mass $m = 1$) [110].

The Liouville equation is very important because it allows (at least formally) to build a bridge between N-particle dynamics, kinetic theory, and hydrodynamics. These three levels of description of physical systems will be briefly discussed in Chapter 7.

1.3 BBGKY-hierarchy

It was assumed in Section 1.2 that $f_N(z_1,\dots,z_N;t)$ is integrable over the whole phase space $\mathbb{R}^{6N}$. This assumption is sometimes too strong. For example, it does not allow to consider translationally invariant (in physical space $\mathbb{R}^3$) systems with initial data like $f_2^{(0)}(x_1 - x_2; v_1, v_2)$. Therefore, it is more convenient to consider an N-particle system confined to a bounded domain $\Omega \subset \mathbb{R}^3$, say, a box or a sphere with "large" diameter L. The volume of Ω is denoted by $|\Omega|$. Then the phase state of the ith particle is described by the point

$$z_i = (x_i, v_i) \in \Omega \times \mathbb{R}^3, \quad i = 1,\dots,N. \tag{1.3.1}$$

We can assume for simplicity that the walls of Ω are specularly reflecting. Another possibility is to assume that Ω is a periodic box. What is important is to keep unchanged the total number N of particles inside Ω.

For the sake of brevity we will use below the symbolic notations

$$F_N(x_1, v_1; \dots; x_N, v_N; t) = F_N(1, 2, \dots, N; t),$$
$$\int\limits_{\Omega} dx_i \int\limits_{\mathbb{R}^3} dv_i \dots = \int\limits_{G} di \dots ; \quad G = \Omega \times \mathbb{R}^3, \quad i = 1, \dots, N, \tag{1.3.2}$$

where F_N is normalized by the equality

$$\int\limits_{G_N} d1 d2 \dots dN F_N(1, \dots, N; t) = 1, \quad G_N = G^N, \tag{1.3.3}$$

and satisfies the Liouville equation (1.2.3)

$$\left(\partial_t + \sum_{i=1}^{N} A_i - \sum_{1 \le i < j \le N} B_{ij}\right) F_N = 0, \quad A_i = v_i \cdot \partial_{x_i},$$
$$B_{ij} = \frac{\partial \Phi(|x_i - x_j|)}{\partial x_i} \cdot (\partial_{v_i} - \partial_{v_j}); \quad i, j = 1, \dots, N; \quad i \ne j. \tag{1.3.4}$$

The function $F_N(1, \dots, N; t)$ is assumed to be invariant under permutations of its arguments $(1, \dots, N)$ because all N particles are identical. This property is preserved by the Liouville equation if it is valid at $t = 0$.

We introduce k-particle probability distributions by equalities

$$F_k^{(N)}(1, 2, \dots, k) = \int\limits_{G_{N-k}} d(k+1) \dots dN \, F_N(1, 2, \dots, N),$$
$$1 \le k \le N - 1. \tag{1.3.5}$$

Here and below, the argument t is omitted in such cases when it does not cause any confusion. The next step is to obtain a set of evolution equations for $F_k^{(N)}(1, 2, \dots, k; t)$. We take any $1 \le k \le N - 1$ and integrate equation (1.3.4) over $d(k+1) \dots dN$. The result reads

$$(\partial_t + L_k) F_k^{(N)} = \Gamma_k^{(N)}, \quad L_k = \sum_{i=1}^{k} A_i - \sum_{1 \le i < j \le k} B_{ij}, \tag{1.3.6}$$

$$\Gamma_k^{(N)} = \int\limits_{G_{N-k}} d(k+1) \dots dN \left[- \sum_{i=k+1}^{N} A_i + \sum_{i=1}^{k} \sum_{j=k+1}^{N} B_{ij} \right.$$
$$\left. + \sum_{i=k+1}^{N-1} \sum_{j=i+1}^{N} B_{ij} \right] F_N(1, \dots, N). \tag{1.3.7}$$

We assume that the boundary conditions guarantee that for any $1 \le i \le N$

$$\int\limits_{G} di (v_i \cdot \partial_{x_i}) \, F_N(1, \dots, N; t) = 0. \tag{1.3.8}$$

Then the first sum in (1.3.7) disappears. The third sum in (1.3.7) also disappears under the natural assumption that

$$F_N(1,\dots,N;\,t) \to 0 \quad \text{if} \quad |v_i| \to \infty, \quad \text{for some} \quad 1 \le i \le N. \tag{1.3.9}$$

Then we obtain

$$\Gamma_k^{(N)} = \sum_{i=1}^{k} \Gamma_{ik}^{(N)}, \quad \Gamma_{ik}^{(N)} = \sum_{j=k+1}^{N} \int_{G_{N-k}} d(k+1)\dots dN\, B_{ij}\, F_N(1,\dots,N).$$

Let us consider the first term of the sum for $\Gamma_{ik}^{(N)}$. Then $j = k+1$, and we obtain

$$\begin{aligned}
&\int_G d(k+1) B_{i\,k+1} \int_{G_{N-k-1}} d(k+2)\dots dN F_N(1,\dots,N) \\
&= \int_G d(k+1)\, B_{i\,k+1}\, F_{k+1}^{(N)}(1,\dots,k+1), \quad 1 \le k \le N-1, \quad F_N^{(N)} = F_N.
\end{aligned}$$

If we take another value of j in the sum for $\Gamma_{ik}^{(N)}$, then the result will be the same. It follows from the symmetry of $F_N(1,\dots,N)$ with respect to permutations. Moreover the operator $B_{i\,k+1}$ from (1.3.4) can be replaced by

$$\tilde{B}_{i\,k+1} = \frac{\partial \Phi(|x_i - x_{k+1}|)}{\partial x_i} \cdot \frac{\partial}{\partial v_i}$$

because of conditions (1.3.9). Therefore, we obtain

$$\Gamma_k^{(N)} = (N-k) \sum_{i=1}^{k} \int_G d(k+1) \tilde{B}_{i\,k+1}\, F_{k+1}^{(N)}(1,\dots,k+1).$$

We substitute this formula into equations (1.3.6) and get the following set of equations:

$$(\partial_t + L_k) F_k^{(N)} = (N-k) C_{k+1} F_{k+1}^{(N)}, \quad 1 \le k \le N-1, \tag{1.3.10}$$

where $F_k^{(N)}(1,\dots,k)$ are given in equations (1.3.5),

$$\begin{aligned}
F_N^{(N)}(1,\dots,N) &= F_N(1,\dots,N), \\
L_k &= \sum_{i=1}^{k} v_i \cdot \partial_{x_i} - \sum_{1 \le i < j \le k} \frac{\Phi(|x_i - x_j|)}{\partial x_i} \cdot (\partial_{v_i} - \partial_{v_j}), \\
C_{k+1} &= \sum_{i=1}^{k} C_{i\,k+1}, \\
C_{i\,k+1} F_{k+1}^{(N)} &= \int_G d(k+1) \frac{\Phi(|x_i - x_{k+1}|)}{\partial x_i} \cdot \partial_{v_i} F_{k+1}^{(N)}(1,\dots,k+1).
\end{aligned} \tag{1.3.11}$$

The set of equations (1.3.10)–(1.3.11) is called (if we ignore its trivial modifications) the BBGKY-hierarchy. BBGKY stands for Bogolyubov, Born, Green, Kirkwood, and Yvon, the names of physicists who introduced independently this system of equations (see, e. g., [10, 56]). Equation (1.3.10) with $k = N$ also makes sense if we set $F_N^{(N)} = F_N$, $F_N^{(N+1)} = 0$. Then it will be the Liouville equation for F_N. Note that equations (1.3.10) (the BBGKY-hierarchy for the N-particle dynamical system (1.1.1) in the box Ω) can be formally considered as exact equations. The only relevant assumption is equality (1.3.8), related to interactions with boundaries of the box Ω. This equality is fulfilled, in particular, for a periodic box or the box with specularly reflecting walls.

The BBGKY-hierarchy is a starting point for all classical kinetic equations. For example, let us assume that the interaction between particles is weak and replace $\Phi(r)$ by $\varepsilon\bar{\Phi}(r)$ in equations (1.3.10) in the following form:

$$\left(\partial_t + \sum_{i=1}^{N} A_i - \varepsilon \sum_{1\le<j\le N} B_{ij}\right) F_{k+1}^{(N)} = \varepsilon\,(N-k) C_{k+1} F_{k+1}^{(N)}, \quad 1 \le k \le N-1,$$

in the notation of equations (1.3.4) and (1.3.11). Then it is natural to consider the limit

$$N \to \infty, \quad \varepsilon \to 0, \quad N\varepsilon = \text{const}.$$

and to assume that $F_k^{(N)} \to F_k$ in that limit for all $k = 1, 2, \ldots$. Then we formally obtain the following infinite set of limiting equations for F_k:

$$\begin{aligned}
&\left(\partial_t + \sum_{i=1}^{k} v_i \cdot \partial_{x_i}\right) F_k(z_1, \ldots, z_k; t) \\
&\quad = (N\varepsilon) \sum_{i=1}^{k} \int_G dz_{k+1} \frac{\Phi(|x_i - x_j|)}{\partial x_i} \cdot \partial_{v_i} F_{k+1}(z_1, \ldots, z_{k+1}; t); \\
&z_i = (x_i, v_i) \in G, \quad dz_i = dx_i dv_i, \quad i = 1, \ldots, k+1; \quad k = 1, \ldots.
\end{aligned}$$

It is easy to verify that these equations admit a class of solutions in factorized form

$$F_k(z_1, \ldots, z_k; t) = \prod_{i=1}^{k} F(z_i, t), \quad k = 1, \ldots,$$

where $F(x, v, t)$ satisfies the Vlasov equation [149]

$$F_t + v \cdot F_x - (N\varepsilon) F_v \cdot \partial_x \int_\Omega dy\, dw\, \Phi(|x - y|)\, F(y, w, t) = 0.$$

The most important applications of the Vlasov equations are related to the Coulomb potential $\Phi(r) = \alpha/r$ (gravitational or electrostatic forces).

On the other hand, in the important case of potentials $U(r)$ with compact support (strong interaction at short distances) it is more convenient to use a different approach. This approach will be considered in Section 1.6.

In the next section we consider pair collisions of particles.

1.4 The two-body problem and pair collisions

For the sake of generality we consider in this section a system of two particles with masses m_1 and m_2 interacting via potential $\Phi(r)$. Then we obtain the following equations of motion:

$$\partial_t x_i = v_i, \quad m_i \partial_t v_i = -\partial_{x_i}\Phi(|x_1 - x_2|); \quad x_i \in \mathbb{R}^3, \quad v_i \in \mathbb{R}^3; \quad i = 1, 2. \tag{1.4.1}$$

These equations coincide with (1.1.1) if $m_1 = m_2 = 1$.

We denote

$$X = \frac{m_1 x_1 + m_2 x_2}{m_1 + m_2}, \quad U = \frac{m_1 v_1 + m_2 v_2}{m_1 + m_2},$$
$$x = x_1 - x_2, \quad v = v_1 - v_2, \quad m = \frac{m_1 m_2}{m_1 + m_2}. \tag{1.4.2}$$

Then we obtain

$$X_t = U, \quad x_t = v; \quad U_t = 0, \quad m v_t = -\partial_x \Phi(|x|). \tag{1.4.3}$$

Hence, the two-body problem reduces to equations of motion of a particle with mass $m > 0$ in the radial field $\Phi(r)$, where $r > 0$ is a distance to origin. We note that there are two conservation laws: energy $E \in \mathbb{R}$ and angular momentum $M \in \mathbb{R}^3$, i. e.,

$$E = \frac{m|v|^2}{2} + \Phi(|x|) = \text{const.}, \quad M = mx \times v = \text{const.}, \tag{1.4.4}$$

where the cross denotes the vector product in $\mathbb{R}^3$.

The conservation laws (1.4.4) follow directly from equations (1.4.3). Note that both $x(t)$ and $v(t)$ are for all $t \in \mathbb{R}$ on the same plane Π_M perpendicular to M. For simplicity we assume that the potential $\Phi(r)$ is repulsive, i. e., $\Phi' < 0$, and $\Phi(r) \to 0$ as $r \to \infty$. Then obviously $|x(t)| \to \infty$ as $t \to \pm\infty$. We denote

$$u = \lim_{t \to -\infty} v(t), \quad u' = \lim_{t \to +\infty} v(t), \tag{1.4.5}$$

and introduce in the plane Π_M the Cartesian coordinates (x^1, x^2) such that the particle trajectory has the form shown in Fig. 1.1.

It is assumed there that $x(t) \cdot v(t) < 0$ if $t \to -\infty$. The angle $\theta \in [0, \pi]$ between vectors u' and u is called the scattering angle. We briefly explain below how to evaluate θ for given values of E and $|M|$ from (1.4.4) (see, e. g., [110] for details). Note that $|u'| = |u|$ because of the conservation of energy E. Introducing polar coordinates by transformation

$$x^1 = r \cos\varphi, \quad x^2 = r \sin\varphi,$$

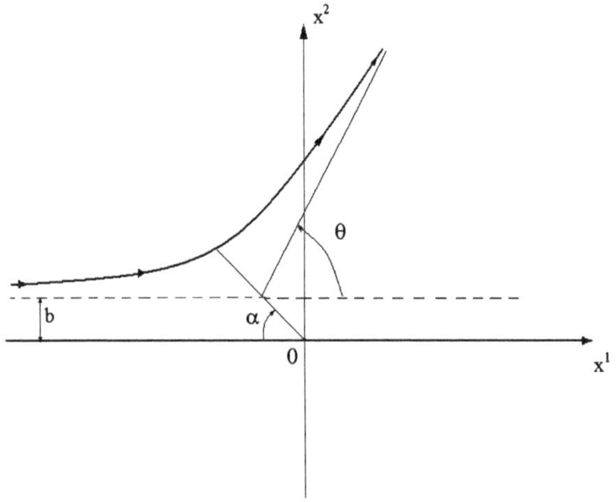

Figure 1.1: Typical trajectory of particle moving in central field $\Phi(r)$.

we obtain for components of the velocity $v = (v^1, v^2)$

$$v^1 = x_t^1 = r \cos\varphi - r\varphi_t \sin\varphi, \quad v^2 = x_t^2 = r_t \sin\varphi + r\varphi_t \cos\varphi.$$

Therefore, it follows from (1.4.4) that

$$E = \frac{m}{2}(r_t^2 + r^2\varphi_t^2) + \Phi(r), \quad |M| = m\,|x^1v^2 - x^2v^1| = mr^2|\varphi_t|.$$

Hence, the equation for $r(t)$ reads

$$r_t^2 = \frac{2}{m}\left[E - \Phi(r) - \frac{|M|^2}{2mr^2}\right]. \tag{1.4.6}$$

We introduce the impact parameter

$$b = \frac{|M|}{m|u|}, \tag{1.4.7}$$

whose geometrical meaning is clear from Fig. 1.1. We also note that $E = m|u|^2/2$ and transform (1.4.6) to the following equation:

$$r_t^2 = |u|^2 - \frac{2\Phi(r)}{m} - \frac{b^2|u|^2}{r^2}. \tag{1.4.8}$$

It is clear from Fig. 1.1 that $r_t < 0$ for $t \in (-\infty, t_*)$ and $r_t > 0$ for $t \in (t_*, \infty)$, where t_* is defined by equation $r_t(t_*) = 0$. Thus $r_{\min} = r(t_*)$ is the minimal distance of the particle to the origin. It satisfies the equation

$$\frac{2\Phi(r_{\min})}{m} + \frac{b^2|u|^2}{r_{\min}^2} = |u|^2. \tag{1.4.9}$$

If $\Phi'(r) < 0$ and $\Phi(r) \underset{r\to\infty}{\longrightarrow} 0$, then this equation has a unique root $r_{\min} > 0$. The polar angle $\varphi(t)$ decreases from $\varphi(-\infty) = \pi$ to $\varphi(+\infty) = \theta$. The equation for $\varphi(t)$ reads

$$\varphi_t = -\frac{b|u|}{r^2(t)}.$$

It is clear from Fig. 1.1 that

$$\varphi(+\infty) = \theta = \pi - 2\alpha, \quad \alpha = \pi - \varphi(t_*).$$

Hence, we obtain

$$\theta = -\pi + 2\varphi(t_*) = -\pi + 2\left[\pi + \int_{-\infty}^{t_*} dt\varphi_t(t)\right] = \pi - 2b|u| \int_{-\infty}^{t_*} \frac{dt}{r^2(t)}.$$

Then we change the variable t to r in the integrand and use equation (1.4.8) to obtain the final result

$$\theta(b,|u|) = \pi - 2b|u| \int_{r_{\min}}^{\infty} \frac{dr}{r^2}\left[|u|^2 - \frac{2\Phi(r)}{m} - \frac{b^2}{r^2}|u|^2\right]^{-1/2}, \tag{1.4.10}$$

where $r_{\min}$ is defined by equation (1.4.9). This formula is very important for kinetic theory, as we shall see below.

1.5 Scattering cross-section

Coming back to the two-particle problem (1.4.1), we can invert the transformation (1.4.2) for velocities and obtain

$$v_1(t) = U + \frac{m}{m_1}v(t), \quad v_2(t) = U - \frac{m}{m_2}v(t), \quad U = \text{const.}, \tag{1.5.1}$$

in the notation of equations (1.4.2). Similarly to (1.4.5), we denote

$$\begin{aligned} v_1 &= \lim_{t\to-\infty} v_1(t) = U + \frac{m}{m_1}u, \quad v_2 = \lim_{t\to-\infty} v_2(t) = U - \frac{m}{m_2}u, \\ v_1' &= \lim_{t\to\infty} v_1(t) = U + \frac{m}{m_1}u', \quad v_2' = \lim_{t\to\infty} v_2(t) = U - \frac{m}{m_2}u'. \end{aligned} \tag{1.5.2}$$

It is assumed that $\Phi'(r) < 0$, $\Phi(r) \to 0$ if $r \to \infty$ and

$$[x_1(t) - x_2(t)] \cdot [v_1(t) - v_2(t)] < 0 \quad \text{if} \quad t \to -\infty. \tag{1.5.3}$$

Then the velocities $v_{1,2}$ and $v_{1,2}'$ are called pre- and post-collisional velocities, respectively. The corresponding transformation of relative velocity $u \to u'$ was studied in Section 1.4. In particular, it was shown that $|u'| = |u|$. Hence,

$$v_1' = U + \frac{m}{m_1}|u|\omega, \quad v_2' = U - \frac{m}{m_2}|u|\omega,$$
$$m = \frac{m_1 m_2}{m_1 + m_2}, \quad U = \frac{m_1 v_1 + m_2 v_2}{m_1 + m_2}, \quad u = v_1 - v_2, \quad \omega \in S^2, \tag{1.5.4}$$

where S^2 denotes the unit sphere in $\mathbb{R}^3$. This transformation $(v_1, v_2) \to (v_1', v_2')$ is very important in kinetic theory (especially for the Boltzmann equation). For any $\omega \in S^2$ the transformation satisfies the conservation laws of energy and momentum:

$$m_1|v_1|^2 + m_2|v_2|^2 = m_1|v_1'|^2 + m_2|v_2'|^2,$$
$$m_1 v_1 + m_2 v_2 = m_1 v_1' + m_2 v_2'. \tag{1.5.5}$$

If pre-collisional velocities $v_{1,2} \in \mathbb{R}^3$ are fixed, then the resulting velocities $v_{1,2}' \in \mathbb{R}^3$ depend on the unit vector $\omega \in S^2$ (see equations (1.5.4)). Generally speaking, the vector ω depends on the pre-collisional relative velocity $u = v_1 - v_2$, the direction of angular momentum M (1.4.3), and the impact parameter $b > 0$ shown in Fig. 1.1. The scattering angle $\theta \in [0, \pi]$ between u and $\omega = u'/|u|$ depends on two positive parameters $|u|$ and b in accordance with the integral formula (1.4.10).

Our goal in this section is to introduce an important for kinetic theory function $\sigma(|u|, \theta)$ called "the differential scattering cross-section." It is assumed below that the scattering angle $\theta(b, |u|)$ defined by (1.4.10) is for any $|u| > 0$ a monotone decreasing function of $b \in [0, R_{\max}]$, where $R_{\max} > 0$ is the radius of action of the potential $\Phi(r)$ in (1.4.10) (the case $R_{\max} = \infty$ is also included). The assumption guarantees the existence of an inverse function $b(|u|, \theta)$, such that $\theta[b(|u|, \theta), |u|] = \theta$ for any $\theta \in [0, \pi]$ and $|u| > 0$. Then the classical definition (see, e. g., [110]) of $\sigma(|u|, \theta)$ is the following:

$$\sigma(|u|, \theta) = \left|\frac{1}{2\sin\theta}\partial_\theta b^2(|u|, \theta)\right|, \quad 0 < \theta < \pi, \quad |u| > 0. \tag{1.5.6}$$

The physical meaning of this function will be explained below. First we consider a few examples of typical intermolecular potentials $\Phi(r)$ in equation (1.4.10).

Example 1 (Potentials with finite radius $R_{\max} = d$ of action and hard spheres). We assume that $\Phi(r) \equiv 0$ if $r > d > 0$. Then we consider (1.4.10) and note that

$$2b \int_d^\infty \frac{dr}{r^2}\left[1 - \left(\frac{b}{r}\right)^2\right]^{-1/2} = 2\arcsin\frac{b}{d}, \quad 0 < b < d.$$

Therefore, we obtain

$$\theta(|u|, b) = \begin{cases} 0 & \text{if } b \ge d; \\ \pi - 2\arcsin\frac{b}{d} - 2b\int_{r_{\min}}^d \frac{dr}{r^2}\left[1 - \left(\frac{b}{r}\right)^2 - \frac{2\Phi(r)}{m|u|^2}\right]^{-1/2} & \text{if } 0 \le b \le d. \end{cases} \tag{1.5.7}$$

In case (1.1.4) of hard spheres with diameter d the integral term disappears, and therefore we get

$$\theta(|u|, b) = \pi - 2\arcsin\frac{b}{d} \Rightarrow b(|u|, \theta) = d\cos\frac{\theta}{2} \leq d, \quad 0 \leq \theta \leq \pi. \tag{1.5.8}$$

It follows from (1.5.6) that

$$\sigma(|u|, \theta) = \frac{1}{4}d^2 \tag{1.5.9}$$

for the hard sphere potential (1.1.4).

Example 2 (Power-like potentials $\Phi_n(r) = \alpha r^{-n}$, $\alpha > 0$, $n > 0$). In this case it is convenient to denote $b = a_n\,(\alpha/mu^2)^{1/n}$. Then we obtain from (1.4.10)

$$\theta(|u|, b) = \tilde{\theta}(a) = \pi - 2\int_0^{x_{\max}} dx\left[1 - x^2 - (x/a_n)^n\right]^{-1/2},$$

where $x_{\max}$ corresponds to zero of the denominator in the integrand. For brevity we just postulate the existence of the inverse function $a_n(\theta)$ such that $\tilde{\theta}[a_n(\theta)] = \theta$. Hence, the differential cross-section for the potential $\Phi(r) = \alpha\, r^{-n}$ reads

$$\sigma(|u|, \theta) = \left(\frac{\alpha}{m\,|u|^2}\right)^{2/n} A_n(\theta), \quad A_n(\theta) = \frac{a_n(\theta)|a_n'(\theta)|}{\sin\theta}, \quad 0 < \theta < \pi. \tag{1.5.10}$$

It is easy to show that $A_1(\theta)$ and $A_2(\theta)$ are elementary functions. The hard sphere potential can be formally considered as a limiting case of power-like potentials $\Phi_n(r)$ for $n \to \infty$.

In order to illustrate a typical application of the function $\sigma(|u|, \theta)$ we consider once more the above discussed collision of two particles. To be more precise, we consider the transformation of velocities $(v_1, v_2) \to (v_1', v_2')$ given in equations (1.5.4); basically it is a transformation from $u = v_1 - v_2$ to $u' = v_1' - v_2'$. Moreover, $|u'| = |u|$. We denote

$$u = |u|\omega_0, \quad u' = |u|\omega, \quad \mu = \cos\theta = \omega \cdot \omega_0, \quad \theta \in [0, \pi],$$

and consider again a special coordinate system, where the trajectory $x(t) = x_1(t) - x_2(t)$ looks like in Fig. 1.1.

Then we introduce a vector $b_\perp$ with components $b_\perp^1 = 0$, $b_\perp^2 = b > 0$. It follows from Fig. 1.1 that

$$\omega = \omega(|u|, b_\perp) = \mu\omega_0 + \sqrt{1 - \mu^2}e_\perp, \quad e_\perp = \frac{b_\perp}{|b_\perp|}, \tag{1.5.11}$$

where

$$\mu = \cos\theta(|u|, b_\perp), \quad |b_\perp| = b,$$

in the notation of equation (1.4.10). Hence, we obtain a transition from $\omega_0 \in S^2$ to $\omega \in S^2$ parameterized by vector $b_\perp \in \mathbb{R}^2$. Note that $\omega = \omega_0$ if $|b_\perp| > d$, where d is the radius of action of the potential $\Phi(r)$ (see Example 1).

In real gases d is the microscopic length of order 10^{-8} cm. Usually we do not know the initial positions of particles with such high accuracy. Therefore, it seems to be a reasonable approximation to consider the vector $b_\perp$ as a random vector in $\mathbb{R}^3$ uniformly distributed inside the disc

$$D = \{b_\perp \in \mathbb{R}^3 : u \cdot b_\perp = 0,\ 0 \le |b_\perp| \le d\}. \tag{1.5.12}$$

Then $\omega \in S^2$ from (1.5.11) becomes also a random unit vector. For arbitrary function $\psi(\omega)$ its average value (with respect to uniform distribution of $b_\perp$ in D) is given by the equality

$$\overline{\Psi(\omega)} = c \int_D db_\perp \Psi[\omega(|u|, b_\perp)], \quad c^{-1} = \int_D db_\perp = \pi d^2, \tag{1.5.13}$$

in the notation of equations (1.5.11). We take any orthonormal basis (e_1, e_2, e_3) such that $e_3 = \omega_0$ and introduce polar coordinates (b, φ) in the plane spanned by (e_1, e_2). Then we obtain

$$\overline{\Psi(\omega)} = c \int_0^d d\rho\rho \int_0^{2\pi} d\varphi \Psi[\mu e_3 + \sqrt{1-\mu^2}(e_1 \cos\varphi + e_2 \sin\varphi)], \quad \mu = \cos\theta(\rho|u|, b),$$

in the notation of equation (1.4.9). Below we temporarily omit the constant argument $|u|$ of $\theta(|u|, b)$ and change the variables from b to $\mu(b) = \cos\theta(b)$. More precisely, we assume the existence of inverse function $b(\mu)$ such that $b(-1) = 0$, $b(1) = d$, and $b'(\mu) > 0$ on $(-1, 1)$. Then we finally obtain

$$\overline{\Psi(\omega)} = c \int_{-1}^{1} d\mu \tilde{\sigma}(|u|, \mu) \int_0^{2\pi} d\varphi \Psi[\mu\omega_0 + \sqrt{1-\mu^2}(e_1 \cos\varphi + e_2 \sin\varphi)], \tag{1.5.14}$$

$$\tilde{\sigma}(|u|, \mu) = \frac{1}{2}\frac{d}{d\mu} b^2(|u|, \mu), \quad c^{-1} = 2\pi \int_{-1}^{1} d\mu \sigma(|u|, \mu) = \pi d^2. \tag{1.5.15}$$

Note that

$$\tilde{\sigma}(|u|, \cos\theta) = \sigma(|u|, \theta), \tag{1.5.16}$$

in the notation of equation (1.5.6). Hence, $\tilde{\sigma}(|u|, \cos\theta)$ is the differential scattering cross-section expressed as a function of $\cos\theta$, not $\theta \in [0, \pi]$. This variable is more convenient for applications. In particular, we note that $\mu = \cos\theta = \omega \cdot \omega_0$ and introduce spherical coordinates in $\mathbb{R}^3$ with the polar axis along ω_0. Then we obtain from (1.5.12) the following formula, which does not depend on the coordinate system:

$$\overline{\Psi(\omega)} = c \int_{S^2} d\omega \tilde{\sigma}(|u|, \omega \cdot \omega_0) \Psi(\omega),$$

$$c^{-1} = \sigma_{\text{tot}} = \int_{S^2} d\omega \tilde{\sigma}(|u|, \omega \cdot \omega_0) = 2\pi \int_{-1}^{1} d\mu \sigma(|u|, \mu), \tag{1.5.17}$$

where $\int_{S^2} d\omega \ldots$ denotes the usual integral over the unit sphere in $\mathbb{R}^3$. Hence, the function $\tilde{\sigma}(|u|, \omega \cdot \omega_0)/\sigma_{\text{tot}}$ is the probability density for distribution of the unit vector $\omega(|u|, b_\perp) = u'/|u|$ (1.5.11) provided (a) the relative velocity before collision $u = |u|\omega_0$ is fixed; (b) the "impact vector" $b_\perp$ is uniformly distributed in the disc D (1.5.12); and (c) the radius of action $d = R_{\max}$ of the potential $\Phi(r)$ is finite. This explains the physical (or probabilistic) meaning of the differential cross-section (1.5.16). Equations (1.5.17) make it clear why we prefer to use in the rest of the book the function $\tilde{\sigma}(|u|, \cos\theta)$ instead of the more conventional $\sigma(|u|, \theta)$. Of course, these two functions are equal. The tildes are therefore usually omitted below. We shall see a connection between the function $\sigma(|u|, \cos\theta)$ and the Boltzmann equation in Chapter 2.

In the next section we consider the so-called "Boltzmann–Grad limit" of an N-particle system.

1.6 Hard spheres and the Boltzmann–Grad limit

We begin with the case of N hard spheres of diameter d. Then the Liouville equation (1.2.13) cannot be used directly because the potential $\Phi_{\text{HS}}(r)$ (1.1.4) is too singular. The function $F_N(x_1, v_1; \ldots; x_N, v_N; t)$ is defined in this case by the free flow equation

$$\partial_t F_N + \sum_{i=1}^{N} v_i \cdot \partial_{x_i} F_N = 0, \tag{1.6.1}$$

valid in the domain

$$B_N = \{x_i \in \mathbb{R}^3 : |x_i - x_j| > d;\ i, j = 1, \ldots, N,\ i \neq j\}. \tag{1.6.2}$$

We add to this equation the boundary conditions on each of $N(N-1)/2$ boundary surfaces $|x_i - x_j| = d$ $(i, j = 1, \ldots, N; i \neq j)$ in $\mathbb{R}^{3N}$. Taking, for example, $i = 1$, $j = 2$, we obtain, assuming the specular reflection law,

$$F_N(x_1, v_1; x_2, v_2; \ldots; t)|_{|x_1 - x_2| = d,\, x \cdot u > 0} = F_N(x_1, v_1'; x_2, v_2'; \ldots; t), \tag{1.6.3}$$

where

$$\begin{aligned} &x = x_1 - x_2 = d\,n, \quad n \in S^2, \quad u = v_1 - v_2, \\ &v_1' = v_1 - n(u \cdot n), \quad v_2' = v_2 + n(u \cdot n). \end{aligned} \tag{1.6.4}$$

The surfaces that correspond to multiple collisions of $k \geq 3$ particles are described by at least $(k-1) \geq 2$ equalities. For example, for $k = 3$ we need to satisfy simultaneously two conditions like $|x_1 - x_2| = d$ and $|x_2 - x_3| = d$. These surfaces have a zero measure as compared with the case $k = 2$ of pair collisions. Therefore, we ignore multiple collisions.

For brevity we assume that

$$
\begin{aligned}
F_N \in L(G_N), \quad G_N = G^N, \quad G = \mathbb{R}^3 \times \mathbb{R}^3, \\
F_N(1, \dots, N; t) = 0 \quad \text{if} \quad |x_i - x_j| < d
\end{aligned}
\tag{1.6.5}
$$

for at least one pair of indices $1 \leq i < j \leq N$. Here and below we use symbolic notations (1.3.2) from Section 1.3, when it does not cause any confusion. We also assume that $F_N \geq 0$ is the probability density in G_N with usual normalization condition (1.3.3). Our aim is to construct the equation for one-particle probability density $F_1^{(N)}(1)$ given in (1.3.5) with $k = 1$. Note that this function can be equally defined by the equality

$$
\begin{aligned}
F_1^{(N)}(1) &= \int\limits_{G_{N-1}} d2 \dots dN \, \Psi(1|2, \dots, N) \, F_N(1, \dots, N), \\
\Psi(1|2, \dots, N) &= \prod_{k=2}^{N} \eta[d^2 - |x_1 - x_k|^2],
\end{aligned}
\tag{1.6.6}
$$

where $\eta(y)$ is the unit function

$$
\eta(y) = \begin{cases} 1, & \text{if } y > 0, \\ 0, & \text{otherwise.} \end{cases}
\tag{1.6.7}
$$

We multiply equation (1.6.1) by $\Psi(1|2, \dots, N)$ and integrate over G_{N-1}. The result reads

$$
\partial_t F_1^{(N)} + I_1 + \sum_{k=2}^{N} I_k = 0,
\tag{1.6.8}
$$

where

$$
I_k = \int\limits_{G_{N-1}} d2 \dots dN \, \Psi(1|2, \dots, N) \, A_k \, F_N(1, \dots, N), \quad A_k = v_k \cdot \partial_{x_k}.
\tag{1.6.9}
$$

We separated the term with $k = 1$ in (1.6.8) because all other terms in the sum are equal to I_2. Indeed, we always assume that all particles are identical and therefore $F_N(1, \dots, N)$ is symmetric with respect to permutations of arguments $(2, \dots, N)$. Hence, $I_k = I_2$ for any $3 \leq k \leq N$ and therefore

$$
\sum_{k=2}^{N} I_k = (N-1) I_2.
\tag{1.6.10}
$$

The integral I_2 can be written as

$$I_2 = \int_{|x_1-x_2|>d} dx_2 \operatorname{div}_{x_2} \int_{\mathbb{R}^3} dv_2\, v_2 \tilde{F}_2^{(N)}(x_1, v_1; x_2, v_2), \tag{1.6.11}$$

where

$$\begin{aligned} \tilde{F}_2^{(N)}(x_1, v_1; x_2, v_2) &= \tilde{F}_2^{(N)}(1, 2) \\ &= \int_{G_{N-2}} d3 \dots dN F_N(1, , 2, \dots, N) \prod_{k=3}^{N} \eta[d^2 - |x_1 - x_k|^2]. \end{aligned} \tag{1.6.12}$$

We use the notation $\tilde{F}_2^{(N)}$ because this function coincides with $F_2^{(N)}$ (1.3.5) only for $d = 0$. We apply the Gauss theorem to integral (1.6.11) and obtain after simple transformations

$$\begin{aligned} I_2 &= -\int_{|y|>d} dy \operatorname{div}_y \int_{\mathbb{R}^3} dv_2\, v_2 \tilde{F}_2^{(N)}(x_1, v_1;\, x_1 - y, v_2) \\ &= d^2 \int_{\mathbb{R}^3 \times S^2} dv_2\, dn\, (v_2 \cdot n) \tilde{F}_2^{(N)}(x_1, v_1;\, x_1 - dn, v_2), \end{aligned}$$

where n denotes the outward unit normal vector to the unit sphere S^2. It remains to evaluate the integral I_1 in (1.6.9). We note that

$$\Psi(1|2, \dots, N) A_1 F_N(1, \dots, N) = A_1 \Psi F_N - F_N A_1 \Psi, \quad A_1 = v_1 \cdot \partial_{x_i}.$$

Since $\eta'(y) = \delta(y)$, where $\delta(y)$ denotes the Dirac delta function, we obtain

$$\begin{aligned} A\Psi &= v_1 \cdot \partial_{x_1} \prod_{i=2}^{N} \eta[d^2 - |x_1 - x_i|^2] \\ &= 2 \sum_{i=2}^{N} [v_1 \cdot (x_1 - x_i)]\, \delta[|x_1 - x_i|^2 - d^2] \prod_{j=2}^{N\ (j \neq i)} \eta[d^2 - |x_1 - x_j|^2]. \end{aligned}$$

Then we perform the integration in (1.6.9) and use again symmetry of F_N and Ψ. The result reads

$$\begin{aligned} I_1 &= v \cdot \partial_{x_1} F_1^{(N)}(x_1, v_1) \\ &\quad - 2(N-1) \int_{\mathbb{R}^3 \times \mathbb{R}^3} dx_2 dv_2\, \delta[|x_1 - x_2|^2 - d^2] v_1 \cdot (x_1 - x_2) \tilde{F}_2^{(N)}(x_1, v_1; x_2, v_2), \end{aligned}$$

in the notation of equation (1.6.12). Note that

$$2 \int_{\mathbb{R}^3} dy \delta[(x - y)^2 - d^2] F(y) = d^2 \int_{S^2} dn F(x - dn).$$

Therefore, we obtain equation (1.6.8) in the following form:

$$\begin{aligned}&(\partial_t + v_1 \cdot \partial_{x_1})F_1^{(N)}(x_1, v_1)\\&\quad = Q^{(N)} = (N-1)\,d^2 \int\limits_{\mathbb{R}^3\times S^2} dv_2 dn\,[(v_1 - v_2)\cdot n]\,\tilde{F}_2^{(N)}(x_1, v_1;\, x_1 - dn, v_2),\end{aligned} \tag{1.6.13}$$

where $\tilde{F}_2^{(N)}(x_1, v_1; x_2, v_2)$ is given in (1.6.12). We can split the integral over S^2 into two parts in the following way:

$$\int\limits_{S^2} dn(u\cdot n)\Psi(n) = \int\limits_{S^2_+} dn|u\cdot n|\Psi(n) - \int\limits_{S^2_-} dn|u\cdot n|\Psi(n)\,,$$

$$S^2_+ = \{n \in S^2 : u\cdot n > 0\}, \quad S^2_- = \{n \in S^2 : u\cdot n < 0\},$$

where $u = v_1 - v_2$, $\Psi(n)$ is an arbitrary integrable function. It is clear from Equations (1.6.3) and (1.6.4) that $\Psi(n) = \tilde{F}_2^{(N)}(x_1, v_1; x_1 - dn, v_2)$ in the integral over S^2_+ can be expressed through $\Psi(n)$ in the integral over S^2_-. Then we obtain

$$\begin{aligned}Q^{(N)} = (N-1)\,d^2 \int\limits_{\mathbb{R}^3\times S^2} & dv_2 dn|u\cdot n|\,[\tilde{F}_2^{(N)}(x_1, v_1';\, x_1 - dn,\, v_2')\\&- \tilde{F}_2^{(N)}(x_1, v_1;\, x_1 + dn,\, v_2)]\,,\end{aligned} \tag{1.6.14}$$

$$v_1' = v_1 - n(u\cdot n), \quad u = v_1 - v_2, \quad v_2' = v_2 + n(u\cdot n),$$

in the notation of equation (1.6.13). Note that equations (1.6.13)–(1.6.14) are formally exact for hard spheres. To our knowledge, they were first published by Harold Grad not later than in 1957 [92]. These equations are very important as a starting point for mathematically rigorous derivation of the Boltzmann equation.

For our goals it is sufficient to introduce the "chaotic" initial data

$$F_N(1,2,\ldots,N)|_{t=0} = c_N \left\{\prod_{k=1}^{N} F_0(k)\right\} \prod_{1\le i<j\le N} \eta[d^2 - |x_i - x_j|^2]\,,$$

$$\int\limits_G d1\, F_0(1) = 1, \tag{1.6.15}$$

where c_N is the normalization constant, and to consider the formal limit of equations (1.6.13)–(1.6.14) under the conditions that

$$N \to \infty, \quad d \to 0, \quad N\,d^2 = \text{const.} \tag{1.6.16}$$

This is the so-called Boltzmann–Grad limit [70, 93]. To be more precise we assume that

$$F_1^{(N)}(x_1, v_1, t) \to F(x, v, t), \quad F_1^{(N)}(x_1, v_1; x_2, v_2; t) \to \prod_{i=1}^{2} F(x_i, v_i, t) \tag{1.6.17}$$

under conditions (1.6.16).

Then we formally obtain from (1.6.13) and (1.6.14) the Boltzmann equation for hard spheres. In simplified notations $x = x_1$, $v = v_1$, $w = v_2$, this equation reads

$$(\partial_t + v \cdot \partial_x)F(x, v, t) = (Nd^2)\tilde{Q}(F, F), \tag{1.6.18}$$

$$\tilde{Q}(F,F) = \frac{1}{2} \int_{\mathbb{R}^3 \times S^2} dw dn |u \cdot n| [F(x, v', t)F(x, w', t) - F(x, v, t)F(x, w, t)],$$

$$u = v - w, \quad n \in S^2; \quad v' = v - (u \cdot n)n, \quad w' = w + (u \cdot n), \tag{1.6.19}$$

where the domain of integration S^2_+ in (1.6.14) is extended to the whole unit sphere S^2 in an obvious way. The limiting initial conditions formally follow from (1.6.15):

$$F(x, v)|_{t=0} = F_0(x, v), \quad \int_G dx dv F_0(x, v) = 1. \tag{1.6.20}$$

We consider in this section the case of the whole space $G = \mathbb{R}^3 \times \mathbb{R}^3$ in order to simplify formal calculations. In fact the same equations (1.6.18)–(1.6.20) can be derived rigorously from N-particle dynamics in the case when all particles are confined to the bounded domain Ω with reflecting walls (see [72] and references therein for details). Therefore, we assume that equations (1.6.18)–(1.6.20) are also valid if $x \in \Omega$, $|\Omega| < \infty$, with appropriate boundary conditions.

Remarks on Chapter 1

1. The material of Chapter 1 is more or less standard. It should be pointed out that the first mathematically rigorous proof of validity of the Boltzmann equation for hard spheres was done by O. Lanford in 1975 [111].
2. In our formal derivation of the Boltzmann equation in Section 1.6 we follow the Grad scheme from [92].
3. A deep and rigorous presentation of the validation problem for the Boltzmann equation can be found in the book [72]. It also contains a detailed discussion of the famous Di Perna–Lions global existence theorem for the Boltzmann equation.

2 The Boltzmann equation

2.1 The Boltzmann equation for hard spheres and its generalizations

The classical Boltzmann equation is usually considered not for the probability density $F(x,v,t)$ from equations (1.6.18)–(1.6.20), but for the distribution function $f(x,v,t)$,

$$f(x,v,t) = N\,F(x,v,t). \tag{2.1.1}$$

The equation for $f(x,v,t)$ reads

$$\begin{aligned} f_t + v\cdot f_x &= Q(f,f) \\ &= \frac{d^2}{2}\int\limits_{\mathbb{R}^3\times S^2} dwdn\,|u\cdot n|\,[f(v')f(w') - f(v)f(w)], \\ u = v-w,\quad n\in S^2;&\quad v' = v-(u\cdot n)n,\quad w' = w+(u\cdot n), \end{aligned} \tag{2.1.2}$$

where irrelevant arguments (x,t) of $f(x,v,t)$ are omitted in the so-called Boltzmann collision integral $Q(f,f)$. Note that $Q(f,f)$ is a quadratic with respect to f operator acting only on variable $v\in\mathbb{R}^3$. The advantage of equation (2.1.2) is that it does not contain the number of particles N. The Boltzmann collision integral in (2.1.2) can be presented in different forms. In particular, the following form of $Q(f,f)$ is very useful:

$$\begin{aligned} Q(f,f) &= \frac{d^2}{4}\int\limits_{\mathbb{R}^3\times S^2} dwd\omega|u|\,[f(v')f(w') - f(v)f(w)], \\ u = v-w,\quad \omega\in S^2;&\quad v' = \frac{1}{2}(v+w+|u|\omega),\quad w' = \frac{1}{2}(v+w-|u|\omega). \end{aligned} \tag{2.1.3}$$

In order to prove that this integral coincides with $Q(f,f)$, from (2.1.2) we consider a simpler integral

$$I(F) = \int\limits_{\mathbb{R}^3} dk\,\delta\Big(k\cdot u + \frac{|k|^2}{2}\Big)F(k), \tag{2.1.4}$$

where $u\in\mathbb{R}^3$ and $F(u)$ is a continuous function. Then the following identity can be easily proved.

Lemma 2.1.1. *We have*

$$I(F) = 2\int\limits_{\mathbb{S}^2} dn|u\cdot n|\,F[-2(u\cdot n)n] = |u|\int\limits_{\mathbb{S}^2} d\omega F(|u|\omega - u). \tag{2.1.5}$$

https://doi.org/10.1515/9783110550986-003

Proof. We evaluate the integral $I(F)$ in spherical coordinates with the polar axis directed along $u \in \mathbb{R}^3$. Thus we denote $k = rn$, $n \in \mathbb{S}^2$ in (2.1.4) and obtain

$$I(F) = \int_0^\infty dr\, r^2 \int_{\mathbb{S}^2} dn\, \delta\left[rn \cdot u + \frac{r^2}{2}\right] F(rn).$$

Since

$$\delta(\alpha x) = \frac{\delta(x)}{\alpha}, \quad \alpha > 0, \quad x \in \mathbb{R}, \tag{2.1.6}$$

we change the order of integration and obtain

$$\begin{aligned} I(F) &= \int_{\mathbb{S}^2} dn \int_0^\infty dr\, r\delta\left[\frac{1}{2}(2n \cdot u + r)\right] F(rn) \\ &= 4 \int_{\mathbb{S}_-^2} dn\, |n \cdot u|\, F[-2(u \cdot n)n], \quad \mathbb{S}_-^2 = \{n \in \mathbb{S}^2 : u \cdot n < 0\}. \end{aligned}$$

The integrand is an even function of $n \in S^2$ and hence the first equality (2.1.5) follows. The second equality is based on the change of variables $k = \tilde{k} - u$ in the integral (2.1.4). Then we obtain

$$I(F) = \int_{\mathbb{R}^3} dk\, \delta\left(\frac{|k|^2 - |u|^2}{2}\right) F(k - u)$$

and evaluate this integral in the same way as above. This completes the proof of Lemma 2.1.1. □

Now we can prove the transformation of $Q(f,f)$ from (2.1.2) to (2.1.3). We consider (2.1.2) and denote

$$F(k) = f(v + k/2)f(w - k/2) - f(v)f(w), \tag{2.1.7}$$

considering v and w as fixed parameters. Then we obtain from (2.1.2)

$$Q(f,f) = \frac{d^2}{2} \int_{\mathbb{R}^3 \times S^2} dw\, dn\, |u \cdot n|\, F[-2(u \cdot n)n].$$

It remains to use the identity (2.1.5) and get the following result:

$$Q(f,f) = \frac{d^2}{4} \int_{\mathbb{R}^3 \times S^2} dw\, d\omega\, |u|\, F(|u|\omega - u),$$

in the notation of equation (2.1.7). It is easy to check that this formula for $Q(f,f)$ coincides with (2.1.3). Hence, the equivalence of (2.1.2) and (2.1.3) is proved. Note also that the same identity (2.1.5) leads to the third useful representation of the collision integral for hard spheres:

$$Q(f,f) = \frac{d^2}{4} \int_{\mathbb{R}^3\times\mathbb{R}^3} dwdk\, \delta(k\cdot u + |k|^2/2) \times [f(v + k/2)f(w - k/2) - f(v)f(w)]. \tag{2.1.8}$$

The physical meaning of the Boltzmann equation can be better understood by considering $Q(f,f)$ in the form (2.1.3). We denote

$$\langle f,\psi\rangle = \int_{\mathbb{R}^3} dv\, f(v)\psi(v), \tag{2.1.9}$$

where $\psi(v)$ is an arbitrary function of velocity $v \in \mathbb{R}^3$ for which the integral exists. The reader can formally consider this integral as the generalized scalar product of functions $f(v)$ and $g(v)$. Then we formally obtain from (2.1.2)

$$\partial_t\langle f,\psi\rangle + \partial_x\cdot\langle f, v\psi\rangle = \langle\psi, Q(f,f)\rangle. \tag{2.1.10}$$

The right hand side with $Q(f,f)$ from (2.1.3) reads

$$\langle\psi, Q(f,f)\rangle = \frac{d^2}{4} \int_{\mathbb{R}^3\times\mathbb{R}^3\times\mathbb{S}^2} dvdwd\omega\, |u|\psi(v)[f(v')f(w') - f(v)f(w)],$$

in the notation of (2.1.3). We denote the center of mass variables (see Section 1.4) by

$$U = \frac{v+w}{2}, \quad u = v - w \Leftrightarrow v = U + \frac{u}{2}, \quad w = U - \frac{u}{2}. \tag{2.1.11}$$

Hence, $dvdw = dUdu$. Therefore, we obtain

$$\langle\psi, Q(f,f)\rangle = \frac{d^2}{4} \int_{\mathbb{R}^3\times\mathbb{R}^3\times\mathbb{S}^2} dUdud\omega\, |u|\psi\left(U + \frac{u}{2}\right)[F(U,|u|\omega) - F(U,u)],$$
$$F(U,u) = f(v)f(w).$$

If we denote $u = r\omega_0$, $\omega_0 \in S^2$ and write down the integral over du as

$$\int_{\mathbb{R}^3} du\varphi(u) = \int_0^\infty dr\, r^2 \int_{\mathbb{S}^2} d\omega_0\varphi(r\omega_0),$$

then the internal integral over $d\omega_0 d\omega$ reads

$$I = \int_{S^2\times S^2} d\omega_0 d\omega\psi\left(U + \frac{r}{2}\omega_0\right)[F(U,r\omega) - F(U,r\omega_0)],$$

where $r = |u|$. Obviously, we can exchange variables ω and ω_0 in the first term and obtain

$$I = \int_{S^2 \times S^2} d\omega_0 d\omega F(U, r\omega_0) \left[\psi\left(U + \frac{r}{2}\omega\right) - \psi\left(U + \frac{r}{2}\omega_0\right)\right].$$

Coming back to initial variables, we get

$$\langle \psi, Q(f,f) \rangle = \int_{\mathbb{R}^3 \times \mathbb{R}^3} dv dw f(v) f(w) |u| \, \sigma_{\text{tot}} \, \overline{[\psi(v') - \psi(v)]}, \tag{2.1.12}$$

where

$$\sigma_{\text{tot}} = \int_{\mathbb{S}^2} d\omega \sigma_{\text{diff}} = \pi d^2,$$

$$\overline{\psi(v') - \psi(v)} = \frac{1}{\sigma_{\text{tot}}} \int_{\mathbb{S}^2} d\omega \sigma_{\text{diff}} \left[\psi\left(U + \frac{|u|}{2}\omega\right) - \psi\left(U + \frac{u}{2}\right)\right], \tag{2.1.13}$$

where $\sigma_{\text{diff}} = d^2/4$ denotes the differential scattering cross-section discussed in detail in Section 1.5. The bar in (2.1.12) means actually an averaging over random impact parameters (see equation (1.5.13)). The physical meaning of equation (2.1.10) becomes clear if we write it as

$$\partial_t \langle f, \psi \rangle + \partial_x \cdot \langle f, v\psi \rangle = \int_{\mathbb{R}^3 \times \mathbb{R}^3} dv dw f(v) f(w) |u| \sigma_{\text{tot}} \overline{[\psi(v') - \psi(v)]}, \tag{2.1.14}$$

in the notation of equation (2.1.13). Indeed, the average total number of collisions per unit time is given by the integral

$$\nu_{\text{tot}}(f,f) = \int_{\mathbb{R}^3 \times \mathbb{R}^3} dv dw f(v) f(w) |u| \sigma_{\text{tot}} . \tag{2.1.15}$$

On the other hand, the average change of ψ in the collision of particles with velocities v and w is equal to the average difference $\overline{[\psi(v') - \psi(v)]}$ given in equation (2.1.13). Hence, the right hand side of equation (2.1.14) defines correctly (at the intuitive level) the rate of change of $\langle f, \psi \rangle$ due to collisions.

These considerations allow us to generalize equations (2.1.12)–(2.1.13) to the case of the general (repulsive) potential $\Phi(r)$ with finite radius of action $R_{\max} = d$. In that case we have the same total cross-section $\sigma_{\text{tot}} = \pi d^2$ as for hard spheres with diameter d. However, the differential cross-section of scattering $(v, w) \to (v', w')$, such that

$$v - w = u, \quad v' - w' = u' = |u|\omega$$

is given for the general potential $\Phi(r)$ by the function

$$\sigma_{\text{diff}} = \sigma(|u|, \omega \cdot u/|u|), \quad \sigma_{\text{tot}} = 2\pi \int_{-1}^{1} d\mu \sigma(|u|, \mu), \tag{2.1.16}$$

discussed in detail in Section 1.5. If we fix the intermolecular potential $\Phi(r)$ and the corresponding differential scattering cross-section $\sigma(|u|, \mu)$ (see equations (1.5.16)–(1.5.17)), then we obtain the same equation (2.1.12), where

$$\begin{aligned} \overline{\psi(v') - \psi(v)} &= \frac{1}{\sigma_{\text{tot}}} \int_{S^2} \sigma(|u|, \omega \cdot u/|u|)[\psi(v') - \psi(v)], \\ v' &= \frac{1}{2}(v + w + |u|\omega), \quad u = v - w. \end{aligned} \tag{2.1.17}$$

Finally, we note that the total cross-section σ_{tot} disappears after substitution of (2.1.17) into (2.1.12). In addition, the differential cross-section $\sigma(|u|, \mu)$ is always multiplied by $|u|$ in the integrand in (2.1.12). Therefore, it is more convenient to introduce a new function

$$g(|u|, \mu) = |u|\sigma(|u|, \mu), \quad \mu \in [-1, 1]. \tag{2.1.18}$$

Then the generalized equation (2.1.12) reads

$$\begin{aligned} &\langle \psi, Q(f,f) \rangle \\ &\quad = \int_{\mathbb{R}^3 \times \mathbb{R}^3 \times S^2} dv dw d\omega f(v) f(w) g(|u|, \omega \cdot u/|u|) \, [\psi(v') - \psi(v)], \\ &u = v - w, \quad v' = \frac{1}{2}[v + w + |u|\omega], \quad \omega \in S^2. \end{aligned} \tag{2.1.19}$$

Note that $\psi(v)$ is an arbitrary test function. Therefore, this expression is sometimes called "a weak form of the Boltzmann collision integral." The corresponding strong form of $Q(f,f)$ is

$$\begin{aligned} Q(f,f) &= \int_{\mathbb{R}^3 \times S^2} dw \, d\omega \, g(|u|, \omega \cdot u/|u|)[f(v')f(w') - f(v)f(w)], \\ \omega \in S^2, \quad u &= v - w, \quad v' = \frac{1}{2}(v + w + |u|\omega), \quad w' = \frac{1}{2}(v + w - |u|\omega), \end{aligned} \tag{2.1.20}$$

where $g(|u|, \mu)$ is given in (2.1.18). We consider below the Boltzmann equation

$$f_t + v \cdot f_x = Q(f,f), \tag{2.1.21}$$

in the notation of equations (2.1.18) and (2.1.20). The connection of the kernel $g(|u|, \mu)$ (2.1.18) with intermolecular potential $\Phi(r)$ is explained in Section 1.5. We shall usually consider $g(|u|, \mu)$ as a given function.

2.2 Basic properties of the Boltzmann equation

In applications to rarefied gas dynamics we are mainly interested not in the distribution function $f(x, v, t)$ itself, but in the (macroscopic) characteristics of the gas averaged over the velocity space. In accordance with the physical meaning of $f(x, v, t)$, the density of the gas or equivalently the number of particles per unit volume is defined by the equality

$$\rho(x,t) = \langle f, 1\rangle = \int_{\mathbb{R}^3} dv f(x,v,t), \quad x \in \mathbb{R}^3, \quad t \geq 0. \tag{2.2.1}$$

Other important macroscopic characteristics of the gas are the bulk (or mean) velocity $u(x, t)$ (not to be confused with the notation u for relative velocity in the collision integral (2.1.20)) and the absolute temperature $T(x, t)$. These functions are defined by the equalities

$$u(x,t) = \frac{1}{\rho}\langle f, v\rangle, \quad T(x,t) = \frac{m}{3\rho}\langle f, |v-u|^2\rangle, \tag{2.2.2}$$

in the notation of (2.1.9). Here m stands for molecular mass, whereas T is expressed in energy units. Usually we assume in this book that $m = 1$ unless the mixture of different gases is considered.

For given values of $\rho(x, t)$, $u(x, t)$, and $T(x, t)$, the following distribution function will be called a local Maxwell distribution (or Maxwellian):

$$f_M(x,v,t) = \rho(2\pi T)^{-3/2} \exp\left[-\frac{|v-u|^2}{2T}\right], \tag{2.2.3}$$

where it is assumed that $m = 1$ in (2.2.2). The same function is called "absolute Maxwellian" if the parameters ρ, u, T are independent of x and t.

Coming back to the Boltzmann equation (2.1.21) we can easily understand the importance of the Maxwellian distribution (2.2.3). Indeed, it follows from (2.1.20) that

$$v' + w' = v + w, \quad |v'|^2 + |w'|^2 = |v|^2 + |w|^2, \tag{2.2.4}$$

i. e., the conservation laws for momentum and energy in each pair collision. Hence, any function of the form

$$f(v) = \exp(\alpha + \beta \cdot v - \gamma|v|^2), \quad \gamma > 0, \tag{2.2.5}$$

with constant parameters (α, β, γ) satisfies the equations

$$f(v')f(w') = f(v)f(w), \quad v \in \mathbb{R}^3, \quad w \in \mathbb{R}^3, \quad \omega \in S^2, \tag{2.2.6}$$

in the notation of equations (2.1.19). Therefore,

$$Q(f_M, f_M) = 0 \tag{2.2.7}$$

for any local Maxwellian (2.2.3).

Another important property of the Boltzmann equation is connected with conservation laws for mass, momentum, and energy. We consider the identity (2.1.18) for a given test function $\psi(v)$ and transform the integral by exchanging variables v and w. Then we easily obtain

$$\langle \psi, Q(f,f)\rangle = \frac{1}{2} \int_{\mathbb{R}^3\times\mathbb{R}^3\times S^2} dv dw d\omega f(v) f(w) g(|u|, \omega\cdot u/|u|)[\psi(v') + \psi(w') - \psi(v) - \psi(w)], \quad (2.2.8)$$

in the notation of (2.1.19). Hence,

$$\langle \psi, Q(f,f)\rangle = 0 \quad \text{if} \quad \psi(v) = a + b\cdot v + c|v|^2 \quad (2.2.9)$$

with any constant parameters a, b, c. This identity leads to conservation laws for mass, momentum, and energy. Indeed, we consider equation (2.1.10) with $\psi = 1$, $\psi = v$, and $\psi = |v|^2$, respectively. Then we obtain

$$\begin{aligned} &\partial_t \rho + \operatorname{div} \rho u = 0, \\ &\partial_t \rho u_\alpha + \partial_{x_\beta}\langle f, v_\alpha v_\beta\rangle = 0, \quad \alpha, \beta = 1, 2, 3; \\ &\partial_t \langle f, |v|^2\rangle + \operatorname{div}\langle f, |v|^2 v\rangle = 0. \end{aligned} \quad (2.2.10)$$

These equations are very basic for the Boltzmann equation. We shall use them for different problems below.

Finally, we shall prove Boltzmann's famous H-theorem. The theorem is based on the following inequality:

$$\langle \log f, Q(f,f)\rangle \le 0. \quad (2.2.11)$$

To prove this we need one more identity for $\langle \psi, Q(f,f)\rangle$, namely,

$$\langle \psi, Q(f,f)\rangle = -\frac{1}{4} \int_{\mathbb{R}^3\times\mathbb{R}^3\times S^2} dv dw d\omega g(|u|, \omega\cdot u/|u|)[f(v')f(w') - f(v)f(w)][\psi(v') + \psi(w') - \psi(v) - \psi(w)]. \quad (2.2.12)$$

The identity (2.2.12) follows from equation (2.2.7) and another general equality

$$\int_{\mathbb{R}^3\times\mathbb{R}^3\times S^2} dv dw d\omega [\Psi(v, w; v', w') - \Psi(v', w'; v, w)] = 0, \quad (2.2.13)$$

that is valid for any function $\Psi(v_1, v_2; v_3, v_4)$ such that integral (2.2.13) is convergent. For the proof it is sufficient to pass to variables U, u (2.1.11) in the integrand and to repeat the considerations used for the proof of identity (2.1.19). For brevity we omit these straightforward calculations.

To complete the proof of (2.2.12) we consider equation (2.2.8) and denote

$$\Psi(v,w;v',w') = \frac{1}{2} f(v)f(w)g(|u|,u'\cdot u/|u|^2)[\psi(v')+\psi(w')-\psi(v)-\psi(w)],$$
$$u' = v' - w' = |u|\omega, \quad u = v - w.$$

Then we apply (2.2.13) and obtain

$$\langle \psi, Q(f,f)\rangle = \frac{1}{2} \int\limits_{\mathbb{R}^3\times\mathbb{R}^3\times S^2} dv dw d\omega [\Psi(v,w;v',w') - \Psi(v',w';v,w)],$$

i. e., the identity (2.2.12). Now we can prove inequality (2.2.11) by substitution of $\psi = \log f(v)$ into (2.2.12). We obtain

$$\begin{aligned}\langle \log f, Q(f,f)\rangle = -\frac{1}{4} \int\limits_{\mathbb{R}^3\times\mathbb{R}^3\times S^2} & dv dw d\omega g(|u|,\omega\cdot u/|u|)[f(v')f(w') \\ & - f(v)f(w)] \log\frac{f(v')f(w')}{f(v)f(w)} \\ \le 0\,. \end{aligned} \tag{2.2.14}$$

This completes the proof of inequality (2.2.11).

The main application of (2.2.11) is the proof of Boltzmann's H-theorem. We introduce the Boltzmann H-functional

$$H(f)(x,t) = \langle f, \log f\rangle = \int\limits_{\mathbb{R}^3} dv f(x,v,t)\log f(x,v,t)\,, \tag{2.2.15}$$

where $f(x,v,t)$ is a solution of equation (2.1.21).

Note that

$$(\partial_t + v\cdot\partial_x) f\log f = (1+\log f)(f_t + v\cdot f_x) = (1+\log f)Q(f,f)\,.$$

Hence, we obtain by integration in v

$$\partial_t\langle f,\log f\rangle + \operatorname{div}\langle f, v\log f\rangle = \langle \log f, Q(f,f)\rangle \le 0. \tag{2.2.16}$$

This inequality is known as "Boltzmann's H-theorem." Its importance can be easily understood in the spatially homogeneous case, considered in the next section.

2.3 Spatially homogeneous problem

The Boltzmann equation (2.1.21) admits a class of spatially homogeneous solutions $f(v,t)$. We usually consider the initial value problem

$$f_t = Q(f,f), \quad f|_{t=0} = f_0(v)\,. \tag{2.3.1}$$

The conservation laws (2.2.10) show that

$$\rho = \langle f, 1 \rangle = \text{const.}, \quad u = \frac{1}{\rho}\langle f, v \rangle = \text{const.},$$
$$T = \frac{1}{3\rho}\langle f, |v-u|^2 \rangle = \frac{1}{3\rho}[\langle f, |v|^2 \rangle - \rho |u|^2] = \text{const.} \tag{2.3.2}$$

We also note that the operator $Q(f,f)$ is invariant under shifting $v \to v + v_0$, $v_0 \in \mathbb{R}^3$, in v-space. Therefore, if $f(v,t)$ is a solution of the equation from (2.3.1), then $f(v+v_0,t)$ is also a solution for any $v_0 \in \mathbb{R}^3$. Hence, we can always reduce the problem (2.3.1) to the case

$$u = \frac{1}{\rho}\langle f, v \rangle = 0, \quad t \geq 0. \tag{2.3.3}$$

Moreover, if $f(v,t)$ is a solution of the spatially homogeneous Boltzmann equation, then so is the function $\tilde{f}(v,t) = \alpha f(v, \alpha t)$ with any $\alpha > 0$. This transformation allows to reduce the general problem (2.3.1) to the case

$$\rho = \langle f, 1 \rangle = 1. \tag{2.3.4}$$

The corresponding Maxwell distribution (2.2.3) reads

$$f_M(v) = (2\pi T)^{3/2} \exp\left(-\frac{|v|^2}{2T}\right), \tag{2.3.5}$$

where

$$T = \frac{1}{3}\langle f_0, |v|^2 \rangle. \tag{2.3.6}$$

The H-theorem (2.2.16) shows that the functional

$$H(f) = \langle f, \log f \rangle = \int_{\mathbb{R}^3} dv f(v,t) \log f(v,t) \tag{2.3.7}$$

cannot increase with time on the solution of (2.3.1) because

$$\partial_t H(f)(t) = \langle \log f, Q(f,f) \rangle \leq 0. \tag{2.3.8}$$

If we consider the explicit formula (2.2.14), then it becomes clear that $\langle \log f, Q(f,f) \rangle = 0$ if and only if

$$f(v')f(w') = f(v)f(w)$$

for almost all values $(v, w, \omega) \in \mathbb{R}^3 \times \mathbb{R}^3 \times S^2$ provided the kernel $g(|u|, \mu)$ in (2.1.20) is positive almost everywhere. This functional equation was studied (in various classes of functions) by many authors, beginning with L. Boltzmann (see [72] and references therein). They proved the uniqueness of its well-known solution (2.2.5). On the other hand, the only function (2.2.5) which satisfies the above discussed conservation laws is the Maxwellian f_M (2.3.5). Hence, we can conclude at the formal level that $H(f)$ decreases monotonically in time unless $f = f_M$. This conclusion can be confirmed by the general inequality

$$\langle f_M, \log f_M \rangle \le \langle f, \log f \rangle, \tag{2.3.9}$$

in the notation of equations (2.2.1)–(2.2.3). Its proof is very simple. Note that $\langle f - f_M, \log f_M \rangle = 0$. Therefore, it is sufficient to prove that $\langle f, (\log f - \log f_M) \rangle \ge 0$. This follows from the elementary inequality

$$G(z, y) = z(\log z - \log y) + y - z = zG_1\left(\frac{z}{y}\right) \ge 0, \quad z > 0, \quad y > 0,$$

where $G_1(t) = \log t + t^{-1} - 1 \ge 0$. We set $z = f(v)$, $y = f_M(v)$ and integrate the inequality $G(f, f_M) \ge 0$ over $v \in \mathbb{R}^3$. This completes the proof of (2.3.9). Inequality (2.3.9) shows that the Maxwellian (2.2.1)–(2.2.3) is the minimizer of the H-functional $H(f) = \langle f, \log f \rangle$ in the class of distribution functions with fixed lower moments (ρ, u, T).

Coming back to the initial value problem (2.3.1) and assuming conditions (2.3.3) and (2.3.4), we know that there is a unique positive stationary solution f_M, given in (2.3.5)–(2.3.6), such that all conservation laws are satisfied. This stationary solution f_M minimizes the H-functional and therefore we expect that the solution $f(v, t)$ converges (in some precise sense) to f_M for large values of time.

This is a qualitative behavior of solutions of the problem (2.3.1) that we expect on the basis of the above formal considerations. The corresponding physical process is called “the relaxation to equilibrium.”

Rigorous mathematical theory of the problem (2.3.1) is not simple. The first steps in its development were made by T. Carleman in the 1930s [64] for the model of hard spheres. A more general and detailed theory was presented by L. Arkeryd [4] in the early 1970s (see also [3]). To understand these and more recent results in this area we need to introduce a sort of classification of collisional kernels $g(|u|, \mu)$ in the Boltzmann integral (2.1.20). This is done in the next section.

2.4 Collisional kernels

We remind the reader that the kernel $g(|u|, \mu)$ is equal to $|u|$ multiplied by the differential cross-section $\sigma(|u|, \mu)$ expressed as a function of $\mu = \cos\theta$. The scattering angle $\theta \in [0, \pi]$ is given in the form (see (1.4.9))

$$\theta(b, |u|) = \pi - 2b|u| \int_{r_{\min}}^{\infty} \frac{r}{r^2}\left[|u|^2\left(1 - \frac{b^2}{|u|^2}\right) - \frac{2\Phi(r)}{m}\right]^{-1/2}, \tag{2.4.1}$$

where b is the impact parameter, $\Phi(r)$ is the intermolecular potential, and m is the reduced mass of colliding particles ($m = 1/2$ for particles with unit mass). To find $\sigma(|u|,\mu)$ we need to construct the inverse function $b = (|u|,\theta)$. Then we express this function as $b = \tilde{b}(|u|,\cos\theta)$ and finally obtain (see (1.5.6))

$$\sigma(|u|,\mu) = \left|\frac{1}{2}\partial_\mu \tilde{b}^2(|u|,\mu)\right|. \tag{2.4.2}$$

Generally speaking, this is a rather complicated calculation. Fortunately it leads to a simple explicit formula $\sigma = d^2/4$ in the important case, when particles are hard spheres with diameter d. It was also shown in Section 1.5 that for power-like potentials $\Phi(r) = \alpha r^{-n}$, $\alpha > 0$, we obtain

$$\sigma(|u|,\mu) = \left(\frac{\alpha}{m|u|^2}\right)^{2/n} \tilde{A}_n(\mu), \quad \mu = \cos\theta, \quad n \geq 1,$$

where all functions from (1.5.10) are expressed as functions of μ. Hence, in the case of power-like potentials $\Phi(r) = \alpha r^{-n}$ the collisional kernel in (2.1.20) reads

$$g(|u|,\mu) = |u|^{\gamma_n} g_n(\mu), \quad \gamma_n = 1 - 4/n\,. \tag{2.4.3}$$

We can use the same formula for hard spheres assuming that $n = \infty$, $g_\infty = d^2/4$. There is, however, an important difference between hard spheres and power-like potentials. We consider again the collision integral (2.1.20) and split it formally into two parts:

$$Q(f,f) = Q^{\text{gain}}(f,f) - Q^{\text{loss}}(f,f)\,, \tag{2.4.4}$$

where

$$\begin{gathered} Q^{\text{loss}}(f,f) = f(v)\nu(v), \quad \nu(v) = \int_{\mathbb{R}^3} dw f(w) g_{\text{tot}}(|v-w|)\,, \\ g_{\text{tot}}(|u|) = |u|\sigma_{\text{tot}}(|u|) = 2\pi|u| \int_{-1}^{1} d\mu\, \sigma(|u|,\mu). \end{gathered} \tag{2.4.5}$$

It was already discussed in Section 1.5 that $\sigma_{\text{tot}} = \pi R_{\text{max}}^2$, where R_{max} denotes the radius of action of the potential. In the case of hard spheres of diameter d or any potential with $R_{\text{max}} = d$, we obtain the following universal formula for the collision frequency:

$$\nu(v) = \pi d^2 \int_{\mathbb{R}^3} dw f(w)|v-w|\,. \tag{2.4.6}$$

However, if we consider the power-like potential with any $n > 0$, then $R_{\text{max}} = \infty$ and therefore the integral $\nu(v)$ diverges. Hence, the splitting (2.4.4) is impossible, though the "whole" collision integral (2.1.20) can be convergent. The matter is that the kernel $g(|u|, \mu)$ has a non-integrable singularity at $\mu = 1$, i. e., $\theta = 0$, for long range potentials with $R_{\max} = \infty$. At the same time, $v' = v$ and $w' = w$ if $\mu = 1$. Therefore, the second factor in the integrand is equal to zero at that point. It is easy to see that the integral (2.1.20) is convergent for a large class of differentiable functions $f(v)$ provided

$$\int_{-1}^{1} d\mu g_n(\mu)(1-\mu) < \infty \tag{2.4.7}$$

in the case of power-like potentials. It can be shown that this condition is satisfied for all $n > 1$. The Coulomb case $n = 1$ is always considered separately.

There are many publications from the last two decades related to the Boltzmann equation with long range potentials (see, e. g., [2, 119] and references therein). We do not consider related questions in this book, though some of our results (in particular, from Chapters 3–5) are valid also for long range potentials.

Note that the splitting (2.4.4) has a clear physical meaning and simplifies some mathematical considerations. Unfortunately, the power-like potentials with cut-off at some distance $r = R_{\max}$ do not lead to a factorized kernel (2.4.3). Grad proposed the so-called "angular cut-off." The idea is to fix some $n > 1$ in (2.4.3) and some number $0 < \varepsilon \ll 1$. Then we can replace $g_n(\mu)$ in (2.4.3) by another function,

$$\tilde{g}_n(\mu) = \begin{cases} g_n(\mu), & \text{if } -1 \le \mu \le 1-\varepsilon, \\ g_n(1-\varepsilon), & \text{otherwise.} \end{cases} \tag{2.4.8}$$

Then the new kernel has both advantages: boundedness for all $\mu \in [-1, 1]$ and a factorized structure (2.4.3). Such kernels are used in many mathematical works. The authors of these works usually distinguish between (a) hard potentials, i. e., $n > 4$; (b) pseudo-Maxwell molecules, i. e., $n = 4$; and (c) soft potentials, i. e., $-1 < n < 4$. For example, a statement that "something is proved for hard potentials with Grad's cut-off" normally means that it is proved for collision integral (2.1.20) with kernel of the form (2.4.3), (2.4.8).

2.5 Boltzmann equations for gas mixtures

Formal generalization of the Boltzmann equation (2.1.21) to the case of gas mixtures is straightforward. We consider a mixture of $n \ge 1$ gases and describe it by distribution functions $\{f_i(x, v, t), i = 1, \dots, n\}$, where $x \in \mathbb{R}^3$, $v \in \mathbb{R}^3$, $t \in \mathbb{R}_+$. It is assumed that particles of the ith sort have a mass $m_i > 0$ and interact with particles of the kth sort with given potential $\Phi_{ik}(r) = \Phi_{ki}(r)$, where $r > 0$ is a distance between interacting particles. For the sake of generality we assume also the presence of an external force field $\Phi_{\text{ext}}(x,t)$ such that

$$a_i(x,t) = -\frac{1}{m_i}\,\partial_x\,\Phi_{\text{ext}}(x,t), \quad 1 \le i \le n\,, \tag{2.5.1}$$

denoting the acceleration caused by external force. Then the set of kinetic equations for the mixture reads

$$(\partial_t + v\cdot\partial_x + a_i\cdot\partial_v)f_i = Q_i, \quad 1 \le i \le n\,, \tag{2.5.2}$$

where

$$Q_i = \sum_{k=1}^{n} Q_{ik}(f_i, f_k)\,, \tag{2.5.3}$$

$$Q_{ik}(f_i, f_k) = \int_{\mathbb{R}^3\times S^2} dw d\omega g_{ik}(|u|, \omega\cdot u/|u|)[f_i(v')f_k(w') - f_i(v)f_k(w)], \tag{2.5.4}$$

$$u = v - w, \quad \omega \in S^2, \quad v' = U + \frac{m_{ik}}{m_i}|u|\omega, \quad w' = U - \frac{m_{ik}}{m_k}|u|\omega,$$

$$U = \frac{m_i v + m_k w}{m_i + m_k}, \quad m_{ik} = \frac{m_i m_k}{m_i + m_k}; \quad i,k = 1,\dots,n.$$

Here we also denote

$$g_{ik}(|u|,\mu) = |u|\sigma_{ik}(|u|,\mu),$$

where the differential cross-section $\sigma_{ik}(|u|, \cos)$, $\theta \in [0,\pi]$, is calculated from the scattering problem for a particle with reduced mass m_{ik} on the potential $\Phi_{ik}(r) = \Phi_{ki}(r)$. Formulas for v' and w' in (2.5.4) follow from general consideration of pair collisions for particles with different masses (see Section 1.4).

Basic properties of equations (2.5.3) are (A) conservation laws; (B) the H-theorem; and (C) uniqueness of equilibrium solutions. These properties are related to the following properties of collision terms Q_i:

(A) conservation laws:

$$\langle Q_i, 1\rangle = 0, \quad i = 1,\dots,n; \quad \sum_{i=1}^{n} m_i\langle Q_i, v\rangle = 0; \quad \sum_{i=1}^{n} m_i\langle Q_i, |v|^2\rangle = 0;$$

(B) the basic inequality for the H-theorem:

$$\sum_{i=1}^{n}\langle \log f_i, Q_i\rangle \le 0;$$

(C) uniqueness of equilibrium solutions;

$$Q_i = 0 \Rightarrow f_i = n_i(2\pi T/m_i)^{-3/2}\exp\left[-\frac{m_i(v-u)^2}{2T}\right], \quad i = 1,\dots,n,$$

where n_i, u, T do not depend on v. The proof of these properties is mainly a repetition of above considerations for the case $n = 1$. Therefore, we omit it.

Remarks on Chapter 2

1. The material of Chapter 2 is also standard. It is contained in almost all books on the Boltzmann equation; see, e. g., [68, 70, 74, 104] and others.
2. The general form (2.1.20) of the Boltzmann collision integral $Q(f,f)$ does not coincide with its more conventional form (see equation (2.1.2) for the case of hard spheres) used in the literature. Therefore, we explain in detail in Section 2.1 the equivalence of these two forms. The form (2.1.20) has some advantages and therefore we use it in this book. The weak form (2.2.8) was in fact considered in almost explicit form in Maxwell's paper [115]. However, the introduction by Boltzmann of the strong form of his equation in [57] was extremely important because Boltzmann had proved in that paper his famous H-theorem.

3 Maxwell molecules and the Fourier transform

3.1 Maxwell molecules

The history of so-called Maxwell molecules begins with the famous Maxwell paper "On the Dynamical Theory of Gases" published in 1867 [115]. He considers in that paper a gas of particles interacting with power-like potential $\Phi(r) = \alpha/r^n$, $\alpha > 0$. Maxwell introduces a sort of weak formulation of the Boltzmann equation (see equations (2.1.10) and (2.1.18)) for that molecular model. He finds that all calculations become simpler for the special case $n = 4$ and evaluates the coefficients of diffusion, viscosity, and heat conductivity for some real gases. In the beginning of the paper Maxwell claims that his experiments on viscosity of the air confirm that the gas molecules interact with this specific potential

$$\Phi(r) = \frac{\alpha}{r^4}, \quad \alpha > 0\,. \tag{3.1.1}$$

In fact it was a mistake. Now it is believed that the hard sphere model is more close to reality, though some more sophisticated models are also used in applied works. However, the Maxwell molecules, i. e., the particles interacting with potential (3.1.1) and their generalizations, remain to be a very popular and useful mathematical model in kinetic theory.

It was already mentioned in Chapters 1 and 2 that the differential scattering cross-section $\sigma(|u|, \theta)$ for the potential (3.1.1) is inversely proportional to $|u|$ (see (1.5.10)). Hence, the kernel of the Boltzmann collision integral

$$g(|u|, \cos\theta) = |u|\sigma(|u|, \theta) = g(\cos\theta) \tag{3.1.2}$$

does not depend on velocities of colliding particles. For completeness we present below some explicit formulas for potential (3.1.1). We consider a collision of two particles with masses m_1 and m_2. Then the scattering angle $0 \le \theta \le \pi$ reads (see equation (1.4.10))

$$\theta(b, |u|) = \pi - 2b|u| \int_{r_{\min}}^{\infty} \frac{dr}{r^2} \left[|u|^2 \left(1 - \frac{b^2}{r^2}\right) - \frac{2\alpha}{mr^4} \right]^{-1/2},$$

where $m = m_1 m_2/(m_1 + m_2)$ and b and $|u|$ denote the impact parameter and the modulus of relative velocity, respectively. The value $r_{\min}$ corresponds to zero of the denominator. After simple transformations, we obtain

$$\theta = \pi - 2I(a), \quad a^4 = \frac{b^4 m|u|^2}{8\alpha}, \quad I(a) = \int_0^{x_{\max}} dx\, \Delta^{-1/2}(x),$$

$$\Delta(x) = 1 - x^2 - \frac{1}{4}\left(\frac{x}{a}\right)^2, \quad \Delta(x_{\max}) = 0.$$

https://doi.org/10.1515/9783110550986-004

The substitution

$$a^4 = \cot^2 2\varphi, \quad x = \sqrt{1 - \tan^2\varphi} \cos t, \quad 0 < \varphi \le \pi/4,$$

allows to express the result in terms of elliptic integrals,

$$\theta = \pi - 2\sqrt{\cos 2\varphi} K(\sin\varphi), \quad K(z) = \int_0^{\pi/2} dt\, (1 - z^2 \sin^2 t)^{-1/2}.$$

This calculation was performed in the original Maxwell paper [115]. Hence, we obtain a relatively simple formula for $\theta = \theta(a)$. Unfortunately the inverse function

$$a = b\left(\frac{m|u|^2}{8\alpha}\right)^{1/4} = a(\theta)$$

is known only in implicit form. The differential cross-section $\sigma(|u|, \theta)$ is defined by equations (1.4.10) and (1.5.6). Hence, we obtain the kernel (3.1.2) in the form

$$g(\mu) = -\sqrt{\frac{2\alpha}{m}} \frac{1}{\sin\theta} \frac{d}{d\theta} a^2(\theta), \quad \mu = \cos\theta, \quad \theta \in (0, \pi), \tag{3.1.3}$$

since $a'(\theta) < 0$. It can be shown that

$$0 < \sqrt{\frac{m}{\alpha}}\, g(\mu)(1-\mu)^{5/4} \le c_1, \quad \lim_{\mu \to 1} \sqrt{\frac{m}{\alpha}}\, g(\mu)(1-\mu)^{5/4} = c_2, \tag{3.1.4}$$

where c_1, c_2 are absolute constants.

We do not use below any specific properties of the kernel (3.1.3). The main restriction used in Chapter 3 is that the kernel $g(\mu)$ does not depend on velocities. This simplifies some calculations with the collisional integral. The best way to use these simplifications is to pass to Fourier representation, as we shall see in the next section.

Before doing this it is convenient to extend the notion of Maxwell molecules or Maxwell models. We shall use below the same terminology for any Boltzmann-type integral operators with velocity-independent kernels. The particles interacting with potential (3.1.1) will be called "true Maxwell molecules." Otherwise any reasonable non-negative function $g(\mu)$, $\mu \in [-1, 1]$ can be used as the kernel of the Boltzmann integral. For simplicity we prefer to avoid terms like "Maxwell pseudo-molecules" [144] or "pseudo-Maxwell molecules" [93].

Of course, the true Maxwell molecules are the only dynamical model formally consistent with the corresponding Boltzmann equation. On the other hand, the Boltzmann equation itself can be equally considered as a limiting case of certain stochastic particle models (for example, the models by Kac [101]). Such models can be constructed for each reasonable kernel $g(\mu)$ in the collision integral.

All results of this chapter are valid for essentially bounded kernels $g \in L^\infty_+(M)$, $M = [-1,1]$. Most of them remain valid for integrable kernels $g \in L^1_+(M)$. A significant part of the results is valid under a less restrictive condition

$$g(\mu)(1-\mu) \in L^1_+(M), \quad M = [-1,1]. \tag{3.1.5}$$

This condition is always assumed below. In particular, this is satisfied for true Maxwell molecules.

3.2 Fourier transform of the Boltzmann equation

We consider the Boltzmann equation in its general form (2.1.20) and introduce the characteristic function (in terminology of the probability theory, see, e. g., [85]),

$$\varphi(x,k,t) = \langle f, \exp(-ik \cdot v)\rangle, \quad k \in \mathbb{R}^3, \tag{3.2.1}$$

in the notation of equation (2.1.9). Then we use (2.1.19) and obtain, omitting arguments x and t,

$$\begin{aligned}
&\varphi_t + i\partial_k \cdot \partial_x \varphi \\
&\quad = \int\limits_{\mathbb{R}^3\times\mathbb{R}^3\times S^2} dv dw d\omega\, g(|u|, \omega\cdot\hat{u})\, f(v) f(w)(e^{-ik\cdot v'} - e^{-ik\cdot v}), \\
&u = v - w, \quad \omega \in S^2, \quad \hat{u} = \frac{u}{|u|}, \quad v' = \frac{v + w + |u|\omega}{2}.
\end{aligned} \tag{3.2.2}$$

The integral on the right hand side can be written as

$$\begin{aligned}
&\langle e^{-ik\cdot v}, Q(f,f)\rangle \\
&\quad = \int\limits_{\mathbb{R}^3\times\mathbb{R}^3} dv dw\, f(v) f(w)\, e^{-ik\cdot(v+w)/2} I(u,k), \quad k \in \mathbb{R}^3, \quad u \in \mathbb{R}^3,
\end{aligned} \tag{3.2.3}$$

where

$$I(u,k) = \int\limits_{S^2} d\omega\, g(|u|, \omega\cdot\hat{u})\, (e^{-ik\cdot|u|\omega/2} - e^{-ik\cdot u/2}).$$

We assume that the kernel $g(|u|,\mu)$ can be factorized, i. e.,

$$g(|u|,\mu) = a(|u|) b(\mu). \tag{3.2.4}$$

Then

$$\begin{aligned}
I(u,k) &= a(|u|) I_1(u,k), \\
I_1(u,k) &= \int\limits_{S^2} d\omega\, b(\omega\cdot\hat{u})\, (e^{-ik\cdot|u|\omega/2} - e^{-ik\cdot u/2}).
\end{aligned} \tag{3.2.5}$$

A key observation is given in the following lemma.

Lemma 3.2.1. *We have*

$$I_1(u,k) = I_1(k,u)\,. \tag{3.2.6}$$

Proof. (1)

It follows from the fact that $I_1(u,k)$ is an isotropic function (invariant under rotation) of two vectors $u \in \mathbb{R}^3$ and $k \in \mathbb{R}^3$. Therefore,

$$I_1(u,k) = I_2(|u|, |k|, u \cdot k)\,,$$

i. e., it depends only on three scalar arguments. Obviously, $u \cdot k = k \cdot u$. Hence, it is sufficient to prove that it is possible to interchange $|u|$ and $|k|$. The proof follows immediately from equation (3.2.5) for $I_1(u,k)$. □

For completeness we also present a direct proof of identity (3.2.6), which does not use "symmetry considerations."

Proof. (2)

The integral (3.2.5) can be evaluated in spherical coordinates $(\theta, \varphi) \in [0,\pi]\times[0,2\pi)$ with the z-axis directed along u. We can always choose the x-axis located in the plane spanned by vectors u and k and such that the projection $k_x \geq 0$. Then the Cartesian coordinates of vectors ω, u, k read as

$$\omega = (\sin\theta\cos\varphi,\ \sin\theta\sin\varphi,\ \cos\theta),$$
$$u = |u|(0,0,1), \quad k = |k|(\sin\theta_0, 0, \cos\theta_0)\,,$$

where $0 \leq \theta_0 \leq \pi$ is the angle between u and k. Therefore, we obtain

$$\omega \cdot \hat{u} = \cos\theta, \quad k \cdot u = |k||u|\cos\theta_0,$$
$$k \cdot |u|\omega = |k||u|(\sin\theta\sin\theta_0\cos\varphi + \cos\theta\cos\theta_0).$$

Note that $\sin\theta_0 = (1-\cos^2\theta_0)^{1/2} \geq 0$ if $0 \leq \theta_0 \leq \pi$. Hence,

$$I_1(u,k) = \int_0^\pi d\theta \sin\theta b(\cos\theta)\, I_3(|k||u|, \theta, \cos\theta_0)\,,$$

where

$$I_3(z,\theta,\zeta) = \int_0^{2\pi} d\varphi (e^{-z\tilde{\zeta}/2} - e^{-z\zeta/2}), \quad \tilde{\zeta} = \zeta\cos\theta + \sqrt{1-\zeta^2}\sin\theta\cos\varphi\,.$$

Hence, $I_1(u,k)$ is presented in the form

$$I_1(u,k) = I_4\left(|k||u|, \frac{k \cdot u}{|k||u|}\right),$$

which makes the identity (3.2.6) obvious. This completes the proof. □

Substituting equations (3.2.4)–(3.2.6) into equation (3.2.3), we obtain after change of order of integration

$$\begin{aligned}\langle e^{-ik\cdot v}, Q(f,f)\rangle &= \int_{S^2} d\omega b(\omega\cdot\hat{k}) \\ &\times \int_{\mathbb{R}^3\times\mathbb{R}^3} dv dw f(v) f(w) a(|u|) e^{-ik\cdot\frac{v+w}{2}} (e^{-i|k|u\cdot\omega/2} - e^{-ik\cdot\frac{u}{2}}) \\ &= \int_{S^2} d\omega b(\omega\cdot\hat{k}) \int_{\mathbb{R}^3\times\mathbb{R}^3} dv dw f(v) f(w) a(|u|) [e^{-i(k_+\cdot v + k_-\cdot w)} - e^{-ik\cdot v}], \\ \hat{k} = \frac{k}{|k|}, \quad k_\pm &= \frac{k \pm |k|\omega}{2}. \end{aligned} \tag{3.2.7}$$

This is the general formula for the Fourier transform of the Boltzmann collision integral with factorized kernel (3.2.4). In the case of Maxwell molecules, the kernel $g(\mu)$ does not depend on $|u|$ and there is no need to introduce a new notation $b(\mu)$ in (3.2.4). In this case, we obtain

$$\begin{aligned}\langle e^{-ik\cdot v}, Q(f,f)\rangle = \hat{Q}(\varphi,\varphi) &= \int_{S^2} d\omega g(\hat{k}\cdot\omega)\,[\varphi(k_+)\varphi(k_-) - \varphi(0)\varphi(k)], \\ \hat{k} = \frac{k}{|k|}, \quad k_\pm &= \frac{k \pm |k|\omega}{2}. \end{aligned} \tag{3.2.8}$$

This formula was firstly obtained in [17]. Hence, the Fourier transform in the velocity space leads to a significant simplification of the Boltzmann collision integral for Maxwell molecules. Unfortunately the left hand side of the transformed equation (3.2.2) becomes more complicated in the spatially inhomogeneous case. Therefore, we will mainly consider applications to the spatially homogeneous problem in this book (except for Chapter 7).

3.3 The spatially homogeneous Boltzmann equation for Maxwell molecules

We consider the Cauchy problem for the distribution function $f(v,t)$ satisfying the equation

$$\begin{aligned} f_t &= Q(f,f) \\ &= \int_{\mathbb{R}^3\times S^2} dw d\omega\, g(\omega\cdot\hat{u})\,[f(v')f(w') - f(v)f(w)], \quad \omega\in S^2, \quad t>0, \\ u &= v-w, \quad \hat{u} = u/|u|, \quad v' = \frac{1}{2}(v+w+|u|\omega), \quad w' = \frac{1}{2}(v+w-|u|\omega) \end{aligned} \tag{3.3.1}$$

and the initial condition

$$f|_{t=0} = f_0(v). \tag{3.3.2}$$

We denote (see equation (2.1.9))

$$\rho_0 = \langle f_0, 1\rangle, \quad u_0 = \frac{1}{\rho_0}\langle f_0, v\rangle, \quad T_0 = \frac{1}{3\rho_0}\langle f_0, |v - u_0|^2\rangle$$

and pass to dimensionless variables such that

$$\begin{aligned} f_0(v) &= \rho_0 T_0^{-3/2} \tilde{f}_0(\tilde{v}), \quad \tilde{v} = \frac{v - u_0}{\sqrt{T_0}}, \\ f(v,t) &= \rho_0 T_0^{-3/2} \tilde{f}(\tilde{v}, \tilde{t}), \quad \tilde{t} = \rho_0 g_* t, \quad g(\mu) = g_* \tilde{g}(\mu), \end{aligned} \tag{3.3.3}$$

where g_* is a typical value of the kernel $g(\mu)$. Substituting these formulas into equations (3.3.1)–(3.3.2) and omitting tildes, we again obtain the same expressions (3.3.1)–(3.3.2). However, the function $f_0(v)$ will now satisfy the conditions

$$\langle f_0, 1\rangle = 1, \quad \langle f_0, v\rangle = 0, \quad \langle f_0, |v|^2\rangle = 3. \tag{3.3.4}$$

General properties of solutions of the spatially homogeneous Boltzmann equation were already discussed in Section 2.3. In particular, we have classical conservation laws (2.3.2) for mass, momentum, and energy. Hence, we formally obtain the same equalities (3.3.4) for $f(v,t)$,

$$\langle f, 1\rangle = 1, \quad \langle f, v\rangle = 0, \quad \langle f, |v|^2\rangle = 3, \quad t \geq 0. \tag{3.3.5}$$

We will usually consider below (except for Sections 5.6 and 5.7) the distribution functions with bounded second moment (energy) in the velocity space. Then, in the spatially homogeneous case, we can assume without loss of generality the normalization conditions (3.3.5). It is useful to bear in mind from the very beginning that these conditions correspond to the standard Maxwellian

$$M(v) = \left(\frac{1}{2\pi}\right)^{3/2} \exp\left(-\frac{|v|^2}{2}\right). \tag{3.3.6}$$

The classical approach to the problem (3.3.1)–(3.3.2) is the following (see [117] and references therein). We assume that the kernel $g(\mu)$ is non-negative and integrable on $M = [-1, 1]$. Then the constant g_* in equations (3.3.3) can be chosen in such a way that

$$g_{tot} = 2\pi \int_{-1}^{1} d\mu g(\mu) = 1 \tag{3.3.7}$$

for the dimensionless kernel $g(\mu)$. Then

$$Q(f,f) = Q_+(f,f) - f, \quad Q_+(f_1,f_2) = \int\limits_{\mathbb{R}^3\times S^2} dwd\omega\, g(\omega\cdot\hat{u}) f(v_1') f(w_2'),$$

in the notation of equation (3.3.1). We change variables

$$f(v,t) = e^{-t}F(v,\tau), \quad \tau = 1 - e^{-t},$$

and obtain

$$F_\tau = Q_+(F,F), \quad F_{\tau=0} = f_0(v).$$

Then we look for the solution in the form

$$F(v,\tau) = \sum_{n=0}^{\infty} f_n(v)\tau^n$$

and find the following recurrence formulas for coefficients:

$$f_{n+1}(v) = \frac{1}{n+1}\sum_{k=0}^{n} Q_+(f_k, f_{n-k}), \quad n = 0,1,\dots, \tag{3.3.8}$$

$$Q_+(\varphi,\psi) = \int\limits_{\mathbb{R}^3\times S^2} dwd\omega g(\omega\cdot\bar{u})\varphi(v')\psi(w'). \tag{3.3.9}$$

Hence, we obtain the distribution function in the form of series

$$f(v,t) = e^{-t}\sum_{n=0}^{\infty}(1-e^{-t})^n f_n(v), \tag{3.3.10}$$

in the notation of equations (3.3.8). It is the so-called Wild sum [116, 155]. Its convergence is easy to prove. Note that

$$\langle 1, Q_+(f_k, f_{n-k})\rangle = \langle f_k, 1\rangle\,\langle f_{n-k}, 1\rangle, \quad 0 \le k \le n, \quad n = 0,1,\dots.$$

Then by induction we obtain $\langle f_n, 1\rangle = 1$, $n = 1,2,\dots$. Therefore, $\langle f, 1\rangle = 1$, as expected. The convergence of series (3.3.10) almost everywhere in $\mathbb{R}^3$ follows from B. Levi's theorem (see, e. g., [106]). It is convenient to introduce the notation

$$f(v) \in L_k^p$$
$$\Leftrightarrow \|f\|_{L_k^p} = \Big[\int\limits_{\mathbb{R}^3} dv |f(v)|^p (1+|v|^2)^{kp/2}\Big]^{1/p} < \infty, \quad p \ge 1,\ k \ge 0, \tag{3.3.11}$$

such that $L_0{}^p = L^p$, where L^p is the usual Lebesgue space. Since $\langle f, 1\rangle = \|f\|_{L^1}$ for any non-negative $f(v)$, we can conclude that the problem (3.3.1)–(3.3.2) has a non-negative solution $f(v,t) \in L^1$ for any $f_0 \in L^1$, $f_0 \geq 0$. It is also easy to prove that the solution is unique in L^1. We just give an idea of the proof.

Consider a symmetric bilinear operator

$$Q(f_1,f_2) = \frac{1}{2} \int\limits_{\mathbb{R}^3\times S^2} dw d\omega g(\omega \cdot \hat{u})[f_1(v')f_2(w') + f_1(w')f_2(v') \\ - f_1(v)f_2(w) - f_1(w)f_2(v)], \tag{3.3.12}$$

in the notation of equation (3.3.1). Note that

$$\begin{aligned} Q(f_1,f_1) - Q(f_2,f_2) &= Q(f_1 + f_2, f_1 - f_2), \\ \|Q(f_1,f_2)\|_{L^1} &= \langle 1, |Q(f_1,f_2)|\rangle \leq 2\|f_1\|_{L^1}\|f_2\|_{L^1}, \end{aligned} \tag{3.3.13}$$

under condition (3.3.7). Then we can assume that $f_{1,2} \in L^1$ are two different solutions of the problem (3.3.1)–(3.3.2) and conclude that $f_1 = f_2$ by contradiction.

Thus we obtain the following theorem, which was firstly proved by E. Wild [155] and D. Morgenstern [117, 118] in the 1950s.

Theorem 3.3.1. *For any non-negative initial data $f_0 \in L^1$, the Cauchy problem* (3.3.1)–(3.3.2) *has a unique in L^1 solution $f(v,t) \geq 0$ given by the Wild sum* (3.3.10).

The Fourier transformed Boltzmann equation (3.3.2), (3.3.8) [17] made it natural to consider the problem (3.3.1)–(3.3.2) in the Fourier representation. We introduce the characteristic function

$$\varphi(k,t) = \langle f(v,t), e^{-ik\cdot v}\rangle \tag{3.3.14}$$

and consider the following equation:

$$\begin{aligned} \varphi_t = J(\varphi,\varphi) &= \int\limits_{S^2} d\omega g(\omega \cdot \hat{k})[\varphi(k_+)\varphi(k_-) - \varphi(0)\varphi(k)], \\ \hat{k} = k/|k|, \quad &\omega \in S^2, \quad k_\pm = (k \pm |k|\omega)/2, \end{aligned} \tag{3.3.15}$$

with initial condition

$$\varphi|_{t=0} = \varphi_0(k) = \langle f_0(v), e^{-ik\cdot v}\rangle. \tag{3.3.16}$$

Note that the normalization conditions (3.3.4) become the boundary conditions at $k = 0$

$$\varphi_0|_{k=0} = 1, \quad \partial_k\varphi_0|_{k=0} = 0, \quad \Delta_k\varphi_0|_{k=0} = -3. \tag{3.3.17}$$

The Maxwellian (3.3.6) is transformed to

$$\varphi_M(k) = \exp\left(-\frac{|k|^2}{2}\right). \tag{3.3.18}$$

A natural class of initial data (3.3.16) for equation (3.3.15) is a set of all characteristic functions, i. e., Fourier transforms of probability measures in $\mathbb{R}^3$. Then we can define a generalized solution of the Boltzmann equation (3.3.1) as a time-dependent probability measure in $\mathbb{R}^3$, whose characteristic function $\varphi(k,t)$ satisfies (3.3.15). This is what is called in the literature the measure solution of the Boltzmann equation (see, e. g., [99]). We do not introduce any special notation for the measure solutions, considering in such case $f(v,t)$ as a generalized density of a probability measure. The classical solution of equation (3.3.1) corresponds to an absolutely continuous measure with density $f(v,t) \in L^1$.

Properties of characteristic functions are well known (see, e. g., [85, 114]). If we consider the problem (3.3.15)–(3.3.16) and construct its formal solution $\varphi(k,t)$ in a similar way (see the Wild sum (3.3.10)), then it is easy to prove that the corresponding series is absolutely convergent for any $k \in \mathbb{R}^3$ and its sum $\varphi(k,t)$ is the characteristic function for any $t \geq 0$. Indeed, we can assume the same condition (3.3.7) on the kernel $g(\mu)$ in equation (3.3.15) and represent this equation as

$$\varphi_t = \hat{Q}_+(\varphi,\varphi) - \varphi, \quad \varphi|_{t=0} = \varphi_0(k), \tag{3.3.19}$$

where $\varphi_0(k)$ satisfies equations (3.3.18) and the bilinear operator $\hat{Q}_+(\varphi,\psi)$ reads

$$\hat{Q}_+(\varphi,\psi) = \int_{S^2} d\omega\, g(\omega\cdot\hat{k})\,\varphi(k_+)\psi(k_-), \tag{3.3.20}$$

in the notation of equations (3.3.15). We assume that $\varphi_0(k)$ is the characteristic function given in equation (3.3.16). Therefore, $|\varphi_0(k)| \leq \varphi_0(0) = 1$. Obviously, we can construct a formal solution of the problem (3.3.19) in the form of the Wild sum

$$\begin{aligned} \varphi(k,t) &= e^{-t}\sum_{n=0}^{\infty}(1-e^{-t})^n\,\varphi_n(k), \\ \varphi_{n+1}(k) &= \frac{1}{n+1}\sum_{k=0}^{n}\hat{Q}_+(\varphi_k,\varphi_{n-k}), \quad n = 0,1,\dots, \end{aligned} \tag{3.3.21}$$

similarly to the construction from Theorem 3.3.1. It follows from elementary properties of characteristic functions [85, 114] that all $\varphi_n(k)$, $n \geq 0$, are characteristic functions and so is $|\varphi(k,t)|$ for any $t > 0$. The convergence of series (3.3.21) at any $k \in \mathbb{R}^3$ and $t > 0$ follows from inequalities $|\varphi_n(k)| \leq 1$, $n \geq 0$. Actually the Wild sum (3.3.21) is the solution of the problem (3.3.19) for initial data $\varphi_0(k)$ such that

$$\|\varphi_0(k)\|_\infty = \sup_{k\in\mathbb{R}^3}|\varphi_0(k)| \leq 1,$$

since this condition is sufficient for absolute and uniform convergence of the series (3.3.19). We shall later need the following uniqueness result.

Lemma 3.3.1. *If $\varphi_0(k)$ is a characteristic function, then the solution $\varphi(k,t)$ (3.3.21) of the problem (3.3.19) is unique in the class of functions satisfying the inequality*

$$\|\varphi\|_{R,T} = \sup_{|k|\le R,\, 0\le t\le T} |\varphi(k,t)| < \infty \tag{3.3.22}$$

for any numbers $R > 0$ and $T > 0$.

Proof. Note that $\varphi(k,t)$ given in equations (3.3.21) is a characteristic function and therefore

$$|\varphi(k,t)| \le \varphi_0(0) = 1, \quad k \in \mathbb{R}^3, \quad t > 0. \tag{3.3.23}$$

We assume that there exists another solution $\psi(k,t)$ of the problem (3.3.19) satisfying condition (3.3.16). Then the function

$$h(k,t) = \varphi(k,t) - \psi(k,t) \tag{3.3.24}$$

is a solution of the problem

$$h_t + h = \hat{P}h, \quad h|_{t=0} = 0, \tag{3.3.25}$$

where the linear operator $\hat{P} = \hat{P}(t)$ is given by

$$\hat{P}h = \int_{S^2} d\omega\, g_1(\omega\cdot\hat{k})[\varphi(k_+,t) + \psi(k_+,t)]\, h(k_-,t),$$

where

$$g_1(\mu) = \frac{1}{2}[g(\mu) + g(-\mu)],$$

in the notation of equations (3.3.15). Then equations (3.3.25) lead to

$$h(k,t) = \int_0^t d\tau\, e^{-(t-\tau)}[\hat{P}h](k,\tau). \tag{3.3.26}$$

We denote for an arbitrary $R > 0$

$$\|h\|_R(t) = \sup_{|k|\le R} |h(k,t)|. \tag{3.3.27}$$

Then we use equations (3.3.7) and (3.3.23) and obtain for any $0 \le t \le T$

$$\sup_{|k|\le R} |\hat{P}h(k,t)| \le (1 + \|\psi\|_{R,T})\, \|h\|_R(t).$$

Therefore, equation (3.3.26) leads to

$$0 \le \|h\|_R(t) \le (1 + \|\psi\|_{R,T}) \int_0^t d\tau \|h\|_R(\tau), \quad 0 \le t \le T.$$

It is well known (Gronwall's lemma) that this inequality has only trivial solution $\|h\|_R(t) = 0$. Hence, $h(k,t) = 0$ for any $|k| \le R$ and $0 \le t \le T$. However, R and T are arbitrary positive numbers and therefore $h = \varphi(k,t) - \psi(k,t) = 0$ on $\mathbb{R}^3 \times \mathbb{R}_+$. This completes the proof. □

A simple and important inequality for solutions of the problem (3.3.15)–(3.3.17) was proved by Toscani and Villani [143]. Let $h(k,t)$ be given by equation (3.3.24), where both $\varphi(k,t)$ and $\psi(k,t)$ are characteristic functions satisfying equations (3.3.15)–(3.3.17). Then $h(k,t)$ is a solution of equation (3.3.25) such that

$$h(k,0) = \varphi(k,0) - \psi(k,0) = O(|k|^2), \quad k \in \mathbb{R}^3.$$

It follows from conservation laws and general properties of characteristic functions that the similar estimate $h(k,t) = O(|k|^2)$ holds for all $t > 0$. On the other hand,

$$|\varphi(k,t)| \le 1, \quad |\psi(k,t)| \le 1, \quad t \ge 0,$$

and therefore

$$\begin{aligned} |\hat{P}h(k,t)| &\le \int_{S^2} d\omega [g(\omega \cdot \hat{k}) + g(-\omega \cdot \hat{k})]\, |h(k_-,t)| \\ &= \int_{S^2} d\omega g(\omega \cdot \hat{k}) [\, |h(k_-,t)| + |h(k_+,t)|\,], \quad k_\pm = \frac{1}{2}(k \pm |k|\omega), \end{aligned} \tag{3.3.28}$$

in the notation of equation (3.3.25). We denote

$$h(k,t) = |k|^2 u(k,t) \tag{3.3.29}$$

and obtain from equation (3.3.25) the following estimate.

Lemma 3.3.2 ([143]). *Let $\varphi(k,t)$ and $\psi(k,t)$ be two characteristic functions satisfying equations (3.3.15)–(3.3.17) with initial data $\varphi_0(k)$ and $\psi_0(k)$, respectively. Then the functional*

$$[d(\varphi,\psi)](t) = \sup_{k \in \mathbb{R}^3} \frac{|\varphi(k,t) - \psi(k,t)|}{|k|^2} \tag{3.3.30}$$

is non-increasing in time.

Proof. We substitute equation (3.3.29) into equation (3.3.25) and obtain

$$u_t + u = \frac{1}{|k|^2}\,\hat{P}u(k,t)\,|k|^2\,. \tag{3.3.31}$$

Then we consider estimate (3.3.28) and note that

$$\left|\hat{P}u(k,t)|k|^2\right| \le \int_{S^2} d\omega g(\omega\cdot\hat{k})(|k_+|^2 + |k_-|^2)\,\|u\|_\infty(t),$$

in the notation of equation (3.3.27) for $R = \infty$. Since $|k_+|^2 + |k_-|^2 = |k|^2$, we obtain from equation (3.3.31) (by $S(k,t)$ we denote the right hand side of the equation)

$$\left|u_t + u(k,t)\right| \le S(k,t), \quad \left\|S(\cdot,t)\right\| \le \|u\|_\infty(t)\,.$$

It remains to show that $\|u\|_\infty(t) \le \|u\|_\infty(0)$. We multiply this equation by $\exp(t)$ and obtain after integration

$$u(k,t) = u(k,0)e^{-t} + \int_0^t d\tau S(k,\tau)e^{-(t-\tau)}\,.$$

Hence, we get the estimate

$$r(t) \le r(0)e^{-t} + \int_0^t d\tau e^{-(t-\tau)}\,r(\tau)\,, \quad r(t) = \|u\|_\infty(t)\,.$$

It is easy to verify that this inequality is equivalent to

$$I'(t) \le I(t), \quad I(t) = \int_0^t d\tau e^{-\tau}\left[r(\tau) - r(0)\right].$$

Then it becomes obvious that $I(t) \le 0$ and therefore $I'(t) \le 0$. Hence, $r(t) \le r(0)$ and this completes the proof. □

The uniqueness of solution to the problem (3.3.1)–(3.3.4) follows from Lemma 3.3.2 even for true Maxwell molecules; see [143] for details. It is also shown in [143] that the functional (3.3.30) can be considered as a distance between corresponding solutions of the Boltzmann equation.

Thus we can prove the existence and uniqueness (at least in the class of probability measures) of generalized solutions of the Cauchy problem (3.3.1)–(3.3.2). Note that the construction of the Wild sum (3.3.10) or the similar sum for $\varphi(k,t)$ assumes only integrability of $f_0(v) \ge 0$. If the second moment of f_0 is bounded, then we expect convergence of $f(v,t)$ to the Maxwellian for large t. Otherwise, the asymptotic behavior will be different (see Sections 5.6 and 5.7), but still the solution of the problem (3.3.1)–(3.3.2) exists for all $t > 0$.

Our main task in this chapter is the detailed study of the solution $f(v,t)$ and related properties of the Boltzmann equation.

3.4 Invariant transformations

We consider equation (3.3.15) and study some transformations of $\varphi(k,t)$ keeping the equation invariant. In particular, these are the following transformations, which map the set of characteristic functions to itself:

1. a time shift,
$$\tilde{\varphi}(k,t;\tau) = e^{\tau L^{(1)}}\varphi = \varphi(k,t+\tau), \quad L^{(1)} = \partial_t;$$
2. multiplication by $\exp(ik\cdot v_0)$, $v_0 \in \mathbb{R}^3$,
$$\tilde{\varphi}(k,t;v_0) = e^{v_0\cdot L^{(2)}}\varphi, \quad L^{(2)} = ik;$$
3. rotation in $\mathbb{R}^3$,
$$\tilde{\varphi}(k,t;\beta) = e^{i\beta\cdot L^{(3)}}\varphi, \quad L^{(3)} = k\times\partial_k, \quad \beta\in\mathbb{R}^3;$$
4. scaling transformation,
$$\tilde{\varphi}(k,t;y) = e^{yL^{(4)}}\varphi = \varphi(e^y k,t), \quad L^{(4)} = k\cdot\partial_k\,;$$
5. multiplication by $\exp(-\theta|k|^2/2)$, $\theta\geq 0$,
$$\tilde{\varphi}(k,t;\theta) = e^{\theta L^{(5)}}\varphi, \quad L^{(5)} = -|k|^2/2. \tag{3.4.1}$$

Using the terminology of group analysis (see, e. g., [124]) we can say that the operators $L^{(i)}$, $i=1,\dots,5$, are "admissible" for equation (3.3.15). The list of corresponding operators, admissible for the Boltzmann equation (3.3.1), reads

$$L^{(1)} = \partial_t, \quad L^{(2)} = \partial_v, \quad L^{(3)} = v\times\partial_v, \quad L^{(4)} = \partial_v\cdot v, \quad L^{(5)} = \frac{1}{2}\Delta_v. \tag{3.4.2}$$

The first four operators are almost obvious for the Boltzmann equation. The operators $L^{(1)}$, $L^{(2)}$, and $L^{(3)}$ correspond, respectively, to shifts in t, shifts in v, and rotations in v. The operator $L^{(4)}$ corresponds to the transformation

$$\tilde{f}(v,t;y) = e^{yL^{(4)}}f = e^{3y}f(e^y v,t), \tag{3.4.3}$$

which preserves the L^1-norm. These are standard one-parameter Lie groups of transformations of independent variables.

However, the operator $L^{(5)}$ is unusual. The explicit transformation of $f(v,t)$ reads

$$\tilde{f}(v,t;\theta) = e^{\frac{\theta}{2}\Delta_v}f(v,t) = f(v,t)*M_\theta(v), \quad \theta\geq 0,$$
$$f_1(v)*f_2(v) = \int_{\mathbb{R}^3} dw f_1(w)f_2(v-w), \quad M_\theta(v) = \frac{\exp(-|v|^2/2\theta)}{(2\pi\theta)^{3/2}}. \tag{3.4.4}$$

These transformations form a one-parameter semi-group, where stars denote the convolution in $\mathbb{R}^3$. They were firstly discovered by Morgenstern [117, 118]. He found the proof of corresponding invariance of the Boltzmann equation (3.3.1) in the two-dimensional case and only for the constant kernel $g = \text{const.}$ in three dimensions. The transition to the Fourier representation makes this invariance obvious for arbitrary kernel $g(\mu)$ in (3.3.1).

The transformation (3.4.4) with any $\theta > 0$ allows to construct a monotone functional

$$H_\theta(f) = H[\tilde{f}(v,t;\theta)] = \langle \tilde{f}, \log \tilde{f} \rangle, \quad \theta > 0, \tag{3.4.5}$$

on any non-negative solution $f(v,t)$ of the Boltzmann equation (3.3.1). These functionals are defined also on generalized solutions of equation (3.3.1).

It is also possible to construct a class of invariant transformations which preserve the normalization conditions (3.3.4)–(3.3.5). We obtain for characteristic functions (equation (3.3.15))

$$\begin{aligned} \varphi_\delta(k,t) &= e^{\delta L_*}\varphi(k,t) = \varphi(ke^{-\delta},t)e^{-\frac{|k|^2}{2}(1-e^{-2\delta})}, \\ L_* &= -|k|^2 - k\cdot\partial_k, \quad 0 \le \delta < \infty. \end{aligned} \tag{3.4.6}$$

The related transformation for distribution functions (equation (3.3.1)) reads

$$\begin{aligned} f_\delta(v,t) &= e^{\delta L}f(v,t) = \int_{\mathbb{R}^3} \frac{dw}{[2\pi T(\delta)]^{3/2}} \exp\left[-\frac{|v-we^{-\delta}|^2}{2T(\delta)}\right], \\ T(\delta) &= 1-e^{-2\delta}, \quad L = \Delta_v + \partial_v\cdot v, \quad 0 \le \delta < \infty. \end{aligned} \tag{3.4.7}$$

The invariance of (3.3.15) under transformation (3.4.6) can easily be proved by direct substitution. These transformations will be used below for construction of a class of invariant solutions of the Boltzmann equation. A detailed analysis of its group properties for general interaction potentials can be found in [28, 46].

3.5 Linearized collision operator

We consider the problem (3.3.1)–(3.3.2) with initial data $f_0(v)$ satisfying the conditions (3.3.5). In addition, we assume that

$$f_0(v) = M(v)[1+\varepsilon F_0(v)], \quad 0 < \varepsilon \ll 1, \tag{3.5.1}$$

in the notation of equation (3.3.6). Then we look for the solution in the form

$$f(v,t) = M(v)[1+\varepsilon F(v,t)] \tag{3.5.2}$$

and, neglecting quadratic in ε terms, obtain the linearized equation

$$F_t = LF = \int_{\mathbb{R}^3\times S^2} dw d\omega\, M(w) g(\hat{u}\cdot\omega)\, [F(v') + F(w') - F(v) - F(w)], \quad F|_{t=0} = F_0(v), \tag{3.5.3}$$

in the notation of equations (3.3.1) and (3.3.6). Note that

$$\langle F_0, 1\rangle = \langle F_0, |v|^2\rangle = 0, \quad \langle F_0, v\rangle = 0. \tag{3.5.4}$$

Our aim in this section is to study some properties of the linearized collision operator L from equations (3.5.3). We can repeat the same steps in the Fourier representation and consider the solution

$$\varphi(k,t) = e^{-\frac{|k|^2}{2}}[1 + \varepsilon\Phi(k,t)], \quad 0 < \varepsilon \ll 1, \tag{3.5.5}$$

of equation (3.3.15). Then the linearized equation for $\Phi(k,t)$ reads

$$\Phi_t = \hat{L}\Phi = \int_{S^2} d\omega g(\omega\cdot\hat{k})[\Phi(k_+) + \Phi(k_-) - \Phi(k) - \Phi(0)],$$

$$\hat{k} = \frac{k}{|k|}, \quad k_\pm = \frac{k \pm |k|\omega}{2}, \quad \omega\in S^2; \quad \Phi|_{t=0} = \Phi_0(k), \tag{3.5.6}$$

where $\Phi_0(k)$ satisfies conditions

$$\Phi|_{k=0} = \Delta\Phi|_{k=0} = 0, \quad \partial_k\Phi|_{k=0} = 0. \tag{3.5.7}$$

Assuming that $g(\mu)\ge 0$ is integrable on $[-1,1]$, we represent $\hat{L}\Phi$ in the form

$$\hat{L}\Phi(k) = \hat{L_+}\Phi(k) - \|g\|_{L^1}[\Phi(k) + \Phi(0)], \quad \|g\|_{L^1} = 2\pi\int_{-1}^{1} d\mu g(\mu),$$

$$\hat{L}^+\Phi(k) = \int_{S^2} d\omega G(\omega\cdot\hat{k})\Phi\left(\frac{k - |k|\omega}{2}\right), \quad \hat{k} = \frac{k}{|k|},$$

$$G(\mu) = g(\mu) + g(-\mu), \quad \mu\in[-1,1], \tag{3.5.8}$$

where we do not assume that $\|g\|_{L^1} = 1$. We will need the following identity (a generalization of equation (2.1.5)).

Lemma 3.5.1. *The identity*

$$\int_{S^2} d\omega\, G(\omega\cdot\hat{k})\Phi\left(\frac{k - |k|\omega}{2}\right) = 2\int_{S^2} dn\, |\hat{k}\cdot n|\, G[1 - 2(\hat{k}\cdot n)^2]\, \Phi[(k\cdot n)n] \tag{3.5.9}$$

holds for any pair of functions $G(\mu)$, $|\mu|\le 1$, and $\Phi(k)$, $k\in\mathbb{R}^3$, provided both integrals are convergent.

Proof. Consider identity (2.1.5) for a function $\tilde{F}(v) = F(-v/2)$, $v \in \mathbb{R}^3$, and assume that $u = \omega_0 \in S^2$. Then $|u| = 1$ and we obtain

$$\int_{S^2} d\omega\, F\left(\frac{\omega_0 - \omega}{2}\right) = 2\int_{S^2} dn\, |\omega_0 \cdot n|\, F[(\omega_0 \cdot n)n],$$

where tildes are omitted. We apply this identity to the function

$$F(v) = G(1 - 2|v|^2)\Phi(|k|v).$$

Noting that

$$1 - 2\left|\frac{\omega_0 - \omega}{2}\right|^2 = \omega_0 \cdot \omega,$$

we obtain

$$\int_{S^2} d\omega G(\omega_0 \cdot \omega)\Phi\left(|k|\frac{\omega_0 - \omega}{2}\right)$$
$$= 2\int_{S^2} dn\, |\omega_0 \cdot n|\, G[1 - 2(\omega_0 \cdot n)^2]\, \Phi[|k|(\omega_0 \cdot n)n].$$

This is precisely the identity (3.5.9) for $k = |k|\omega_0$. Hence, the proof is completed. □

By using this lemma we can transform the operator $\hat{L}^+$ in the following way:

$$\hat{L}^+\Phi(k) = 4\int_{S_+^2} dn(\hat{k} \cdot n)\, \{g[1 - 2(\hat{k} \cdot n)^2]$$
$$+ g[2(\hat{k} \cdot n)^2 - 1]\}\, \Phi[(k \cdot n)n], \quad S_+^2 = \{n \in S^2 : \hat{k} \cdot n \geq 0\}. \tag{3.5.10}$$

We assume that $\Phi(k)$ is bounded on any sphere $|k| = R$. Then we can use the series

$$\Phi(k) = \sum_{l=0}^{\infty} \sum_{m=-l}^{l} \Phi_{lm}(|k|)\, Y_{lm}(\hat{k}), \tag{3.5.11}$$

where $Y_{lm}(\hat{k})$ are spherical functions (see, e. g., [94]). We use only a few properties of these functions:

(A)

$$\int_{S^2} d\omega Y_{l_1 m_1}(\omega) Y^*_{l_2 m_2}(\omega) = \delta_{l_1 l_2}\delta_{m_1 m_2},$$

where $l_{1,2} = 0, 1, \ldots,$ $|m_{1,2}| \leq l_{1,2}$, and the star denotes the complex conjugate;

(B)

$$P_l(\omega_1 \cdot \omega_2) = \frac{4\pi}{2l+1} \sum_{m=-l}^{l} Y_{lm}(\omega_1) Y^*_{lm}(\omega_2), \quad \omega_{1,2} \in S^2, \quad l = 0, 1, \dots,$$

where $P_l(\mu)$ are Legendre polynomials [94];

(C)

$$\int_{-1}^{1} d\mu P_{l_1}(\mu) P_{l_2}(\mu) = \frac{2}{2l_1+1} \delta_{l_1 l_2}, \quad l_{1,2} = 0, 1, \dots.$$

We introduce the linear integral operator B acting on variable $\omega \in S^2$:

$$Bf(\omega) = \int_{S^2} dn\, b(\omega \cdot n) f(n),$$

where $b(\mu)$, $|\mu| \le 1$, is a given function. Then it is easy to prove the following identity.

Lemma 3.5.2. *We have*

$$BY_{lm}(\omega) = \int_{S^2} dn b(\omega \cdot n) Y_{lm}(n) = b_l Y_{lm}(\omega),$$

$$b_l = 2\pi \int_{-1}^{1} b(\mu) P_l(\mu), \quad l = 0, 1, \dots, \quad |m| \le l, \quad \omega \in S^2, \tag{3.5.12}$$

provided

$$\|b\|^2_{L^2([-1,1])} = \int_{-1}^{1} d\mu \, |b(\mu)|^2 < \infty. \tag{3.5.13}$$

Proof. Since $b \in L^2([-1, 1])$, we can expand $b(\mu)$ in the series

$$b(\mu) = \sum_{l=0}^{\infty} c_l P_l(\mu), \quad c_l = \frac{2l+1}{2} \int_{-1}^{1} d\mu \, b(\mu) P_l(\mu).$$

Then we obtain by using (B) and (A)

$$\begin{aligned} BY_{lm}(\omega) &= \sum_{l_1=0}^{\infty} c_{l_1} \int_{S^2} dn P_{l_1}(\omega \cdot n) Y_{lm}(n) \\ &= 4\pi \sum_{l_1=0} \frac{c_{l_1}}{2l_1+1} Y_{l_1 m}(\omega) \delta_{l_1 l}(n) = \frac{4\pi c_l}{2l+1} Y_{lm}(\omega), \quad l = 0, 1, \dots, \quad |m| \le l. \end{aligned}$$

This completes the proof. □

The identity (3.5.12) plays an important role in linear transport theory [68]. The condition (3.5.13) can be weakened. In fact the identity (3.5.12) holds for any $b \in L^1([-1,1])$.

Coming back to equations (3.5.10)–(3.5.11), we obtain

$$\begin{aligned} \hat{L}^+\Phi(k) &= \sum_{l=0}^{\infty}\sum_{m=-l}^{l} \hat{A}_{lm}(|k|)Y_{lm}(\hat{k}), \\ \hat{A}_{lm}(|k|)Y_{lm}(\hat{k}) &= 4\int_{S_+^2} dn\,(\hat{k}\cdot n)b(\hat{k}\cdot n)\Phi_{lm}[|k|(\hat{k}\cdot n)]Y_{lm}(n), \\ b(\mu) &= g(1-2\mu^2) + g(2\mu^2-1). \end{aligned} \tag{3.5.14}$$

Then we apply Lemma 3.5.2 with $\omega = \hat{k}$ and get the following series:

$$\hat{L}^+\Phi(k) = \sum_{l=0}^{\infty}\sum_{m=-l}^{l} Y_{lm}(\hat{k})\hat{L}_l^+\Phi_{lm}(|k|), \tag{3.5.15}$$

where operators $\hat{L}_l^+$ are given by equalities

$$\hat{L}_l^+\Psi_{lm}(|k|) = 8\pi\int_0^1 d\mu\mu b(\mu)P_l(\mu)\Psi(\mu|k|), \quad l = 0,1,\dots. \tag{3.5.16}$$

Note that

$$8\pi\int_0^1 d\mu\mu b(\mu) = 8\pi\int_0^1 d\mu\mu[g(1-2\mu^2)+g(2\mu^2-1)] = 4\pi\int_{-1}^1 d\mu g(\mu) = 2\|g\|_{L^1},$$

in the notation of equations (3.5.8). Finally, we obtain the following representation of the linearized operator $\hat{L}$ (3.5.8). Since $\hat{L}1 = 0$, it is enough to consider such functions $\Phi(k)$ so that $\Phi(0) = 0$.

Theorem 3.5.1. *If $g \in L^1([-1,1])$, $\Phi(k)$ is bounded in any ball $|k| \le R$ in $\mathbb{R}^3$, and $\Phi(0) = 0$, then the operator $\hat{L}$ reads as*

$$\hat{L} = \sum_{l=0}^{\infty} \hat{L}_l \Pi_l, \tag{3.5.17}$$

where Π_l are projectors

$$\Pi_l\Phi(k) = \frac{2l+1}{4\pi}\int_{S^2} dn\,P_l(\hat{k}\cdot n)\,\Phi(|k|n), \quad l = 0,1,\dots, \tag{3.5.18}$$

and $\hat{L}_l$ are the operators acting only on variable $|k|$:

$$\hat{L}_l\Psi(|k|) = 2\pi\int_0^\pi d\theta\,\sin\theta\, g(\cos\theta)\Big[P_l\Big(\cos\frac{\theta}{2}\Big)\Psi\Big(|k|\cos\frac{\theta}{2}\Big) + P_l\Big(\sin\frac{\theta}{2}\Big)\Psi\Big(|k|\sin\frac{\theta}{2}\Big) - \Psi(|k|)\Big]. \quad (3.5.19)$$

Proof. The proof follows directly from equations (3.5.8) and (3.5.14)–(3.5.16). Two different parts of the operator $\hat{L}_l$ are obtained by substitution of $\mu = \cos\theta/2$ or $\mu = \sin\theta/2$ into corresponding parts of $\hat{L}_l^+$ (3.5.16). The formula for Π_l follows from the property (B) of spherical functions. This completes the proof. □

3.6 Eigenfunctions and eigenvalues

Coming back to the Cauchy problem (3.5.6), we can look for its solution in the form

$$\Phi(k,t) = \sum_{l=0}^{\infty}\sum_{m=-l}^{l}\Phi_{lm}(|k|,t)Y_{lm}(\hat{k}),$$
$$\Phi_{lm}(|k|,0) = \Phi(|k|)_{lm}^{(0)} = \int_{S^2} d\omega\Phi_0(|k|\omega)Y_{lm}^*(\omega), \quad l = 0,1,\ldots, \quad |m| \le l. \quad (3.6.1)$$

Assuming that $g(\mu)$ and $\Phi_0(|k|)$ satisfy conditions of Theorem 3.5.1, we can reduce the problem (3.5.6) to an infinite set of independent one-dimensional problems for $\Phi_{lm}(|k|,t)$,

$$\partial_t\Phi_{lm} = \hat{L}_l\Phi_{lm}, \quad \Phi(|k|)_{lm}|_{t=0} = \Phi_{lm}^{(0)}(|k|), \quad (3.6.2)$$

in the notation of equations (3.5.19).

We begin our consideration with the simplest class of particular solutions, namely,

$$\Phi_l(r,t)e^{\lambda_l t} = \Psi_l(r), \quad r = |k|, \quad (3.6.3)$$

where $\Psi_l(r)$ satisfies the equation

$$\begin{aligned}\lambda_l\Psi_l(r) &= \hat{L}_l\Psi_l(r)\\ &= 2\pi\int_0^\pi d\theta\,\sin\theta\, g(\cos\theta)\Big[P_l\Big(\cos\frac{\theta}{2}\Big)\Psi_l\Big(r\cos\frac{\theta}{2}\Big) + P_l\Big(\sin\frac{\theta}{2}\Big)\Psi_l\Big(r\sin\frac{\theta}{2}\Big) - \Psi_l(r)\Big],\end{aligned}$$
$$r \ge 0, \quad \Psi_l(0) = 0; \quad l = 0,1,\ldots. \quad (3.6.4)$$

It is easy to guess the following solution of this eigenvalue problem:

$$\Psi_l(r) = r^p,$$

$$\lambda_l(p) = 2\pi \int_0^{\pi} d\theta \, \sin\theta \, g(\cos\theta)$$

$$\times \left[P_l\left(\cos\frac{\theta}{2}\right)\cos^p\frac{\theta}{2} + P_l\left(\sin\frac{\theta}{2}\right)\sin^p\frac{\theta}{2} - 1 \right], \quad p > 0; \quad l = 0, 1, \dots . \tag{3.6.5}$$

The corresponding result for the operator $\hat{L}_l$ from (3.5.6) is formulated below.

Lemma 3.6.1. *The eigenvalue problem*

$$\lambda\Phi(k) = \hat{L}\Phi(k) \tag{3.6.6}$$

for the operator $\hat{L}$ *from* (3.5.6) *has the following solutions:*

$$\lambda = \lambda_l(p), \quad \Phi_{lm}(k; p) = |k|^p \, Y_{lm}\left(\frac{k}{|k|}\right),$$

$$p > 0, \quad l = 0, 1, \dots, \quad m = -l, \dots, l, \tag{3.6.7}$$

where $\lambda_l(p)$ *are given in equations* (3.6.5).

Proof. It is sufficient to combine equations (3.6.5) with the representation equations (3.5.17) of $\hat{L}$. Then the result follows. □

Of course, equations (3.6.5)–(3.6.7) are valid also for complex values of p with $\Re p > 0$. However, for our goals it is sufficient to consider only real $p > 0$.

The next step is to re-formulate the results of Lemma 3.6.1 in terms of the linearized Boltzmann collision operator L given in (3.6.3). It is easy to see that functions $F(v, t)$ and $\Phi(k, t)$, satisfying equations (3.5.3) and (3.5.6), respectively, are connected by transformation

$$e^{-\frac{|k|^2}{2}} \Phi(k, t) = (2\pi)^{3/2} \langle e^{-\frac{|v|^2}{2}} F(v, t), \, e^{-ik\cdot v} \rangle, \quad k \in \mathbb{R}^3, \tag{3.6.8}$$

in the notation of equation (2.1.9).

The same transformation is valid for eigenfunctions of linear operators L and $\hat{L}$. If $F(v)$ is a solution of the eigenvalue problem

$$\lambda F(v) = L\,F(v), \tag{3.6.9}$$

then the function

$$\Phi(k) = (2\pi)^{-3/2} \int_{\mathbb{R}^3} dv F(v) \exp\left[-\frac{|v + ik|^2}{2} \right] \tag{3.6.10}$$

satisfies equation (3.6.6) with the same eigenvalue λ. The usual inverse Fourier transform (see, e. g., [85]) leads to the equality

$$F(v) = (2\pi)^{-3/2} \int_{\mathbb{R}^3} dk\, \Phi(k) \exp\left[-\frac{|k - iv|^2}{2}\right], \tag{3.6.11}$$

provided that the integral converges absolutely. This condition is obviously satisfied for eigenfunctions (3.6.7) of $\hat{L}$. Hence, the corresponding eigenfunctions of L have the following form:

$$F_{lm}(v; p) = (2\pi)^{-3/2} \int_{\mathbb{R}^3} dk\, |k|^p\, Y_{lm}\left(\frac{k}{|k|}\right) \exp\left[-\frac{|k - iv|^2}{2}\right],$$
$$l = 0, 1, \dots, \quad |m| \le l, \quad p > 0.$$

Then we use Lemma 3.5.2 and obtain

$$F_{lm}(v; p) = \tilde{F}_l(|v|; p)\, Y_{lm}\left(\frac{v}{|v|}\right), \tag{3.6.12}$$

where

$$\tilde{F}_l(r; p) = (2\pi)^{-1/2} e^{r^2/2} \int_0^\infty dx x^{p+2} e^{-x^2/2} \int_{-1}^1 d\mu P_l(\mu)\, e^{ixr\mu}.$$

Note that constant factors are unimportant for eigenfunctions $F_{lm}(v, p)$. Therefore, such factors are denoted below by C. We use the tables from [94] and obtain

$$\int_{-1}^1 d\mu e^{ixr\mu}\, P_l(\mu) = C(xr)^{-1/2} J_{l+1/2}(xr),$$

where $J_{l+1/2}(z)$ is the Bessel function. Hence, we obtain a one-dimensional integral for $F_l(r; p)$. We use the tables from [94] once more and obtain, omitting constant factors,

$$F_l(r; p) = r^l\, {}_1F_1\left(\frac{l-p}{2}, l + \frac{3}{2}, \frac{r^2}{2}\right), \quad l = 0, 1, \dots, \tag{3.6.13}$$

where ${}_1F_1(\alpha, \gamma, z)$ is the degenerate or, equivalently, confluent hypergeometric function [94]. We note that

$${}_1F_1(\alpha, \gamma, z) = \sum_{k=0}^\infty \frac{(\alpha)_k}{(\gamma)_k} \frac{z^k}{k!}, \quad (\alpha)_k = \alpha(\alpha + 1) \dots (\alpha + k - 1) = \frac{\Gamma(\alpha + k)}{\Gamma(\alpha)}.$$

Here and below, $\Gamma(\cdot)$ denotes the gamma function [94]. Since $\gamma = l + 3/2$, the series converges for all complex z. If

$$\alpha = \frac{l-p}{2} = -n \Leftrightarrow p = 2n + l, \quad n = 0, 1, \dots,$$

then the series reduces to the Laguerre polynomial $L_n^{l+1/2}(z)$:

$$ {}_1F_1(-n, l+3/2, z) = CL_n^{l+1/2}(z), \quad l \geq 0. $$

We are interested only in real positive values of z. Then the following asymptotic formula is valid [94] for $\alpha \neq 0, -1, \ldots$:

$$ {}_1F_1(\alpha, \gamma, x) = \frac{\Gamma(\gamma)}{\Gamma(\alpha)} e^x x^{\alpha-\gamma} \left[1 + O\left(\frac{1}{x}\right) \right], \quad x \geq 1. $$

Hence,

$$ {}_1F_1\left(\frac{l-p}{2}, l + \frac{3}{2}, \frac{r^2}{2}\right) \approx C\, e^{\frac{r^2}{2}} r^{-l-(p+3)}, \quad r \to \infty, $$

and therefore

$$ F_l(r;p) \approx C e^{\frac{r^2}{2}} r^{-(p+3)}, \quad r \to \infty; \quad p - l \neq 0, 2, 4, \ldots. $$

Hence, the following statement is proved.

Theorem 3.6.1.

(i) *The general solution of the eigenvalue problem* (3.6.9) *reads*

$$ \begin{aligned} F(v) &= F_{lm}(v;p) = |v|^l\, {}_1F_1\left(\frac{l-p}{2}, l + \frac{3}{2}, \frac{|v|^2}{2}\right) Y_{lm}\left(\frac{v}{|v|}\right), \\ \lambda &= \lambda_l(p), \quad l = 0, 1, \ldots, \quad |m| \leq l; \quad p > 0, \end{aligned} \tag{3.6.14} $$

where the eigenvalues $\lambda_l(p)$ are given in equations (3.6.5).

(ii) *There is an exceptional set of discrete values of p, namely,*

$$ p_{nl} = 2n + l; \quad n, l = 0, 1, \ldots, \tag{3.6.15} $$

such that

$$ F_{lm}(v; 2n+l) = C_{n,l} L_n^{l+1/2}\left(\frac{|v|^2}{2}\right) |v|^l\, Y_{lm}\left(\frac{v}{|v|}\right), \tag{3.6.16} $$

where $L_n^{l+1/2}(z)$ are Laguerre polynomials and C_{nl} are irrelevant constant factors. The corresponding eigenvalues are

$$ \begin{aligned} \lambda_{ln} = 2\pi \int_0^\pi d\theta \sin\theta g(\cos\theta) \Big[& P_l\left(\cos\frac{\theta}{2}\right) \cos^{2n+l}\frac{\theta}{2} \\ & + P_l\left(\sin\frac{\theta}{2}\right) \sin^{2n+l}\frac{\theta}{2} - 1 - \delta_{n0}\delta_{l0} \Big], \quad n, l = 0, 1, \ldots. \end{aligned} \tag{3.6.17} $$

(iii) *If $p > 0$ and $p \neq 2n + l$, $n, l = 0, 1, \ldots$, then the following asymptotic formula holds:*

$$F_{lm}(v; p) = C_l(p)\, Y_{lm}\left(\frac{v}{|v|}\right)|v|^{-(p+3)} e^{|v|^2/2}$$
$$\times \left[1 + O\left(\frac{1}{|v|}\right)\right], \quad |v| \to \infty, \tag{3.6.18}$$

where $C_l(p)$ are irrelevant constant factors.

Remark. Eigenfunctions (3.6.13) satisfy the condition $F(0) = 0$. Corresponding eigenvalues $\lambda_l(p)$ are given in equations (3.6.5). However, there is also an eigenfunction $F(v) = 1$, which corresponds to the eigenvalue $\lambda = 0$. Therefore, we need to correct the formula for $\lambda_l(p)$ in the case $l = p = 0$. This obvious correction is made in equations (3.6.17) for discrete eigenvalues.

Properties of functions $\lambda_l(p)$ for complex values of p are discussed in Section 3.7. It should be mentioned that discrete eigenvalues (3.6.17) and polynomial eigenfunctions (3.6.16) were firstly published in 1950s by Wang-Chang and Uhlenbeck [150]. The other eigenvalues and eigenfunctions were found in 1975 by using the Fourier transform in [17]. Our considerations in this section were based on the assumption that the kernel $g(\mu)$ is absolutely integrable in $[-1, 1]$. However, it is clear that Theorem 3.6.1 remains valid under a weaker assumption of absolute integrability of $g(\mu)(1 - \mu)$. For brevity we omit the proof.

3.7 General solution of the linearized equation

Thus far, we found two different classes of eigenfunctions: polynomials (3.6.16) and exponentially increasing at infinity functions (3.6.13) and (3.6.17). How can we explain this difference? We consider the operator L given in equations (3.5.3) with kernel $g(\mu) \in L^1([-1, 1])$. It is convenient to introduce two functional spaces: the weighted L^2-space H (Hilbert space) with norm

$$\|\Phi\|_H = \left(\int_{\mathbb{R}^3} dv e^{-|v|^2/2} |\Phi(v)|^2\right)^{1/2} < \infty \tag{3.7.1}$$

and the weighted L^1-space B (Banach space) with norm

$$\|\Phi\|_B = \int_{\mathbb{R}^3} dv e^{-|v|^2/2} |\Phi(v)| < \infty. \tag{3.7.2}$$

It is easy to verify that the integral operator L is a bounded operator, acting from H to H, or from B to B. On the other hand, we have an obvious estimate

$$\|\Phi\|_B = \int_{\mathbb{R}^3} dv e^{-|v|^2/2} |\Phi(v)| \leq C\|\Phi\|_H, \quad C = \text{const}. \tag{3.7.3}$$

Hence, $H \subset B$. It was proved in Section 3.3 that the non-linear Boltzmann equation (3.3.1) with $g(\mu) \in L^1_+([-1,1])$ has a unique solution for any initial data $f_0(v) \in L^1_+(\mathbb{R}^3)$. Here and below, the index "+" is used for the subset of all non-negative functions. Therefore, we can conclude from the definition (3.5.2) of $F(v,t)$ that the most general and natural assumption for the initial data $F_0(v)$ is $F_0 \in B$. Thus, we consider L as a bounded operator acting in B. We also have in mind that L has an invariant subspace $H \subset B$. Sometimes it is convenient to use "smaller" Banach spaces B_k with norm

$$\|\Phi\|_{B_k} = \int_{\mathbb{R}^3} dv e^{-|v|^2/2}(1+|v|^2)^{k/2}|\Phi(v)|, \quad k > 0, \tag{3.7.4}$$

related to the norm (3.3.11) for distribution functions. It is easy to check that

$$B_0 = B, \quad H \subset B_{k_1} \subset B_{k_2} \quad \text{if} \quad k_1 > k_2 \geq 0. \tag{3.7.5}$$

The linearized Boltzmann collision operator L (3.5.3) is often considered in the literature (not only for Maxwell molecules; see, e. g., [70]) as acting only in Hilbert space H with the scalar product

$$(\varphi, \psi) = \int_{\mathbb{R}^3} dv e^{-|v|^2/2}\varphi(v)\psi(v)\,. \tag{3.7.6}$$

Then we can consider the bilinear form $(L\varphi, \psi)$ and obtain after standard transformations (see Section 2.2) the following identity:

$$\begin{aligned}(L\varphi, \psi) = -\frac{1}{4} \int_{\mathbb{R}^3\times\mathbb{R}^3\times S^2} & dv dw d\omega e^{-\frac{|v|^2+|w|^2}{2}} g(\hat{u}\cdot\omega) \\ & \times [\varphi(v') + \varphi(w') - \varphi(v) - \varphi(w)]\,[\psi(v') + \psi(w') - \psi(v) - \psi(w)],\end{aligned} \tag{3.7.7}$$

valid for any $\varphi \in H$ and $\psi \in H$. Hence, the operator $L : H \to H$ is self-adjoint and non-positive, i. e.,

$$(L\varphi, \psi) = (L\psi, \varphi), \quad (L\varphi, \varphi) \leq 0. \tag{3.7.8}$$

Its null-space $\operatorname{Ker} L$ is also known:

$$L\varphi = 0 \Leftrightarrow \varphi \in \operatorname{Ker} L, \quad \operatorname{Ker} L = \operatorname{Span}(1, v, |v|^2). \tag{3.7.9}$$

In other words, any solution $\varphi \in H$ to equation $L\varphi = 0$ is a linear combination of 1, v, and $|v|^2$. These are the well-known properties of the linearized collision operator acting in the Hilbert space H_l. They are valid for any reasonable kernel $g(|u|, \mu)$, but in this chapter we are interested only in the case of Maxwell molecules. Then we can use Theorem 3.6.1 and conclude that $L : H \to H$ has a discrete set of eigenvalues (3.6.17)

$\{\lambda_{nl};\ n,l = 0,1,\dots\}$. The corresponding eigenfunctions (3.6.15) form an orthogonal basis $\{\varphi_{nlm};\ n,l = 0,1,\dots;\ |m| \le l\}$ in H with respect to the scalar product (3.7.6). Hence, the evolution equation (3.5.6) for L with initial data from H can be easily solved by series in orthogonal polynomial eigenfunctions.

At the same time the norm (3.7.1) is very restrictive. Indeed, equality (3.5.2) with $F(v,t) \in H$ implies that

$$\int_{\mathbb{R}^3} dv f^2(v,t) e^{|v|^2/2} < \infty, \tag{3.7.10}$$

where $f(v,t)$ is a solution of the non-linear Boltzmann equation (3.3.1).

In fact, this inequality can be violated at any given $t_0 > 0$ even if it was satisfied at $t = 0$, as we shall see in Section 5.2. Therefore, our aim is to construct some more or less explicit representation of solution $F(v,t)$ to the problem (3.5.3), where $F_0(v) \in B$. For the sake of generality we do not use the assumption (3.5.4) at the first stage of our consideration.

Thus we consider the problem (3.5.3) and pass to the Fourier representation by transformation

$$\Phi(k,t) = \int_{\mathbb{R}^3} dv F(v,t) \exp[-|v + ik|^2/2)], \quad k \in \mathbb{R}^3. \tag{3.7.11}$$

Note that

$$|\Phi(k,t)| \le \|F(\cdot,\ t)\|_B e^{|k|^2/2}, \tag{3.7.12}$$

in the notation of equation (3.7.2). Since $L : B \to B$ is a bounded operator, we have

$$\|F(\cdot,t)\|_B \le e^{t\|L\|_B}\ \|F_0(\cdot)\|_B,$$

where

$$\|L\|_B = \sup \|L\Phi\|_B / \|\Phi\|_B, \tag{3.7.13}$$

in the notation of equations (3.7.2) and (3.5.3). Hence, $\Phi(k,t)$ is bounded for any $k \in \mathbb{R}^3$ and therefore we can use the expansion (3.5.11) and obtain a sequence of independent Cauchy problems for $\Phi_{lm}(|k|,t)$:

$$\partial_t \Phi_{lm} = \hat{L}_l \Phi_{lm} - \delta_{l0} \Phi_{00}(0,t), \quad l = 0,1,\dots, \quad |m| \le l, \tag{3.7.14}$$

where it is assumed that

$$2\pi \int_{-1}^{1} d\mu g(\mu) = 1, \quad \Phi_{lm}(|k|,0) = \Phi_{lm}^{(0)}(|k|). \tag{3.7.15}$$

Operators $\hat{L}_l$ are defined in equations (3.5.19). It is straightforward to show that

$$\Phi_{00}(r,t) = \Phi_{00}^{(0)}(0) + \Psi_{00}(r,t)\,e^{-t}, \quad r = |k| \geq 0;$$
$$\Phi_{lm}(r,t) = \Psi_{lm}(r,t)\,e^{-t}, \quad l \geq 1, \quad |m| \leq l, \tag{3.7.16}$$

where $\Psi_{lm}(r,t)$ satisfy equations

$$\partial_t \Psi_{lm}(r) = 2\pi \int_0^{\pi} d\theta \sin\theta\, g(\cos\theta) \Big[P_l\Big(\cos\frac{\theta}{2}\Big)\Psi_{lm}\Big(r\cos\frac{\theta}{2}\Big) + P_l\Big(\sin\frac{\theta}{2}\Big)\Psi_{lm}\Big(r\sin\frac{\theta}{2}\Big)\Big] \tag{3.7.17}$$

and initial conditions

$$\Psi_{00}|_{t=0} = \Psi_{00}^{(0)}(r) - \Phi_{00}^{(0)}(0)\,; \quad \Psi_{lm}|_{t=0} = \Psi_{lm}^{(0)}(r), \tag{3.7.18}$$

provided $l \geq 1$. It remains to solve problems (3.7.17)–(3.7.18).

We begin with the simplest case $l = m = 0$. Then we omit indices and obtain after obvious transformation

$$\Psi_t = 2\pi \int_0^{\pi} d\theta \sin\theta [g(\cos\theta) + g(-\cos\theta)] \Psi\Big(r\sin\frac{\theta}{2}\Big) = \int_0^1 ds K(s)\Psi(rs),$$
$$K(s) = 8\pi s[g(1-2s^2) + g(2s^2-1)], \quad \Psi|_{t=0} = \Psi^{(0)}(r). \tag{3.7.19}$$

It is convenient to use the substitution $s = \exp(-\tau)$, $\tau \geq 0$. Then we obtain

$$\hat{K}\Psi(r) = \int_0^1 ds K(s)\Psi(rs) = \int_0^{\infty} d\tau a(\tau)\Psi(re^{-\tau}), \quad a(\tau) = e^{-\tau}K(e^{-\tau}). \tag{3.7.20}$$

Note that formally

$$\Psi(re^{-\tau}) = \exp(-\tau\hat{D})\Psi(r), \quad \hat{D} = r\frac{d}{dr}.$$

Therefore,

$$\hat{K}\Psi(r) = \Big[\int_0^{\infty} d\tau a(\tau) e^{-\tau\hat{D}}\Big]\Psi(r) = A(\hat{D})\Psi(r), \quad A(p) = \mathcal{L}(a) = \int_0^{\infty} d\tau a(\tau) e^{-p\tau},$$

where $\mathcal{L}(a)$ denotes the Laplace transform of $a(\tau)$, $\tau \geq 0$.

In other words, the operator K is the analytic function of the differential operator $\hat{D} = r\frac{d}{dr}$. It is easy to check the rule of multiplication of such operators: if

$$A_i(\hat{D}) = \int_0^{\infty} d\tau a_i(\tau) e^{-\tau\hat{D}}, \quad i = 1, 2,$$

then

$$A_3(\hat{D}) = A_1(\hat{D})A_2(\hat{D}) = \int_0^\infty d\tau a_3(\tau)e^{-\tau\hat{D}},$$

where

$$a_3(\tau) = a_1(\tau) * a_2(\tau) = \int_0^\tau d\tau' a_1(\tau')a_2(\tau - \tau').$$

Hence, it is clear that the solution of the problem (3.7.19) reads

$$\Psi(r,t) = \int_0^\infty d\tau b(t,\tau)\Psi_0(re^{-\tau}),$$

where $b(t,\tau)$ satisfies equations

$$b_t(t,\tau) = b(t,\tau) * a(\tau) = \int_0^\tau d\tau' b(t,\tau')a(\tau - \tau'), \quad b|_{t=0} = \delta(\tau),$$

and $a(\tau)$ is given in equations (3.7.20). It is not difficult to find $b(t,\tau)$ by using the Laplace transform. Indeed, we can take any $p > 0$ and choose the initial condition $\Psi_0(r)$ in the form of the eigenfunction r^p of the operator $\hat{L}_0$ in (3.7.14). Then we obtain

$$\Psi(r,t) = r^p \int_0^\infty d\tau b_0(t,\tau)e^{-pt} = r^p e^{[\Lambda(p)+1]t}, \quad p > 0,$$

or, equivalently,

$$\Phi_{00}(r,t) = r^p \int_0^\infty d\tau b_0(t,\tau)e^{-pt} = r^p e^{\Lambda(p)t},$$

for the solution of equations (3.7.14) with $l = m = 0$. We can repeat similar considerations for arbitrary $l > 0$, $|m| \le l$ and obtain the following result.

Theorem 3.7.1. *The general solution of the problem* (3.7.14)–(3.7.15) *reads*

$$\begin{aligned} \Phi_{00}(r,t) &= \Phi_{00}^{(0)}(0) + \hat{B}_0[\Phi_{00}^{(0)}(r) - \Phi_{00}^{(0)}(0)], \\ \Phi_{lm}(r,t) &= \hat{B}_l\Phi_{lm}^{(0)}(r), \quad l = 1,\dots, \quad |m| \le l, \end{aligned} \tag{3.7.21}$$

where

$$\hat{B}_l\Phi(r) = \int_0^\infty d\tau b_l(t,\tau)\Phi(re^{-\tau}),$$

$$b_l(t,\tau) = \frac{1}{2\pi i}\int_{a-i\infty}^{a+i\infty} d\tau e^{p\tau+\lambda_l(p)t}, \quad l = 0,1,\dots, \quad a > 0, \tag{3.7.22}$$

in the notation of equations (3.6.5).

Proof. The proof is already given above. The formula for $b_l(t,\tau)$ follows from the usual inversion formula for the Laplace transform. This completes the proof. □

Note that eigenvalues $\lambda_l(p)$ (3.6.5) are analytic in the half-plane $\Re p \geq 0$. Here and below, $\Re p$ and $\Im p$ denote, respectively, real and imaginary parts of $p \in \mathbb{C}$. It follows from equations (3.6.5) that

$$\Re \lambda_l(p) = 2\pi \int_0^\pi d\theta \sin\theta\, g(\cos\theta) \Big[P_l\Big(\cos\frac{\theta}{2}\Big) \cos^{p_1}\frac{\theta}{2} + P_l\Big(\sin\frac{\theta}{2}\Big) \sin^{p_1}\frac{\theta}{2} - 1 \Big] \leq \Re \lambda_0(p), \quad p_1 = \Re p,$$

since $P_l(\mu) \leq 1$, $|\mu| \leq 1$. On the other hand,

$$\Re \lambda_0(p) = 2\pi \int_{-1}^{1} d\mu g(\mu) \Big[\Big(\frac{1+\mu}{2}\Big)^{p_1/2} + \Big(\frac{1-\mu}{2}\Big)^{p_1/2} - 1 \Big].$$

It is easy to check that

$$\begin{aligned} \Re\lambda_l(p) \leq \Re \lambda_0(p) < 0 \quad &\text{if} \quad \Re p > 2; \\ \Re \lambda_0(p) > 0 \quad &\text{if} \quad 0 \leq \Re p < 2; \\ \Re \lambda_0(p) = 0 \quad &\text{if} \quad \Re p = 2. \end{aligned}$$

For brevity we consider below only the case $l = 0$ (radial solutions), $p \geq 0$. The corresponding eigenfunctions (3.6.13)

$$F_{00}(v; p) = {}_1F_1(-p/2, 3/2, |v|^2/2)$$

behave for large velocities like

$$F_{00}(v; p) \approx C_0 |v|^{-(p+3)} \exp(|v|^2/2), \quad |v| \to \infty.$$

Therefore, the perturbation $F(v,t)$ of the distribution function in (3.5.2) has an infinite second moment (energy) for $1 \leq p < 2$. Of course, the Maxwellian $M(v)$ is unstable with respect to such perturbations and that is why $\lambda_0(p) > 0$ for $p < 2$. On the other hand, eigenfunctions with $p = 2 + \varepsilon$, $0 < \varepsilon \ll 1$, show the examples of initial perturbation in (3.5.2), whose relaxation to the Maxwellian can be very slow. Such examples will be considered in more detail for the non-linear Boltzmann equation in Section 5.6.

Finally, we perform the inverse Fourier transform and derive the general formula for radial solutions of the Cauchy problem (3.5.3), where

$$F_0(v) = \tilde{F}_0(r), \quad r = |v|.$$

Note that solutions $F(v,t)$ and $\Phi(k,t)$ of equations (3.5.3) and (3.5.6), respectively, are connected by transformation (3.6.10), (3.6.11) at any fixed $t \geq 0$. We use Theorem 3.7.1 and obtain the following formula for $\Phi(k,t)$:

$$\Phi(k,t) = \int_0^\infty d\tau b_0(t,\tau)\Phi_0(|k|e^{-\tau}),$$

provided $\Phi_0(0) = 0$. The kernel $b_0(t,\tau)$ is given in (3.7.22). Note that

$$\Phi_0(|k|) = (2\pi)^{-3/2} \int_{\mathbb{R}^3} dv F_0(|v|) \exp[-|v+ik|^2/2].$$

Then we combine these two formulas and apply to $\Phi(k,t)$ the inverse Fourier transform (3.6.11). The integration can be performed explicitly and we finally obtain

$$F(v,t) = \int_{\mathbb{R}^3} dw F_0(|w|)\, R(v,w,t)\,, \tag{3.7.23}$$

where

$$R(v,w,t) = \left[2\pi(1-e^{-2\tau})\right]^{-3/2} \int_0^\infty d\tau b_0(t,\tau) \exp\left[-\frac{|w-ve^{-\tau}|^2}{2(1-e^{-2\tau})}\right],$$

$$b_0(t,\tau) = \mathcal{L}^{-1}[e^{\lambda(p)t}] = \frac{1}{2\pi i}\int_{a-i\infty}^{a+i\infty} dp\, e^{p\tau+\lambda(p)t},$$

$$\lambda(p) = 2\pi \int_{-1}^{1} d\mu g(\mu)\left[\left(\frac{1+\mu}{2}\right)^{p/2} + \left(\frac{1-\mu}{2}\right)^{p/2} - 1\right].$$

At the end of this section we consider an explicit example. If

$$g(\mu) = \frac{1}{4\pi} = \text{const.},$$

then

$$\lambda(p) = \frac{2-p}{2+p}, \quad b_0(t,\tau) = e^{-(t+2\tau)}\,\mathcal{L}^{-1}_{p\to\tau}(e^{4t/p}).$$

Hence,

$$\begin{aligned} b_0(t,\tau) &= e^{-(t+2\tau)}\left[\delta(\tau) + 4t\sum_{n=0}^{\infty}\frac{(4t/p)^{n+1}}{(n+1)!}\right] \\ &= e^{-(t+2\tau)}\left[\delta(\tau) + 2\sqrt{\frac{t}{\tau}}I_1(2\sqrt{t\tau})\right], \end{aligned}$$

where $I_1(z)$ is the modified Bessel function [94]. Therefore,

$$R(v,w,t) = e^{-t}\left[\delta(w) + \sqrt{\frac{t}{2\pi^3}}\, R_1(v,w,t)\right],$$

$$R_1(v,w,t) = \int_0^\infty \frac{d\tau}{\sqrt{\tau}}\, T^{-3/2}(\tau)\, I_1(2\sqrt{t\tau})\, \exp\left[-\frac{\tau}{2} - \frac{|w - ve^{-\tau}|^2}{2T(\tau)}\right],$$

where $T(\tau) = 1 - \exp(-2\tau)$. The integral (3.7.23) with this formula for $R(v,w,t)$ can be used for various asymptotic expansions of the radial solution $F(|v|,t)$.

Of course, a similar approach can be used for the general case $l \ge 1$ in equations (3.7.21) and (3.7.22) (see [20, 27] for details). We also note that the general solution of the Cauchy problem (3.5.6) can be written in the exponential form as

$$\Phi(k,t) = e^{tL}\,\Phi_0(k), \quad e^{tL} = \sum_{l=0}^{\infty} e^{t\lambda(k\cdot\partial_k)}\Pi_l\,, \tag{3.7.24}$$

in the notation of equations (3.5.18) and (3.6.5). The projectors Π_l, $l \ge 0$, act only on the direction of $k \in \mathbb{R}$, whereas the differential operator $k \cdot \partial_k$ acts only on $|k|$. The analytic for $\Re p > 0$ functions $\lambda_l(p)$ (see (3.6.5)) can be reduced to Laplace-type integrals. Similarly, the corresponding exponential operators in the sum (3.7.24) can be explicitly expressed through inverse Laplace transform (3.7.22). In principle the result of this section can be deduced from general theory of semi-groups of operators (see, e. g., [97]), but we prefer to use a more elementary direct approach.

In the next section we come back to the non-linear Boltzmann equation and consider power moments of its solution.

3.8 Equations for moments

We consider below the Cauchy problem (3.3.1)–(3.3.2) and its Fourier representation (3.3.15)–(3.3.16). The normalization conditions (3.3.4) and (3.3.17) are also assumed. If $f(v,t) \ge 0$ and

$$\langle f(v,t), |v|^n \rangle < \infty$$

for any $n \ge 0$, then we formally obtain the following identity:

$$\varphi(k,t) = \langle f(|v|,t), e^{-ik\cdot v}\rangle = \sum_{n=0}^{\infty} \frac{(-i)^n}{n!} \langle f, (k\cdot v)^n\rangle\,. \tag{3.8.1}$$

Hence,

$$\varphi(k,t) = \sum_{n=0}^{\infty} \frac{(-i)^n}{n!}\, m^{(n)}_{\alpha_1\ldots\alpha_n}(t) \prod_{j=1}^{n} k_{\alpha_j}\,, \tag{3.8.2}$$

where the summation over repeating Greek indices $\alpha_j = 1,2,3$ is assumed and

$$m^{(n)}_{\alpha_1\ldots\alpha_n}(t) = \left\langle f(v,t), \prod_{j=1}^{n} v_{\alpha_j} \right\rangle, \quad n = 1,2,\ldots. \tag{3.8.3}$$

This is the well-known connection between tensor moments of $f(v)$ and coefficients of Taylor series of its Fourier representation $\varphi(k)$.

The remarkable property of Maxwell molecules (or Maxwell models) is that the moments of the collision integral can be explicitly expressed through the moments of the distribution function. The simplest way to derive equations for moments (3.8.3) of the solution $f(v,t)$ of the problem (3.3.1)–(3.3.2) is to consider the Fourier transformed equation (3.3.15). Then we substitute the series (3.8.2) into this equation and collect the terms of orders $n = 0,1,\ldots$ in $|k|$. The conservation laws and the normalization conditions (3.3.4) yield

$$m^{(0)} = 1; \quad m^{(1)}_{\alpha} = 0, \quad \alpha = 1,2,3; \quad m^{(2)}_{\alpha\alpha} = \operatorname{tr}(m^{(2)}) = 3. \tag{3.8.4}$$

For the sake of brevity we denote

$$m^{(n)}_{\alpha_1\ldots\alpha_n} \prod_{j=1}^{n} k_{\alpha_j} = m^{(n)} \cdot k^{(n)}, \quad n \geq 1, \tag{3.8.5}$$

and obtain for $n \geq 2$

$$\frac{d}{dt} m^{(n)} \cdot k^{(n)} = \prod_{l=1}^{n-1} \binom{n}{l} I(l, n-l) - m^{(n)} y^{(n)}, \tag{3.8.6}$$

where

$$I(l,p) = \int_{S^2} d\omega g(\omega \cdot k)(m^{(l)} \cdot k_+^{(l)})\,(m^{(p)} \cdot k_-^{(p)}),$$

$$y^{(n)} = \int_{S^2} d\omega g(\omega \cdot k)[k^{(n)} - k_+^{(n)} - k_-^{(n)}],$$

$$k_{\pm} = \frac{k \pm |k|\omega}{2}, \quad l \geq 1, \quad p \geq 1, \quad n \geq 2. \tag{3.8.7}$$

Since $m^{(1)} = 0$, we have $I(1,p) = I(l,1) = 0$. Hence, we obtain linear equations

$$\frac{d}{dt} m^{(n)} \cdot k^{(n)} = -m^{(n)} \cdot y^{(n)} \tag{3.8.8}$$

for $n = 2,3$. Coming back to usual tensor notations, we obtain after simple calculations

$$y^{(2)}_{\alpha\beta} = \frac{1}{2} \int_{S^2} d\omega g(\omega \cdot \hat{k})\,(k_\alpha k_\beta - |k|^2 \omega_\alpha \omega_\beta).$$

Using the standard representation of isotropic symmetric tensor function of $k \in \mathbb{R}^d$ in the form

$$y^{(2)}_{\alpha\beta} = A(|k|)k_\alpha k_\beta + B(|k|)\delta_{\alpha\beta},$$

we obtain

$$\operatorname{Tr} y^{(2)} = y^{(2)}_{\alpha\alpha} = A|k|^2 + 3B = 0, \quad y^{(2)}_{\alpha\beta}\, k_\alpha k_\beta = A|k|^4 + B|k|^2 = \frac{b|k|^4}{2},$$

where

$$b = 2\pi \int_{-1}^{1} d\mu g(\mu)(1-\mu^2)\,. \tag{3.8.9}$$

Hence,

$$\frac{d}{dt} m^{(2)}_{\alpha\beta} k_\alpha k_\beta = -\frac{3b}{4} m^{(2)}_{\alpha\beta}\left(k_\alpha k_\beta - \frac{|k|^2}{3}\delta_{\alpha\beta}\right).$$

Since $m^{(2)}_{\alpha\alpha} = 3$, we obtain

$$m^{(2)}_{\alpha\beta} = \delta_{\alpha\beta} + g_{\alpha\beta}(t), \quad g_{\alpha\alpha} = 0\,. \tag{3.8.10}$$

Hence,

$$\left[g'_{\alpha\beta}(t) + \frac{3b}{4} g_{\alpha\beta}(t)\right] k_\alpha k_\beta = 0,$$

and therefore

$$g_{\alpha\beta}(t) = g_{\alpha\beta}(0)\exp\left(-\frac{3bt}{4}\right). \tag{3.8.11}$$

Similarly we can solve equation (3.8.8) for $n = 3$. It is easy to see that

$$y^{(3)}_{\alpha\beta\gamma} = \frac{1}{2}[k_\alpha\, y^{(2)}_{\beta\gamma} + k_\beta\, y^{(2)}_{\alpha\gamma} + k_\gamma\, y^{(2)}_{\alpha\beta}], \quad \alpha, \beta, \gamma = 1, 2, 3.$$

Hence,

$$y^{(3)}_{\alpha\beta\gamma} = \frac{b}{8}\,[9\, k_\alpha k_\beta k_\gamma - |k|^2\,(k_\alpha \delta_{\beta\gamma} + k_\beta \delta_{\alpha\gamma} + k_\gamma \delta_{\alpha\beta})]\,.$$

Then we denote

$$M_\alpha = m^{(3)}_{\alpha\beta\beta}, \quad \hat{m}^{(3)}_{\alpha\beta\gamma} = m^{(3)}_{\alpha\beta\gamma} - \frac{1}{5}(M_\alpha \delta_{\beta\gamma} + M_\beta \delta_{\alpha\gamma} + M_\gamma \delta_{\alpha\beta})\,. \tag{3.8.12}$$

The tensor $\hat{m}$ is called an irreducible component of $m^{(3)}$ [92]. Then we obtain from (3.8.8) the following equation:

$$k_\alpha k_\beta k_\gamma \left[\hat{m}'_{\alpha\beta\gamma}(t) + \frac{9b}{8} \hat{m}_{\alpha\beta\gamma}(t) \right] + \frac{3}{5}|k|^2 k_\alpha \left[M'_\alpha(t) + \frac{b}{2} M_\alpha(t) \right] = 0 .$$

Obviously, the functions

$$M_\alpha(t) = M_\alpha(0) \exp\left(-\frac{bt}{2}\right), \quad \hat{m}_{\alpha\beta\gamma}(t) = \hat{m}_{\alpha\beta\gamma}(0) \exp\left(-\frac{9bt}{8}\right) \tag{3.8.13}$$

satisfy the equation. The resulting formula for tensor $m^{(3)}(t)$ follows from (3.8.13). Explicit formulas for moments (3.8.3) of any given order $n \geq 4$ can be found in a similar way.

It is easy to see that the exponents in (3.8.11) and (3.8.13) coincide with corresponding eigenvalues λ_{nl} (3.6.16) of the linearized operator:

$$\lambda_{02} = -\frac{3}{4} b, \quad \lambda_{11} = -\frac{1}{2} b, \quad \lambda_{03} = -\frac{9}{8} b .$$

This is not surprising because the characteristic function $\varphi(k, t)$ (3.8.1) is the solution of equation (3.3.15), satisfying conditions (3.3.16) and (3.3.17). The substitution

$$\varphi(k,t) = \psi(k,t) \exp\left(-\frac{|k|^2}{2}\right),$$

$$\psi(k,t) = 1 + \sum_{n=2}^{\infty} \frac{(-i)^n}{n!} u^{(n)}_{\alpha_1 \cdots \alpha_n}(t) \prod_{j=1}^{n} k_{\alpha_j}, \quad u^{(2)}_{\alpha\alpha} = 0 , \tag{3.8.14}$$

does not change equation (3.3.15). Hence, the equations for tensor coefficients $u^{(n)}(t)$ are formally the same as equations (3.8.6) and (3.8.7) for tensor moments $m^{(n)}(t)$, $n \geq 2$. On the other hand, equations for $u^{(2,3)}(t)$ are linear (see (3.8.8)). Therefore, the evolution of lower moments of order $n < 4$ follows the same law in the fully non-linear case.

The series (3.8.14) is related to the classical Grad moment method [92]. Grad's idea was to look for solutions to the Boltzmann equation in the form (for brevity we consider only the spatially homogeneous case here)

$$f(v,t) = \frac{1}{(2\pi)^{3/2}} e^{-|v|^2/2} \left[1 + \sum_{n=2}^{\infty} a^{(n)}(t) \cdot H^{(n)}(v) \right], \tag{3.8.15}$$

where the notations like in (3.8.5) for symmetric tensors are used. The tensor Hermite polynomials $H^{(n)}(v)$ read as

$$H^{(n)}_{\alpha_1 \cdots \alpha_n}(v) = (-1)^n e^{|v|^2/2} \frac{\partial^n}{\partial v_{\alpha_1} \cdots \partial v_{\alpha_n}} e^{-|v|^2/2},$$

$$n = 1, \ldots; \quad \alpha_1 \cdots \alpha_n = 1, 2, 3. \tag{3.8.16}$$

It is easy to check that

$$\varphi(k,t) = \langle f, e^{-ik\cdot v}\rangle = e^{-|k|^2/2}\left[1 + \sum_{n=2}^{\infty} \frac{(-i)^n}{n!} u^{(n)}(t)\cdot k^{(n)}\right],$$

$$a^{(n)} = n!\, u^{(n)}, \quad n \geq 2.$$

Hence, the series (3.8.14) is formally equivalent to the Grad series (3.8.15).

It is often more convenient to use expansion in eigenfunctions of the linearized collision operator. Then we obtain

$$f(v,t) = \frac{1}{(2\pi)^{3/2}} e^{-|v|^2/2} \sum_{n=0}^{\infty}\sum_{l=0}^{\infty}\sum_{m=-l}^{\infty} b_{nlm}(t)\,|v|^l \mathcal{L}_n^{l+1/2}\left(\frac{|v|^2}{2}\right) Y_{lm}\left(\frac{v}{|v|}\right),$$

in the notation of (3.6.15). The Fourier transform yields

$$\begin{aligned}\varphi(k,t) &= \langle f, e^{-ik\cdot v}\rangle \\ &= \sum_{n=0}^{\infty}\sum_{l=0}^{\infty}\sum_{m=-l}^{\infty} b_{nlm}(t) \frac{(-i|k|)^{2n+l}}{(2n)!!} Y_{lm}\left(\frac{k}{|k|}\right).\end{aligned}$$

Then we can substitute the series into equation (3.3.15) and obtain ordinary differential equations (ODEs) for coefficients. It is always assumed in this section that the initial data $\varphi(k,0)$ satisfy the normalization conditions (3.3.17).

First we use the orthogonality of spherical functions, and get

$$\begin{aligned}&\sum_{n=0}^{\infty} \frac{(-i|k|)^{2n+l}}{(2n)!!} [b'_{nlm}(t) - \lambda_{nl} b_{nlm}(t)] \\ &= \sum_{\substack{n_1,n_2,l_1,l_2=0 \\ (2n_1+l_1)(2n_2+l_2)>0}}^{\infty} \frac{(-1)^{n_1+n_2} i^{l_1+l_2}}{(2n_1)!!\,(2n_2)!!} \sum_{m_1=-l_1}^{l_1}\sum_{m_2=-l_2}^{l_2} b_{n_1 l_1 m_1}(t) \\ &\quad \cdot b_{n_2 l_2 m_2}(t) \int_{S^2} d\omega Y^*_{lm}(\omega)\, F(|k|\omega), \quad n \geq 0, \quad l \geq 0, \quad |m| \leq l,\end{aligned} \tag{3.8.17}$$

where eigenvalues λ_{nl} are given in (3.6.17) and

$$\begin{aligned}F(|k|,\omega) &= \int_{S^2} dn g(\omega\cdot n) |k_+|^{2n_1+l_1} |k_-|^{2n_2+l_2} Y_{l_1 m_1}(\hat{k}_+) Y_{l_2 m_2}(\hat{k}_-), \\ k_\pm &= |k|\frac{\omega \pm n}{2}, \quad \hat{k}_\pm = \frac{k_\pm}{|k_\pm|}.\end{aligned} \tag{3.8.18}$$

Note that

$$|k_\pm|^2 = |k|^2\left(\frac{1 \pm \omega\cdot n}{2}\right).$$

Hence, we can collect terms with the same powers of $|k|$ in equations (3.8.17) and obtain a set of ODEs

$$\begin{aligned} & b'_{nlm}(t) - \lambda_{nl}\, b_{nlm}(t) \\ & \quad = (1 - \delta_{n0}\delta_{l0}) \sum_{n_1 m_1 l_1} \sum_{n_2 m_2 l_2} \delta_{2(n_1+n_2)+l_1+l_2,\, 2n+l} \\ & \qquad \times A(n, l, m | n_1, l_1, m_1;\ n_2, l_2, m_2)\, b_{n_1 l_1 m_1}(t)\, b_{n_2 l_2 m_2}(t)\,, \\ & n \geq 0, \quad l \geq 0, \quad |m| \leq l\,, \end{aligned} \tag{3.8.19}$$

where the limits in sums are the same as in (3.8.17),

$$\begin{aligned} & A(n, l, m | n_1, l_1, m_1;\ n_2, l_2, m_2) \\ & \quad = \frac{(2n)!!(1 - \delta_{2n_1+l_1,\, 2n+l})\,(1 - \delta_{2n_2+l_2,\, 2n+l})}{(2n_1)!!\,(2n_2)!!} \\ & \qquad \times K_{n,n_1,n_2}(l, m | l_1, m_1;\ l_2, m_2)\,, \\ & K_{n,n_1,n_2}(l, m | l_1, m_1;\ l_2, m_2) \\ & \quad = \int_{S^2 \times S^2} d\omega dn\, g_{n_1 n_2}(\omega \cdot n) \left[\frac{1 + \omega \cdot n}{2}\right]^{l_1/2} \left[\frac{1 - \omega \cdot n}{2}\right]^{l_2/2} \\ & \qquad \times Y^*_{lm}(\omega) Y_{l_1 m_1}\left(\frac{\omega + n}{|\omega + n|}\right) Y_{l_2 m_2}\left(\frac{\omega - n}{|\omega - n|}\right), \end{aligned} \tag{3.8.20}$$

$$g_{n_1 n_2}(\mu) = g(\mu) \left[\frac{1 + \mu}{2}\right]^{n_1} \left[\frac{1 - \mu}{2}\right]^{n_2}. \tag{3.8.21}$$

These integrals can be significantly simplified and expressed through Clebsch–Gordan coefficients and some relatively simple integrals. It was firstly done by V. Vedenyapin [145] (see also the paper [95] by E. M. Henriks and T. M. Nieuwenhuizen). We evaluate the integrals (3.8.20) below on the basis of the famous Wigner–Eckart theorem on matrix elements of spherical tensor linear operators (see formula (21.19) in Wigner's book [154]). This is done in Appendix A (see Section 3.8.1). Here we just mention that the integrals $K_{n,n_1,n_2}(l, m \,|\, l_1, m_1;\ l_2, m_2)$ have non-zero values only for such indices that

$$|l_1 - l_2| \leq l \leq l_1 + l_2\,, \quad m = m_1 + m_2\,, \quad l_1 + l_2 + l_3 = 0, 2, 4, \ldots.$$

It is also assumed in (3.8.20) that $|m| \leq l$, $|m_{1,2}| \leq l_{1,2}$.

Finally, we briefly discuss a way of solution of ODEs (3.8.19). Note that

$$\lambda_{00} = \lambda_{01} = \lambda_{02} = 0$$

and $\lambda_{nl} < 0$ for other values of n and l. Moreover, it is assumed that the initial data are such that

$$b_{000} = 1, \quad b_{01\pm1} = b_{010} = 0, \quad b_{100} = 0\,. \tag{3.8.22}$$

We consider an increasing sequence of values $p = 2m + l$. The solution of equations (3.8.1) for $p \leq 1$ is obvious from (3.8.22). For $p = 2$ we have two options: (a) $n = 1$, $l = 0$; (b) $n = 0$, $l = 2$. Since $b_{100} = 0$, we need to consider only the second option. Then it is easy to see that the right hand side of equations (3.8.19) is equal to zero, as it follows from (3.8.22). The same is true for $p = 3$. Hence, equations (3.8.19) for $2n+l \leq 3$ are linear. Their solutions read as

$$b_{nlm}(t) = b_{nlm}(0)e^{\lambda_{nl}t}, \quad |m| \leq l; \quad 2n + l \leq 3.$$

Obviously, we have obtained the same result as before (see (3.8.11) and (3.8.13)), when the tensor form of the moment system had been used. It is clear that the right hand side of ODEs (3.8.19) depends only on such coefficients $b_{n'l'm'}(t)$ that $2n' + l' < 2n + l$. Therefore, we can construct the solution of equations (3.8.19) for $p = 2n + l \geq 4$ in recursive form:

$$b_{nlm}(t) = b_{nlm}(0)e^{\lambda_{nl}t} + \int_0^t d\tau S_{nlm}(\tau)e^{\lambda_{nl}(t-\tau)},$$

where $S_{nlm}(t)$ denote the right hand side of equations (3.8.19). This formula allows in principle to find any coefficient $b_{nlm}(t)$ in explicit form, though the sums become too big and cumbersome for large values of n and l. On the other hand, there is an important class of solutions for which the complete set of moment equations become relatively simple and therefore very useful. These are radial (or, equivalently, isotropic) solutions, which will be considered in the next chapter.

3.8.1 Appendix A. Spherical functions, the Wigner–Eckart theorem, and evaluation of some integrals

We introduce a Cartesian coordinate system in $\mathbb{R}^3$ and use standard notations for vectors $x = (x_1, x_2, x_3) \in \mathbb{R}^3$. We also introduce spherical coordinates (r, θ, φ) such that $x_1 = r\sin\theta\cos\varphi$, $x_2 = r\sin\theta\sin\varphi$, $x_3 = r\cos\theta$; $r \in \mathbb{R}_+$, $\theta \in [0, \pi]$, $\varphi \in [0, 2\pi)$. Unit vectors, or, equivalently, points of the unit sphere S^2, will be denoted below by letters ω, n, e or by their spherical coordinates (θ, φ). Suppose that all points $x \in \mathbb{R}^3$ undergo a rotation $x' = Rx$ about the origin $x = (0, 0, 0)$, which remains at rest. We are interested mainly in the rotations of the unit sphere. Clearly, $|x'| = |Rx|$ and therefore

$$n' = Rn \in S^2 \quad \text{if} \quad n \in S^2.$$

It is natural to introduce corresponding operators G_R acting on functions f:

$$S^2 \to C \quad \text{such that} \quad G_R f(\omega) = f(R\omega).$$

We recall the following well-known property of spherical functions $\{Y_{lm}(\omega),\ \omega \in S^2,\ l = 0,1,\dots;\ |m| \le l\}$ (see, e. g., [94]). For any fixed $l \ge 1$ and any fixed rotation R, there exists a quadratic matrix $D^{(l)}(R)$ of order $(2l+1)$ such that

$$G_R Y_{lm}(\omega) = Y_{lm}(R\omega) = \sum_{m=-l}^{l} D^{(l)}_{mm'}(R)\, Y_{lm'}(\omega), \quad l = 1,\dots;\ |m| \le l.$$

The same is true for $l = 0$ since $Y_{00}(\omega) = \text{const}$. Hence, the $(2l+1)$-dimensional set $\Lambda^{(l)} = \text{Span}\{Y_{lm}(\omega),\ m = -l,\dots,l\}$ is the linear subspace of $L^2(S^2)$ that is invariant under rotations.

We are often interested in properties of the various operators with respect to rotations. For example, let us consider the simplest linear operator L_φ of multiplication acting in $L^2(S^2) : L_\varphi f(\omega) = \varphi(\omega) f(\omega)$. Then we perform a rotation R and obtain

$$G_R L_\varphi f(\omega) = \varphi(R\omega) f(R\omega) = \varphi(R\omega) G_R f(\omega)\,.$$

Since $f(\omega)$ is any function, we can obtain the corresponding equality for operators:

$$G_R L_\varphi = \varphi(R\omega) G_R \implies L' = G_R L_\varphi G_R^{-1} = L_{\varphi'}\,,$$

where $\varphi' = \varphi(R\omega)$. In other words the operator L_φ of multiplication by $\varphi(\omega)$ after the rotation R becomes the operator $L_{\varphi'}$ of multiplication by $\varphi'(\omega) = \varphi(R\omega)$.

Of course we can consider the case of the whole space $\mathbb{R}^3$, i. e., the multiplication by $\varphi(x)$, $x \in \mathbb{R}^3$. Then the same consideration shows that the operator L_φ is isotropic or, equivalently, invariant under rotation if $\varphi(x) = \tilde{\varphi}(|x|)$. This is the simplest example of an isotropic linear operator. Another example of that kind is the linearized Boltzmann collision operator considered in Sections 3.5–3.7.

Coming back to L_φ on the unit sphere we can choose

$$\varphi(\omega) = Y_{lm}(\omega), \quad \omega \in S^2, \quad l \ge 1, \quad |m| \le l\,.$$

Then we obtain

$$G_R L_\varphi f(\omega) = Y_{lm}(R\omega) f(R\omega) = \sum_{m'=-l}^{l} D^{l)}_{mm'}(R) Y_{lm'}(\omega) f(R\omega)\,.$$

If we denote $L_{Y_{lm}(\omega)} = T^{(l)}_m$, $m = -l,\dots,l$, then

$$T'^{(l)}_m = G_R T^{(l)}_m G_R^{-1} = \sum_{m'=-l}^{l} D^{l)}_{mm'}(R) T'^{(l)}_m\,. \tag{3.8.23}$$

The set of $(2l+1)$ linear operators $T^{(l)}_{-l},\dots,T^{(l)}_l$ acting in $L^2(S^2)$ and such that they are transformed under any rotation R in accordance with equation (3.8.23) are called spherical tensor operators. Sometimes in the literature [133] the formulas (3.8.23) are slightly different because they depend on the choice of G_R (it is possible to define it by the equality $G_R f(\omega) = f(R^{-1}\omega)$) and an exact definition of matrix elements $D^l_{mm'}(R)$. This, however, is not important for the Wigner–Eckart theorem. The only important point is that transformations of operators $\{T^{(l)}_m, \ |m| \le l\}$ under rotation must be the same as for almost trivial operators of multiplication by spherical functions $\{Y_{lm}(\omega), \ |m| \le l\}$. The following remarkable property of such operators is valid.

Theorem 3.8.1 (The Wigner–Eckart theorem). *The matrix elements of spherical tensor operators satisfy*

$$\int_{S^2} d\omega Y^*_{lm}(\omega)\, T^{(l_1)}_{m_1}\, Y_{l_2 m_2}(\omega)$$
$$= \langle l_1 l_2;\ m_1 m_2 | l_1 l_2; lm\rangle \frac{\langle l \| T^{(l_1)} \| l_2\rangle}{\sqrt{2l_2+1}}, \qquad m = m_1 + m_2, \tag{3.8.24}$$

where the double bar matrix element is independent of m_1, m_2, and m. The Clebsch–Gordan coefficients $\langle \cdots | \cdots \rangle$ are defined by equalities

$$\int_{S^2} d\omega Y^*_{lm}(\omega)\, Y_{l_1 m_1}(\omega)\, Y_{l_2 m_2}(\omega)$$
$$= \frac{(2l_1+1)(2l_2+1)}{4\pi(2l+1)}$$
$$\times \langle l_1 l_2;\ m_1 m_2 \,|\, l_1 l_2; lm\rangle\, \langle l_1 l_2;\ 00 \,|\, l_1 l_2; l0\rangle\, \delta_{m, m_1+m_2}\,. \tag{3.8.25}$$

The formulation of the theorem is almost identical to the one from the textbook of J. Sakurai [133], who calls it "one of the most important theorems in quantum mechanics." The reader can also find a proof there. The original proof of equation (3.8.24) (in the form of equation (21.19) of Wigner's book [154]) was probably given by E. Wigner in the first edition of that book in 1931. Anyway, we will need this formula only for one technical application. It is related to the set of ODEs (3.8.19) that includes some coefficients expressed through integrals (3.8.20). We shall see below how to apply the Wigner–Eckart theorem to these integrals.

Omitting irrelevant indices in (3.8.20), we fix two numbers $l_{1,2} \ge 0$ and consider integrals

$$K(l, m \,|\, l_1, m_1; l_2, m_2)$$
$$= \int_{S^2\times S^2} d\omega dn Y^*_{lm}(\omega)\, g(\omega\cdot n) |\omega_+|^{l_1} |\omega_-|^{l_2} Y_{l_1 m_1}\Big(\frac{\omega_+}{|\omega_+|}\Big) Y_{l_2 m_2}\Big(\frac{\omega_-}{|\omega_-|}\Big),$$
$$\omega_\pm = \frac{\omega \pm n}{2}; \quad |m_{1,2}| \le l_{1,2}\,. \tag{3.8.26}$$

These integrals for fixed $l_1 \geq 0$ can be considered as matrix elements of the tensor operator $T^{(l_1)}_{m_1} : L^2(S^2) \to L^2(S^2)$, namely,

$$T^{(l_1)}_{m_1}\varphi(\omega) = \int_{S^2} dn g_{l_1,l_2}(\omega \cdot n) Y_{l_1 m_1}\left(\frac{\omega + n}{|\omega + n|}\right)\varphi\left(\frac{\omega - n}{|\omega - n|}\right),$$

$$g_{l_1,l_2}(\omega \cdot n) = g(\mu)\left[\frac{1+\mu}{2}\right]^{l_1/2}\left[\frac{1-\mu}{2}\right]^{l_2/2}, \quad |m_1| \leq l_1 . \tag{3.8.27}$$

Then it is easy to check that $T^{(l_1)}_{m_1}$ is the spherical tensor operator. Indeed,

$$G_R T^{(l_1)}_{m_1}\varphi(\omega) = \int_{S^2} dn g(\omega' \cdot n) Y_{l_1 m_1}\left(\frac{\omega' + n}{|\omega' + n|}\right)\varphi\left(\frac{\omega' - n}{|\omega' - n|}\right),$$

where $\omega' = R\omega$. Then we change the integration variable to $n' = R^{-1}n$ and obtain by using the invariance of the scalar product

$$\begin{aligned} G_R T^{(l_1)}_{m_1}\varphi(\omega) &= \int_{S^2} dn' g(R\omega \cdot Rn') Y_{l_1 m_1}\left(\frac{R\omega + n'}{|\omega' + n'|}\right)\varphi\left(\frac{\omega - n'}{|\omega - n'|}\right) \\ &= \sum_{m_1'=-l_1}^{-l_1} D_{mm_1'}(R) T^{(l_1)}_{m_1'} G_R \varphi(\omega) . \end{aligned}$$

Hence, we can apply equations (3.8.27) and reduce in such a way the integrals (3.8.24) to simpler integrals.

There are several ways to do this. We can introduce standard integrals

$$A(l, m|l_1, m_1;\, l_2, m_2) = \int_{S^2} d\omega Y^*_{lm}(\omega) Y_{l_1 m_1}(\omega) Y_{l_2 m_2}(\omega) . \tag{3.8.28}$$

Then, in accordance with the Wigner–Eckart theorem we obtain the following formula for integral (3.8.2):

$$K(l, m|l_1, m_1;\, l_2, m_2) = \lambda^l_{l_1 l_2} A(l, m|l_1, m_1;\, l_2, m_2) , \tag{3.8.29}$$

where coefficients $\lambda^l_{l_1 l_2}$ do not depend on m_1, m_2, and m. By using the identity

$$\int_{S^2} d\omega F(\omega) = 2 \int_{\mathbb{R}^3} dx \delta(|x|^2 - 1) F(x)$$

we can transform the integral $K(\cdots)$ to

$$\begin{aligned} &K(l, m|l_1, m_1;\, l_2, m_2) \\ &= \int_0^\pi d\theta_0 \sin\theta_0 g(\cos\theta_0)\left(\cos\frac{\theta_0}{2}\right)^{l_1}\left(\sin\frac{\theta_0}{2}\right)^{l_1} R_{\theta_0}(l, m|l_1, m_1;\, l_2, m_2) , \end{aligned}$$

$$R_{\theta_0}(l,m|l_1,m_1;\ l_2,m_2) = \int\limits_{S^2\times S^2} d\omega dn\delta(\omega\cdot n)Y_{l_1m_1}(\omega)\,Y_{l_2m_2}(n)\,Y^*_{lm}(\omega_0)\,,$$
$$\omega_0 = \omega\cos\frac{\theta_0}{2} + n\sin\frac{\theta_0}{2}\,.$$

The straightforward calculations are omitted for the sake of brevity. Then we obtain

$$R_{\theta_0}(l,m|l_1,m_1;\ l_2,m_2) = r^l_{l_1l_2}(\theta_0)A(l,m|l_1,m_1;\ l_2,m_2) \tag{3.8.30}$$

and therefore

$$\lambda^l_{l_1l_2} = \int\limits_0^\pi d\theta_0\sin\theta_0 g(\cos\theta_0)\left(\cos\frac{\theta_0}{2}\right)^{l_1}\left(\sin\frac{\theta_0}{2}\right)^{l_1} r^l_{l_1l_2}(\theta_0)\,. \tag{3.8.31}$$

It remains to find functions $r^l_{l_1l_2}(\theta_0)$ for integers $l_{1,2} \geq 0$ and $|l_1 - l_2| \leq l \leq l_1 + l_2$. Noting that

$$Y_{l0}(\theta,\varphi) = \sqrt{\frac{2l+1}{4\pi}}P_l(\cos\theta)\,,$$

we consider equation (3.8.27) for $m = m_1 = m_2 = 0$ and obtain

$$\left[2\pi\int\limits_{-1}^1 d\mu P_{l_1}(\mu)P_{l_2}(\mu)P_l(\mu)\right] r^l_{l_1l_2}(\theta_0)$$
$$= I^l_{l_1l_2}(\theta_0) = \int\limits_{S^2\times S^2} d\omega dn\delta(\omega\cdot n)P_{l_1}(\omega\cdot n_0)P_{l_2}(n\cdot n_0)P_l(\omega_0\cdot n_0)\,,$$
$$\omega_0 = \omega\cos\frac{\theta_0}{2} + n\sin\frac{\theta_0}{2}\,, \tag{3.8.32}$$

where $n_0 \in S^2$ is an arbitrary unit vector chosen as a polar axis of the spherical coordinate system. Obviously, nothing depends on n_0. Hence, we can integrate this equation over $n_0 \in S^2$ and divide the result by 4π. Then the left hand side remains unchanged, whereas the right hand side reads as

$$I^l_{l_1l_2}(\theta_0) = \int\limits_{S^2\times S^2} d\omega dn\delta(\omega\cdot n)F_{\theta_0}(\omega,n)\,,$$
$$F_{\theta_0}(\omega,n) = \frac{1}{4\pi}\int\limits_{S^2} dn_0P_{l_1}(\omega\cdot n_0)P_{l_2}(n\cdot n_0)P_l(\omega_0\cdot n_0)\,. \tag{3.8.33}$$

It is easy to see that $F_{\theta_0}(\omega,n)$ is invariant under rotations of ω and n (for fixed θ_0) and therefore depends only on the scalar product $\omega{\cdot}n$. On the other hand, these two vectors are orthogonal in equations (3.8.30) because of the delta function. Hence,

$$F_{\theta_0}(\omega,n) = \tilde{F}_{\theta_0}(\omega,n) = \tilde{F}_{\theta_0}(0)$$

does not depend on ω and n. Therefore, we can perform integration in the first integral in (3.8.30) and obtain

$$I^l_{l_1 l_2}(\theta_0) = 8\pi^2 \tilde{F}_{\theta_0}(0) = 2\pi \int_{S^2} dn_0 P_{l_1}(\omega \cdot n_0) P_{l_2}(n \cdot n_0) P_l(\omega_0 \cdot n_0)\,,$$

$$\omega_0 = \omega \cos\frac{\theta_0}{2} + n \sin\frac{\theta_0}{2}\,, \quad \theta_0 \in [0, \pi], \tag{3.8.34}$$

where ω and n form any pair of orthogonal unit vectors. It is easy to check that

$$I^l_{l_1 l_2}(\theta_0) = I^l_{l_1 l_2}(\pi - \theta_0)\,. \tag{3.8.35}$$

Therefore, it is sufficient to consider the case $0 \le l_2 \le l_1$. Then we evaluate the integral in such coordinate system that Cartesian coordinates of ω, n, and ω_0 are

$$\omega = (0, 0, 1), \quad n = (1, 0, 1), \quad \omega_0 = \left(\sin\frac{\theta_0}{2},\, 0,\, \cos\frac{\theta_0}{2}\right).$$

The polar axis of spherical coordinates (θ, φ) is directed along ω; therefore,

$$\omega = (0, 0), \quad n = (\pi/2, 0), \quad \omega_0 = (\theta_0/2, 0)$$

in spherical coordinates.

To evaluate the integral (3.8.34) we apply the addition theorem (see property (B) from Section 3.5) to $P_{l_2}(n_0 \cdot \omega)$ and $P_l(n_0 \cdot \omega_0)$. Then the resulting integral reads

$$I^l_{l_1 l_2}(\theta_0) = \frac{32\pi^3}{(2l_2 + 1)(2l + 1)} \times \sum_{|m| \le l} \sum_{|m_2| \le l_2} Y_{lm}(\omega_0) Y_{l_2 m_2}(n) \int_{S^2} dn_0\, P_{l_1}(\omega \cdot n_0) Y_{l_2 m_2}(n_0) Y^*_{lm}(n_0)\,.$$

Denoting $n_0 = (\theta, \varphi)$ in spherical coordinates and

$$Y_{lm}(n_0) = \Phi_{lm}(\cos\theta)\, e^{im\varphi}\,, \quad \theta = \omega \cdot n_0\,,$$

we obtain

$$I^l_{l_1 l_2}(\theta_0) = \frac{32\pi^3}{(2l_2 + 1)(2l + 1)} \sum_{|m| \le \mathrm{Min}(l_1, l_2)} \Phi_{lm}\left(\cos\frac{\theta_0}{2}\right) \Phi_{l_2 m}(0) \times \sqrt{\frac{4\pi}{2l_1 + 1}} A(l, m | l_1, 0;\, l_2, m)\,,$$

in the notation of equations (3.8.28). On the other hand, in the same notation we obtain from (3.8.32)

$$I^l_{l_1 l_2}(\theta_0) = \sqrt{\frac{4\pi^3}{(2l_1 + 1)(2l_2 + 1)(2l + 1)}} A(l, 0 | l_1, 0;\, l_2, 0) r^l_{l_1 l_2}(\theta_0)\,. \tag{3.8.36}$$

The final formula for $r^l_{l_1 l_2}(\theta_0)$ reads

$$r^l_{l_1 l_2}(\theta_0) = \frac{8\pi^2}{\sqrt{(2l_2+1)(2l+1)}} \sum_{|m|\le \mathrm{Min}(l_1,l_2)} C^l_{l_1,l_2}(m)\Phi_{lm}\Big(\cos\frac{\theta_0}{2}\Big),$$

$$C^l_{l_1,l_2}(m) = \Phi_{lm}(0)\frac{A(l,m|l_1,0;\,l_2,m)}{A(l,0|l_1,0;\,l_2,0)},$$

$$0 \le l_2 \le l_1, \quad l_1 - l_2 \le l \le l_1 + l_2, \quad l_1 + l_2 + l = 0, 2, \dots . \tag{3.8.37}$$

These formulas are valid also for $l_2 > l_1$ (then we in addition assume $l_2 - l_1 \le l \le l_1 + l_2$). We can also use the symmetry relations (3.8.35) and avoid longer sums over m. It follows from (3.8.35) and (3.8.37) that

$$r^l_{l_1 l_2}(\theta_0) = r^l_{l_1 l_2}(\pi - \theta_0). \tag{3.8.38}$$

This formula can be used for $l_2 > l_1$.

The functions $\Phi_{lm}(x)$, $x \in [-1,1]$ for $m \ge 0$ are proportional to the associate Legendre functions [94]

$$P^m_l(x) = (-1)^m (1-x^2)^{m/2} \frac{d^m}{dx^m} P_l(x), \tag{3.8.39}$$

where $P_l(x)$ are the Legendre polynomials, $l \ge 0$, $0 \le m \le l$. It is well known that

$$\int_{-1}^{1} dx P^m_l(x) P^m_r(x) = \delta_{lr} \frac{2}{2l+1} \frac{(l+m)!}{(l-m)!},$$

and therefore

$$\Phi_{lm}(x) = c_{lm}\left[\frac{2l+1}{4\pi}\frac{(l-m)!}{(l+m)!}\right]^{1/2} P^m_l(x), \quad m \ge 0, \quad c_{lm} = \pm 1, \tag{3.8.40}$$

in accordance with the normalization conditions. Of course, the signs of $\Phi_{lm}(x)$ and $\Phi_{l-m}(x)$ can be chosen in a non-unique way. We choose it in agreement with the book [133]. Then $c_{lm} = (-1)^m$ for $m \ge 0$, whereas

$$\Phi_{l-m}(x) = (-1)^m \Phi_{l-m}(x), \quad m = 1, 2, \dots, l. \tag{3.8.41}$$

The value

$$P^m_{l_2}(0) = \frac{2^m \sqrt{\pi}}{\Gamma(\frac{l_2-m}{2}+1)\Gamma(\frac{1}{2}-\frac{l_2+m}{2})}$$

is also known [94]. Obviously, $P^m_l(0) = 0$ if $l + m = 2k+1$, $k = 0, 1, \dots$. Therefore, the summation in (3.8.37) is taken only over $m = l_2 - 2k$, $k \ge 0$. Then we obtain the following formula:

$$P^{l_2-2k}_{l_2}(0) = (-1)^{l_2+k}\frac{[2(l_2-k)-1]!!}{(2k)!!}, \quad 0 \le k \le l_2, \tag{3.8.42}$$

for non-zero terms of sum (3.8.36).

Finally, we combine the above expressions and formulate the result related to the coefficients of the set of ODEs (3.8.19)–(3.8.20). For brevity the formulation will be given only for the case $0 \le l_2 \le l_1$. Then the alternative case $l_2 > l_1$ can be obtained by simple transformation (3.8.37) or by directly using the same formulas as for $l_2 \le l_1$.

Proposition 3.8.2. *The integral* (3.8.20) *for fixed values of indices* (n, l, m) *and* $(n_{1,2}, l_{1,2}, m_{1,2})$, $0 \le l_2 \le l_1$, *can be presented in the form*

$$K_{nn_1n_2}(l, m|l_1, m_1; l_2, m_2) = \delta_{m,m_1+m_2}\, \delta_{2n+l,2n_1+l_1+2n_2+l_2} \times \sum_{|j|\le \mathrm{Min}(l_2,l)} C_j(l, l_1, l_2, m_1, m_2) I^{ljl_1l_2}_{n_1n_2}(g), \quad l_1 + l_2 + l = 0, 2, \dots,$$

where $C_j(\dots)$ *are absolute constants known in explicit form,*

$$I^{ljl_1l_2}_{n_1n_2}(g) = \int_0^\pi d\theta \sin\theta g(\cos\theta) \cos^{2n_1+l_1}\left(\frac{\theta}{2}\right) \sin^{2n_1+l_1}\left(\frac{\theta}{2}\right) P^j_l\left(\cos\frac{\theta}{2}\right),$$

in the notation of equations (3.8.39). *The exact values of coefficients* $C_j(\dots)$ *follow from equations* (3.8.29), (3.8.31), (3.8.36), *and* (3.8.42) *and from well-known values of integrals* (3.8.28).

The proof of Proposition 3.8.2 is given above.

Remarks on Chapter 3

1. Chapters 3–5 are devoted to study of the Boltzmann equation for Maxwell molecules. It is clear that in this case we can get a lot of information about solutions of the Boltzmann equation (at least in the spatially homogeneous case) by using advantages of the Fourier representation. The material of Chapter 3 is mainly based on papers [17, 20, 26].
2. The computation of anisotropic moments of the collision integral (see Sections 3.8 and 3.8.1) was performed by several authors in tensor form (see, in particular, [98, 144] and references therein). Of course, the transition to the Fourier representation simplifies this problem, but still does not make it very simple. It seems more convenient to use coefficients of expansion in the orthogonal series in eigenfunctions of the linearized collision operator. Then we reduce the problem to computation of integrals (3.8.20). The computation was firstly performed in papers [95, 145] (see also [146]). We present in Section 3.8.1 a slightly different way of computation based on the Wigner–Eckart theorem [154] for linear operators.
3. There are many interesting results on solutions of homogeneous Boltzmann equation for hard potentials published in recent decades. We mention just two examples. The structure of moment equations from Section 3.8 shows that the absolute

moment of order $N \geq 3$ is finite for all $t \geq 0$ if and only if it is finite at $t = 0$. This property for Maxwell molecules has been well known for a long time (see, e. g., [117]). However, in the case of hard potentials, all moments of the solution become finite at any $t > 0$ even if they do not exist (for sufficiently large $N \geq 3$) at $t = 0$. This was proved by L. Desvillettes [77] and B. Wennberg [151–153]. This is an example of properties which are different for Maxwell molecules and hard potentials.

The second example shows the coincidence of properties of solutions for both kinds of molecular models. It is proved in [88] (see also [30]) for hard potentials and in [49] for pseudo-Maxwell molecules that the solution $f(v,t)$ is bounded in L_v^∞ by some Maxwellian for all $t > 0$ provided that the initial distribution $f(v,0)$ is bounded by another Maxwellian.

4. Also we mention some interesting publications on solutions of the Boltzmann equation with long range Maxwell potential (see, e. g., [2, 119] and references therein).

4 Radial solutions

4.1 Equation for characteristic function $\varphi(|k|, t)$

In this chapter we continue to study the spatially homogeneous Cauchy problem (3.3.1), (3.3.2) under normalization conditions (3.3.4). In addition, we assume that the distribution function $f_0(v)$ at $t = 0$ depends only on $|v|$. Then this property holds for all $t \geq 0$ and we obtain the so-called radial solution $f(|v|, t)$ of the Boltzmann equation (3.3.1). The corresponding characteristic function

$$\varphi(|k|, t) = \langle f(|v|, t), e^{-ik\cdot v}\rangle , \quad k \in \mathbb{R}^3,$$

is also radial or, equivalently, isotropic in k-space. We can represent $\varphi(|k|, t)$ by the integral

$$\varphi(|k|, t) = 4\pi \int_0^\infty drr^2 f(r, t) \frac{\sin |k|r}{|k|r} , \tag{4.1.1}$$

omitting details of elementary integration in spherical coordinates. If $f(r, t)$ decreases sufficiently fast for large $r > 0$, then we can consider a formal expansion in Taylor series under the integral sign and obtain the series

$$\varphi(|k|, t) = \sum_{n=0}^{\infty} (-1)^n m_{2n}(t) \frac{|k|^{2n}}{(2n+1)!} , \quad m_{2n}(t) = 4\pi \int_0^\infty drr^2 f(r, t) , \tag{4.1.2}$$

which can be considered as a simplified version of the series (3.8.2) for radial solutions.

The expansion (4.1.2) shows that the most convenient variable in the Fourier representation is $|k|^2$, not $|k|$. Having in mind the standard Maxwellian (3.3.6), we denote

$$\varphi(|k|, t) = \tilde{\varphi}(x, t) , \quad x = \frac{|k|^2}{2} . \tag{4.1.3}$$

This notation will be used below in this chapter. Hopefully it will not be confused with the notation $x \in \mathbb{R}^3$ for spatial variables used in other parts of the book.

Let us now consider equation (3.3.15) for radial solutions. Straightforward transformations lead to the following equation for $\varphi(x, t)$:

$$\varphi_t = \int_0^1 ds p(s) \{\varphi(sx)\varphi[(1-s)x] - \varphi(0)\varphi(x)\}, \quad x \geq 0 , \tag{4.1.4}$$

where tildes are omitted,

$$p(s) = 4\pi g(1-2s), \quad s \in [0, 1] . \tag{4.1.5}$$

https://doi.org/10.1515/9783110550986-005

The initial data and the normalization conditions read as

$$\varphi|_{t=0} = \varphi_0(x), \quad \varphi(0,t) = 1, \quad \varphi_x(0,t) = -1. \tag{4.1.6}$$

In this part of the book we will mostly consider solutions of the Boltzmann equation $f(|v|, t)$ having bounded moments of all orders. Then a natural form of representation of the characteristic function $\varphi(x, t)$ is a power series in x. Equations for coefficients of such series will be derived in the next section.

4.2 Equations for moments

We consider a formal solution $\varphi(x, t)$ of equation (4.1.4),

$$\varphi(x,t) = \sum_{n=0}^{\infty} (-1)^n z_n(t) \frac{x^n}{n!} \tag{4.2.1}$$

and substitute the series into the equation. Then we obtain a set of ordinary differential equations (ODEs) for coefficients,

$$\dot{z}_0 = \dot{z}_1 = 0; \quad \dot{z}_n = \sum_{k=0}^{n} \binom{n}{k} h_{k,n-k} (z_k z_{n-k} - z_0 z_n), \tag{4.2.2}$$

$$\binom{n}{k} = \frac{n!}{k!(n-k)!}, \quad h_{k,l} = \int_0^1 ds \rho(s) s^k (1-s)^l; \quad k, l = 0, 1, \ldots, \quad n \geq 2, \tag{4.2.3}$$

where the dot denotes differentiation with respect to time t. Note that we did not assume the condition $\rho(s) \in L_+^1([0,1])$ here. If it is satisfied, then the equations for $z_n(t)$ can be also written as

$$\dot{z}_n + z_n \int_0^1 ds \rho(s) = \sum_{k=0}^{n} \binom{n}{k} h_{k,n-k} z_k z_{n-k}, \tag{4.2.4}$$

since

$$\sum_{k=0}^{n} \binom{n}{k} h_{k,n-k} = \int_0^1 ds \rho(s).$$

On the other hand, terms with $k = 0$ and $k = n$ are actually absent in sums in (4.2.2). Therefore, all coefficients in ODEs (4.2.2) are finite under weaker conditions,

$$\int_0^1 ds \rho(s) s(1-s) < \infty, \quad \rho(s) \geq 0, \tag{4.2.5}$$

which are equivalent to (3.1.5) and fulfilled also for true Maxwell molecules.

Coefficients $\{z_n(t),\ n = 0, 1, \dots\}$ will be called “normalized moments” of the distribution function $f(|v|, t)$. Their connection with usual moments is given by equalities

$$z_n(t) = \frac{m_{2n}(t)}{(2n+1)!!}\,, \quad m_{2n}(t) = 4\pi \int_0^\infty dr r^{2+2n} f(r, t)\,, \quad n = 0, 1, \dots\,. \tag{4.2.6}$$

Another form of representation of a similar class of solutions is the series

$$\varphi(x, t) = e^{-x} \sum_{n=0}^{\infty} u_n(t) \frac{x^n}{n!}\,. \tag{4.2.7}$$

Since the transformation

$$\varphi(x, t) = e^{-x} \psi(x, t)$$

does not change equation (4.1.4), we obtain the same equations as equations (4.2.2) for $u_n(t)$, namely,

$$\dot{u}_0 = \dot{u}_1 = 0, \quad \dot{u}_n = \sum_{k=1}^{n-1} \binom{n}{k} h_{k,n-k} (u_k u_{n-k} - u_0 u_n)\,, \quad n \geq 2\,, \tag{4.2.8}$$

in the notation of equations (4.2.3).

There are some differences between series (4.2.1) and (4.2.7). In particular, the second series has a direct analog in the velocity space. It can be considered as the Fourier transformed Laguerre series

$$f(|v|, t) = M(|v|) \sum_{n=0}^{\infty} u_n(t)\, L_n^{1/2}\left(\frac{|v|^2}{2}\right), \quad M(|v|) = \frac{1}{(2\pi)^{3/2}} e^{-|v|^2/2}. \tag{4.2.9}$$

The corresponding weighted $L^2_{M^{-1}}$-space has a norm

$$\|f\|^2_{L^2_{M^{-1}}} = \int_{\mathbb{R}^3} dv f^2(|v|)\, M^{-1}(|v|)\,. \tag{4.2.10}$$

Then we obtain from (4.2.9)

$$\|f(|v|, t)\|^2_{L^2_{M^{-1}}} = \sum_{n=0}^{\infty} u_n^2(t) \frac{(2n+1)!!}{(2n)!!}\,. \tag{4.2.11}$$

We shall see below that the space with norm (4.2.10) is “too small” for the Cauchy problem (3.3.1), (3.3.2) in certain precise sense (see Section 5.2). On the other hand, the series (4.2.1) and (4.2.7) are absolutely equivalent in the Fourier space. They are connected by a simple transformation (4.2.7). The corresponding formulas for coefficients $z_n(t)$ and $u_n(t)$ read as

$$z_n(t) = \sum_{k=0}^{n} \binom{n}{k} (-1)^k u_k(t)\,, \quad n = 0, 1, \dots\,. \tag{4.2.12}$$

How can we solve equations (4.2.8)? Obviously, it can be done in a recursive way. By using the normalization conditions (4.1.6), we obtain $u_0 = 1$, $u_1 = 0$. Hence,

$$\begin{aligned} \dot{u}_n + \lambda_n u_n &= 0\,, \quad n = 2, 3; \\ \dot{u}_n + \lambda_n u_n &= \sum_{k=2}^{n-2} \binom{n}{k} h_{k,n-k} u_k u_{n-k}\,, \quad n \geq 4; \end{aligned} \tag{4.2.13}$$

where

$$\lambda_n = -\lambda_0(2n) = \int_0^1 ds \rho(s)[1 - s^n - (1-s)^n]\,, \quad n \geq 1, \tag{4.2.14}$$

are familiar (see (3.6.5)) eigenvalues of the linearized collision operator with minus sign. The solution of these ODEs reads

$$\begin{aligned} u_n(t) &= u_n(0)e^{-\lambda_n t}\,, \quad n = 2, 3; \\ u_n(t) &= u_n(0)e^{-\lambda_n t} \\ &+ \sum_{k=2}^{n-2} \binom{n}{k} h_{k,n-k} \int_0^t d\tau e^{-\lambda_n(t-\tau)} u_k(\tau) u_{n-k}(\tau))\,, \quad n \geq 4\,. \end{aligned} \tag{4.2.15}$$

Thus we obtain a significantly simplified version of equations (3.8.19) in slightly different notation. Note that $\lambda_n > 0$ for all $n \geq 2$.

4.3 Distribution functions with Maxwellian tails

For a moment we come back to the initial value problem (3.3.1), (3.3.2) for the Boltzmann equation. Let us assume that $f_0(v) = f(v, 0)$ has a compact support in $\mathbb{R}^3$ such that $f_0(v) = 0$ if $|v| > R$ with some $R > 0$. It is well known that the solution $f(v, t)$ becomes strictly positive for any $t > 0$. Moreover, there are such positive $a(t)$ and $b(t)$ that

$$f(v, t) \geq a(t) \exp(-b(t)|v|^2) \tag{4.3.1}$$

for any $t > 0$ and almost all $v \in \mathbb{R}^3$. This was originally proved by T. Carleman for hard spheres [64] and later by A. Pulverenti and B. Wennberg for pseudo-Maxwell molecules [129]. In informal language we usually call it "the formation of the Maxwellian tail of $f(v, t)$ for large velocities." Of course, the most famous example of the solution with Maxwellian tail is the Maxwellian distribution function itself. It is clear that we cannot expect any faster decay of solutions $f(v, t)$ for large $|v|$. On the other hand, we shall see below that the functions with Maxwellian tails represent an important "closed" class of solutions to the Cauchy problem (3.3.1), (3.3.2). It is convenient for our goal to define such functions $f(v)$, $v \in \mathbb{R}^3$, in the following way.

Definition 4.3.1. We call the distribution function $f(v) \geq 0$ a rapidly decreasing function (as $|v| \to \infty$) if for a certain $r > 0$ the following integral converges:

$$\Psi(r) = \int_{\mathbb{R}^3} dv\, f(v) \exp\left(r\frac{|v|^2}{2}\right) < \infty\,. \tag{4.3.2}$$

Note that Definition 4.3.1 includes not only functions $f(|v|)$. One of the advantages of this class of functions is the fact that it enables us to introduce a convenient asymptotic characteristic as $|v| \to \infty$.

Definition 4.3.2. The quantity $\tau \geq 0$, defined by equation

$$\tau^{-1} = \sup\, r, \tag{4.3.3}$$

will be termed the tail temperature of the function $f(v)$, satisfying Definition 4.3.1, where the *supremum* is taken over those values of $r \geq 0$ for which integral (4.3.2) converges.

In this chapter we consider only radial rapidly decreasing distribution functions. The whole set of such functions $f(|v|)$ satisfying Definition 4.3.1 and normalization conditions

$$4\pi \int_0^\infty dr r^2 f(r) = 1\,, \quad 4\pi \int_0^\infty dr r^4 f(r) = 3 \tag{4.3.4}$$

will be denoted by B_+:

$$B_+ = \{f(|v|) \geq 0,\ v \in \mathbb{R}^3 :\ f(|v|) \text{ satisfies (4.3.4) and Definition 4.3.1}\}. \tag{4.3.5}$$

We will denote by $B_+(\tau_0)$ the subset of B_+ consisting of such functions for which $\tau \leq \tau_0$. If $f(|v|) \in L^2_{M^{-1}}$, i. e., integral (4.2.10) is bounded, and $f(|v|) \geq 0$ satisfies conditions (4.3.4), then elementary application of the Schwartz inequality to (4.3.2) with $r < 1/2$ leads to the estimate

$$\Psi(r) \leq C(1 - 2r)^{-3/2}\, \|f\|^2_{L^2_{M^{-1}}}\,. \tag{4.3.6}$$

Hence, $L^2_{M^{-1}} \subset B_+$ for non-negative functions.

The same letter B was already used in Section 3.7 for another set of functions (see (3.7.2)). We hope that this will not cause any confusion.

We will now consider the Cauchy problem (3.3.1)–(3.3.2) on functions of the class B_+. Reverting to the Fourier representation, i. e., to the problem (4.1.4)–(4.1.6) for the isotropic in k characteristic function $\varphi(x, t)$, $x = |k|^2/2$, it is first necessary to determine what properties the Fourier transforms of the functions of class B_+ possess. This is considered in the next section.

4.4 Analytic properties of isotropic characteristic functions; entire functions of exponential type

The explicit formula for the isotropic characteristic function $\varphi(x,t)$ reads (cf. (4.1.1))

$$\varphi(x,t) = 4\pi \int_0^\infty dr r^2 f(r,t) \frac{\sin r\sqrt{2x}}{r\sqrt{2x}}, \qquad x = \frac{|k|^2}{2}, \tag{4.4.1}$$

where $f(|v|,t)$ is a radial solution of the Boltzmann equation (3.3.1). The formal representation of $\varphi(x,t)$ as the series

$$\varphi(x,t) = \sum_{n=0}^{\infty} (-1)^n z_n(t) \frac{x^n}{n!}, \tag{4.4.2}$$

$$z_n(t) = \frac{4\pi}{(2n+1)!!} \int_0^\infty dr r^{2+2n} f(r,t) \tag{4.4.3}$$

follows directly from (4.4.1). Equations for normalized moments $z_n(t)$ were already obtained in Section 4.2. However, we did not discuss the convergence of corresponding power series (4.4.2). This will be done below.

It is easy to see that solutions $f(|v|,t)$ of the Boltzmann equation from the class B_+ of rapidly decreasing distribution functions are closely connected with a class A of analytic in x solutions $\varphi(x,t)$ of equation (4.1.4). We begin with a lemma that has no relation to the Boltzmann equation. In its formulation we consider the t-variable in equations (4.4.1)–(4.4.2) as a fixed irrelevant parameter and therefore omit it.

Lemma 4.4.1. *Suppose that the functions $f(|v|) \geq 0$ and $\varphi(x)$ are related by transformation (4.4.1). If $f(|v|)$ satisfies Definition 4.3.1 and conditions (4.3.4), then the power series (4.4.2) for $\varphi(x)$ converges absolutely for all $x \geq 0$ and the following relations hold:*

$$z_0 = z_1 = 1; \quad z_n \geq \frac{3^n}{(2n+1)!!}, \qquad n = 0, 1, \dots; \tag{4.4.4}$$

$$\tau = \limsup \, (z_n)^{1/n}, \qquad n \to \infty, \tag{4.4.5}$$

where τ is the tail temperature of $f(|v|)$ (see Definition 4.3.2).

Proof. We consider integral (4.3.2) and expand it in power series,

$$\Psi(r) = \sum_{n=0}^{\infty} \frac{m_{2n}}{(2^n n!)} r^n, \quad m_{2n} = 4\pi \int_0^\infty dr r^{2+2n} f(r).$$

Hence, we obtain from (4.2.6) the series

$$\Psi(r) = \sum_{n=0}^{\infty} \frac{(2n+1)!!}{(2n)!!} z_n r^n, \quad z_n \geq 0,$$

which converges for complex r in the circle $|r| \le \tau^{-1}$, where τ is understood in the sense of Definition 4.3.2. On the other hand, the standard Hadamard formula (see, e. g., [1]) leads to equation (4.4.5) for τ. Inequalities (4.4.4) follow from the elementary properties of the moments of $f(|v|)$ [85]

$$m_{2n} \ge m_2^n, \quad n = 1, 2, \dots .$$

Then absolute convergence of the series (4.4.2) for all $x \ge 0$ obviously follows from the assumption that $\tau < \infty$ and equation (4.4.5). This completes the proof. □

Hence, any function $\varphi(x)$, $x \ge 0$, satisfying the conditions of Lemma 4.4.1 can be analytically continued using equation (4.4.2) over the whole plane of the complex variable x, and is the entire function of the order $\lambda \le 1$.

Indeed

$$\lambda = \limsup_{r\to\infty} \frac{\log\log M(r)}{\log r},$$

where $M(r)$ is the maximum of $|\varphi(x)|$ on $|x| = r$ in the complex plane [1]. Since $z_n > 0$ in (4.4.2), we obtain

$$M(r) = \varphi(-r) = \sum_{n=0}^{\infty} \frac{z_n}{n!} r^n .$$

The sequence $\{z_n^{1/n}, n = 1, \dots\}$ is bounded by some $a \ge 1$, since $\tau < \infty$. Hence, $M(r) \le \exp(ar)$ and therefore $\lambda \le 1$.

We shall see in the next section that the inequality $\lambda < 1$ for solution $\varphi(x,t)$ of equation (4.1.4) can be valid only for initial conditions, i. e., at $t = 0$. It will be shown that at any $t > 0$ the solution $\varphi(x,t)$ has the order $\lambda = 1$. Such functions are sometimes called the entire functions of exponential type.

4.5 Solution of the Cauchy problem

We consider below the initial value problem (4.1.4)–(4.1.6) and assume that the corresponding distribution function $f_0(|v|) \in B_+$, in the notation of equation (4.3.5). Hence, the initial data $\varphi_0(x)$ (4.1.6) is related to $f_0(|v|)$ by transformation (4.4.1) with $t = 0$.

According to Lemma 4.4.1, we obtain

$$\varphi_0(x) = \sum_{n=0}^{\infty} \frac{(-1)^n}{n!} z_n^{(0)} x^n, \tag{4.5.1}$$

$$z_0^{(0)} = z_1^{(0)} = 1; \quad z_n^{(0)} \ge \frac{3^n}{(2n+1)!!}, \quad n = 0, 1, \dots, \tag{4.5.2}$$

$$\tau^{(0)} = \limsup_{n\to\infty} [z_n^{(0)}]^{1/n} < \infty . \tag{4.5.3}$$

We denote

$$A = \left\{ \varphi(x),\ x \geq 0 \;\middle|\; \varphi(x) = \sum_{n=0}^{\infty} (-1)^n z_n \frac{x^n}{n!}, \right.$$

$$\left. z_0 = z_1 = 1,\ \sup |z_n|^{1/n} < \infty, n = 0, 1, \dots \right\}, \tag{4.5.4}$$

where $z_2, z_3, \dots$ are complex numbers. Obviously, $\varphi(x) \in A$. Then it is natural to try to construct the solution of the problem (4.1.4)–(4.1.6) in the form of power series

$$\varphi(x) = \sum_{n=0}^{\infty} (-1)^n z_n \frac{x^n}{n!}, \tag{4.5.5}$$

as we already did in Section 4.2 at the formal level. Thus we obtain ODEs (4.2.2) and solve them by a recursive scheme similar to equation (4.2.15). Now we can make all necessary estimates and prove the following statement.

Theorem 4.5.1. *If*

$$0 \leq \rho(s) \leq C s^{-2+\varepsilon}, \quad \varepsilon > 0, \quad s \in (0, 1], \tag{4.5.6}$$

in equation (4.1.4) *and* $\varphi_0 \in A$ *in equation* (4.1.6), *then the problem* (4.1.4)–(4.1.6) *has a solution* $\varphi(x, t)$ *such that*

(1) $\varphi(x, t)$ *is a unique solution in the class of power series* (4.5.5);
(2) $\varphi(x, t) \in A$ *for any* $t \geq 0$;
(3) $\varphi(x, t) \to e^{-x}$ *as* $t \to \infty$.

Proof. First we note that condition (4.5.6) guarantees that all coefficients $\{\lambda_n,\ h_{k,l};\ n, k, l = 0, 1, \dots\}$ of ODEs (4.2.2) are bounded. The uniqueness of solution in the class of power series follows from uniqueness of solution of ODEs for $\{z_n(t),\ n = 0, 1, \dots\}$. To prove that the series (4.5.5) converges we need to estimate the growth of $|z_n(t)|$. The assumption $\varphi_0 \in A$ means, in particular, that there exists $a \geq 1$ such that

$$z_0^{(0)} = z_0^{(1)} = 1; \quad 1 \leq z_n^{(0)} \leq a^n, \quad n = 2, 3, \dots. \tag{4.5.7}$$

We consider equations (4.2.2) and transform them to

$$\dot{z}_0 = \dot{z}_1 = 0; \quad \dot{z}_n + \lambda_n z_n = \sum_{k=1}^{n-1} \binom{n}{k} h_{k,n-k}\, z_k\, z_{n-k}, \quad n \geq 2, \tag{4.5.8}$$

in the notation of equations (4.2.3), (4.2.14). Then we can obtain the solution in recursive form

$$\begin{aligned} z_0 &= z_1 = 1, \\ z_n &= z_n^{(0)} e^{-\lambda_n t} \\ &\quad + \sum_{k=1}^{n-1} \binom{n}{k} h_{k,n-k} \int_0^t d\tau e^{-\lambda_n (t-\tau)} z_k(\tau) z_{n-k}(\tau), \quad n \geq 2, \end{aligned} \tag{4.5.9}$$

and prove by induction that $|z_n(t)| \le a^n$ for all $t > 0$ and $n = 0, 1, \ldots$. Since $a \ge 1$, it is fulfilled for $n = 0, 1$. We assume that $|z_k(t)| \le a^k$ for $k = 0, 1, \ldots, n-1$, where $n \ge 2$. By using this assumption we obtain from (4.5.9)

$$|z_n(t)| \le K_n(t)a^n, \quad K_n(t) = K^{(n)}(1 - e^{-\lambda_n t}) + e^{-\lambda_n t},$$
$$K^{(n)} = \frac{1}{\lambda_n} \sum_{k=1}^{n-1} \binom{n}{k} h_{k,n-k}, \quad n \ge 2.$$

Then we use explicit formulas (4.2.3) and (4.2.14) for $h_{k,l}$ and λ_n, respectively, and obtain

$$\sum_{k=1}^{n-1} \binom{n}{k} h_{k,n-k} = \int_0^t ds\rho(s) \sum_{k=1}^{n-1} \binom{n}{k} s^k (1-s)^{n-k}$$
$$= \int_0^t ds\rho(s)[1 - s^n - (1-s)^n] = \lambda_n, \quad n \ge 1. \tag{4.5.10}$$

Hence, $K^{(n)} = 1$ and therefore $|z_n(t)| \le a^n$ for all integer $n \ge 1$ and real $t \ge 0$.

It remains to prove that

$$\lim_{t\to\infty} \varphi(x,t) = e^{-x}. \tag{4.5.11}$$

To do it we need to transform the series (4.5.5) to a slightly different form, like it was done in Section 4.2 (cf. (4.2.7)). We represent the solution $\varphi(x,t)$ in the form

$$\varphi(x,t) = e^{-x} \sum_{n=0}^{\infty} u_n(t) \frac{x^n}{n!},$$
$$u_n(t) = \sum_{k=0}^{n} \binom{n}{k} (-1)^k z_k(t), \quad u_n(0) = u_n^{(0)} = \sum_{k=0}^{n} \binom{n}{k} (-1)^k z_k^{(0)}. \tag{4.5.12}$$

Then the solution $\{u_n(t),\ n = 0, 1, \ldots\}$ reads

$$u_0 = 1, \quad u_1 = 0; \quad u_n(t) = u_n^{(0)} e^{-\lambda_n t}, \quad n = 2, 3;$$
$$u_n(t) = u_n^{(0)} e^{-\lambda_n t}$$
$$+ \sum_{k=2}^{n-2} \binom{n}{k} h_{k,n-k} \int_0^t d\tau e^{-\lambda_n (t-\tau)} u_k(\tau) u_{n-k}(\tau), \quad n \ge 4. \tag{4.5.13}$$

Since $|z_n(t)| \le a^n$, it is easy to see from equations (4.5.12) that $|u_n(t)| \le (a+1)^n$, $n = 0, 1, \ldots$.

Hence, for any fixed $x \ge 0$, the series converge uniformly with respect to $t \in (0, \infty)$. Therefore, it is enough to prove that $u_n(t) \to 0$, as $t \to \infty$, $n \ge 4$. In addition, we can prove the following exponential estimate of the rate of convergence for $\{u_n(t),\ n = 2, 3, \dots\}$:

$$|u_n(t)| \le \alpha\, b^n\, e^{-\lambda_2 t}, \quad \alpha = 1 - \frac{\lambda_2}{\lambda_4},$$
$$b = \sup\left\{\left[\frac{|u_n^{(0)}|}{\alpha}\right]^{1/n}, n = 2, \dots\right\}. \tag{4.5.14}$$

We can use for the proof the inductive approach again. Let us assume that

$$|u_k(t)| \le \alpha\, b^k\, e^{-\gamma t}, \quad k = 1, 2, \dots, n-1,$$

for some positive numbers α, b, γ. Moreover we assume that $\lambda_n > \gamma$ for $n \ge 4$ and $|u_m(0)| \le \alpha\, b^m$ for all $m \ge 1$. Then we obtain from equation (4.5.13) for $u_n(t)$

$$|u_n(t)| \le \alpha\, b^n\, e^{-\gamma t}\, R_n(t), \quad n \ge 4,$$
$$R_n(t) = e^{-(\lambda_n - \gamma)t} + \alpha S_n \frac{e^{-\gamma t} - e^{-(\lambda_n - \gamma)t}}{\lambda_n - 2\gamma},$$
$$S_n = \sum_{k=2}^{n-2} \binom{n}{k} h_{k,n-k} = \lambda_n - n(h_{1,n-1} + h_{n-1,1}) \le \lambda_n .$$

Hence, we obtain

$$R_n(t) \le (1-\theta)e^{-\gamma_1 t} + \theta e^{-\gamma t}, \quad \theta = \alpha\frac{\gamma_1 + \gamma}{\gamma_1 - \gamma}, \quad \gamma_1 = \lambda_n - \gamma > 0 . \tag{4.5.15}$$

We need to choose α and γ in such a way that $0 < R(t) \le 1$ for all $t \ge 0$. It is sufficient to choose

$$0 < \alpha \le \frac{\gamma_1}{\gamma_1 + \gamma} . \tag{4.5.16}$$

Then

$$\theta(e^{-\gamma t} - e^{-\gamma_1 t}) \le \gamma_1 \frac{e^{-\gamma t} - e^{-\gamma_1 t}}{\gamma_1 - \gamma} .$$

Therefore, we obtain

$$R_n(t) \le \tilde{R}_n(t) = \frac{e^{-\gamma t} - e^{-\gamma_1 t}}{\gamma_1 - \gamma} .$$

It is easy to check that $0 < \tilde{R}_n(t) \le 1$ for all $t \ge 0$. It remains to check that inequality for some constant α can be satisfied simultaneously for all $n \ge 4$. Coming back to initial notation (see (4.5.15)) we re-write (4.5.16) as

$$\alpha \le 1 - \frac{\gamma}{\lambda_n}, \quad n = 4, \dots .$$

Since $\lambda_{n+1} > \lambda_n$ for all $n \geq 1$, it is sufficient to satisfy this inequality for $n = 4$. Obviously, the estimate (4.5.15) is valid for $n = 2$ only if $\gamma \leq \lambda_2$. Therefore, we can choose the values of parameters α, γ, b given in equations (4.5.14). This completes the proof of estimates (4.5.14). Moreover, the end of the proof of Theorem 4.5.1 obviously follows from these estimates. □

Thus we have proved the existence and uniqueness of solution to the Cauchy problem (4.1.4)–(4.1.6) for initial data from class A defined in equations (4.5.4). The proof is given under general assumptions on kernel $g(\cos\theta)$ in the Boltzmann equation (3.3.1), which include the case of true Maxwell molecules. This follows from equations (4.5.6), (4.1.5), and (3.1.4).

We are interested mainly in positive solutions of the Boltzmann equation. Therefore, we need to show that the solution $\varphi(x,t)$, $x = |k|^2/2$, is the characteristic function, i. e., the Fourier transform with respect to $k \in \mathbb{R}^3$ of the radial distribution function $f(|v|,t)$, provided that this is so at $t = 0$. It is convenient to denote

$$A_* = \{\varphi(x),\ x \geq 0 \mid \exists f(|v|) \in B_+ :\ \varphi(|k|^2/2) = \langle f, e^{-ik\cdot v}\rangle\}, \tag{4.5.17}$$

in the notation of equations (2.1.9). Then $A_* \subset A$ and the following statement is valid.

Lemma 4.5.1. *Let $\varphi(x,t) \in A$ be a solution of the problem* (4.1.4)–(4.1.6), *where $\varphi_0(x) = \varphi(x,0) \in A_*$, constructed in Theorem* 4.5.1. *Then $\varphi(x,t) \in A_*$ for all $t > 0$.*

Proof. The proof of this lemma can be found in [27]. For brevity we explain here just a general idea of the proof. First we consider the cut-off case such that

$$\rho_{\text{tot}} = \int_0^1 ds\, \rho(s) < \infty \tag{4.5.18}$$

and assume without loss of generality that $\rho_{\text{tot}} = 1$. Then we can construct the solution $\varphi_1(x,t)$ of the problem (4.1.4)–(4.1.6) in the form of the Fourier transformed Wild sum (3.3.21). This solution is obviously the characteristic function. On the other hand, the function $\varphi(x,t)$, $x = |k|^2/2$, from Theorem 4.5.1 obviously satisfies conditions of Lemma 3.3.1. Therefore, it is the same solution, i. e.,

$$\varphi(x,t) = \varphi_1(x,t) \in A_*$$

for all $t \geq 0$. This proves Lemma 4.5.1 for any $\rho(s)$ satisfying (4.5.18). Then we can approximate the non-integrable kernel by integrable kernels and make usual transition to the limit on known properties of analytic characteristic functions [85]. It is clear that the limit will be a characteristic function represented by power series from Theorem 4.5.1. This completes the proof of Lemma 4.5.1. □

4.6 Stationary and self-similar solutions

It is easy to guess that equation (4.1.4) admits a class of self-similar solutions having the following form:

$$\varphi(x,t) = y(x\,e^{-\mu t}), \quad \mu = \text{const.} \tag{4.6.1}$$

We shall see below that this kind of self-similar solutions is typical for various Maxwell models in kinetic theory. The equation for $y(x)$ can be obtained by substitution of (4.6.1) into (4.1.4). It reads

$$-\mu x y'(x) = J(y,y) = \int_0^1 ds\,\rho(s)\{y(sx)y[(1-s)x] - y(0)y(x)\}, \quad x \geq 0. \tag{4.6.2}$$

We look for $y(x)$ in the form of power series

$$y(x) = \sum_{n=0}^{\infty} (-1)^n y_n \frac{x^n}{n!}. \tag{4.6.3}$$

Then we obtain algebraic equations for y_n,

$$-n\mu y_n = \sum_{k=0}^{n} \binom{n}{k} h_{k,n-1}(y_k y_{n-k} - y_0 y_n), \quad n = 0, 1, \dots, \tag{4.6.4}$$

in the notation of equations (4.2.3). The equation for $n = 0$ is satisfied for any y_0. We note that μ is a free parameter. Therefore, without any loss of generality we can choose $y_0 = 1$. The equation for $n = 1$ reads

$$-\mu y_1 = 0 \Rightarrow \mu = 0 \quad \text{or/and} \quad y_1 = 0. \tag{4.6.5}$$

If $\mu = 0$, then $y(x)$ in (4.6.1) is a stationary solution of equation (4.1.4). In that case $y_1 = a$ is a free parameter. For $n \geq 2$, we obtain

$$y_0 = 1, \quad y_1 = a, \quad \sum_{k=1}^{n-1} \binom{n}{k} h_{k,n-k}(y_k y_{n-k} - y_0 y_n) = 0. \tag{4.6.6}$$

Therefore, the coefficients $\{y_n,\ n \geq 2\}$ are uniquely defined by the recursive formula

$$y_n = \frac{1}{\lambda_n y_0} \sum_{k=1}^{n-1} \binom{n}{k} h_{k,n-k} y_k y_{n-k}, \quad y_0 = 1, \quad y_1 = a,$$

in the notation of equation (4.5.10). Hence,

$$y_k = a^k, \quad k = 0, 1, \dots, \tag{4.6.7}$$

is a unique solution of equations (4.6.6). Thus, we obtain the following result.

Lemma 4.6.1. *The equation* (4.6.1) *with* $\mu = 0$

$$J(y, y) = 0 \tag{4.6.8}$$

has under conditions

$$y(0) = 1, \quad y'(0) = -a \tag{4.6.9}$$

a unique solution $y(x) = \exp(-a\,x)$ *in the class of analytic at* $x = 0$ *functions.*

Proof. Any analytic at $x = 0$ function $y(x)$ can be represented by Taylor series (4.6.3) that converge for small values of $|x|$. The equations (4.6.6) for coefficients $\{y_n,\ n \geq 2\}$ lead to the unique solution $y(x) = \exp(-ax)$. This completes the proof. □

Thus, everything is clear with equation (4.6.2) for $\mu = 0$. From now on we assume that $\mu \neq 0$. Then it follows from (4.6.5) that $y_1 = 0$. Equations with $n = 2, 3$ read as

$$y_n\,(\lambda_n - n\mu) = 0, \quad n = 2, 3\,. \tag{4.6.10}$$

Equations with $n \geq 4$ have the following form:

$$y_n\,(\lambda_n - n\mu) = \sum_{k=2}^{n-2} \binom{n}{k} h_{k,n-k} y_k y_{n-k}\,, \quad n \geq 4\,. \tag{4.6.11}$$

The general solution of equations (4.6.10)–(4.6.11) is described below.

Proposition 4.6.1. *There are two possibilities:*
(A) *If* $y_2 = 0$ *and* $y_k = 0$ *for all* $2 \leq k \leq p - 1$, *but* $y_p \neq 0$, *where* $p \geq 3$, *then*

$$\mu = \mu_p = \frac{\lambda_p}{p}, \quad y_n = \begin{cases} y_{mp}, & \textit{if } n = mp\,,\ m = 1, 2, \ldots, \\ 0, & \textit{otherwise}\,, \end{cases} \tag{4.6.12}$$

where y_p *can be chosen arbitrarily and*

$$y_{mp} = (\lambda_{mp} - m\lambda_p)^{-1} \sum_{k=1}^{m-1} \binom{np}{kp} h_{kp,(n-k)p}\, y_{kp}\, y_{(m-k)p}\,, \quad m = 2, 3, \ldots. \tag{4.6.13}$$

(B) *If* $y_2 \neq 0$, *then both* y_2 *and* y_3 *can be chosen arbitrarily, and*

$$y_n = (\lambda_n - n\lambda_2/2)^{-1} \sum_{k=2}^{n-2} \binom{n}{k} h_{k,n-k}\, y_k\, y_{n-k}\,, \quad n = 4, 5, \ldots. \tag{4.6.14}$$

Proof. We begin with case (B). If $y_2 \neq 0$, then it follows from (4.6.10) that $\mu = \mu_2 = \lambda_2/2$. The explicit formula for μ_p reads (see equations (4.5.10))

$$\mu_p = \frac{\lambda_p}{p} = \frac{1}{p} \int_0^1 ds p(s)\left[1 - s^p - (1-s)^p\right], \quad p \geq 1\,. \tag{4.6.15}$$

Note that

$$1 - s^2 - (1-s)^2 = 2s(1-s), \quad 1 - s^3 - (1-s)^3 = 3s(1-s).$$

Hence, $\mu_2 = \mu_3$. Therefore, with

$$\mu = \mu_2 = \mu_3 = \int_0^1 ds\rho(s)\, s\,(1-s), \tag{4.6.16}$$

both equations (4.6.10) are satisfied for arbitrary values of two free parameters y_2 and y_3. Then other coefficients $y_4, y_5, \ldots$ are uniquely defined by the recursive formula (4.6.14). An elementary inequality which shows that the denominators in (4.6.14) cannot vanish is proved below.

It remains to consider case (A) with $y_2 = 0$. For simplicity we present the proof for $p = 4$ (the value $p = 3$ is automatically included in case (B) with equations (4.6.14) for $y_2 = 0$). If $p = 4$, we obtain

$$y_2 = y_3 = 0; \quad y_4 \neq 0 \quad \Rightarrow \quad \mu = \mu_4 = \lambda_4/4.$$

Then equations (4.6.11) imply that

$$y_5(\lambda_5 - 5\mu_4) = 0, \quad y_6(\lambda_6 - 6\mu_4) = 0, \quad y_7(\lambda_7 - 7\mu_4) = 0.$$

We shall see below that μ_p is a positive monotone decreasing function of p for $p \geq 3$. Therefore, the equalities $y_5 = y_6 = y_7 = 0$ follow. We obtain for y_8

$$y_8 = (\lambda_8 - 8\mu_4)^{-1} \binom{8}{4} h_{4,4}\, y_4^2,$$

i. e., equation (4.6.13) with $m = 2$. Note that $y_8 = y_8^{(4)}$ and $y_4 = y_4^{(4)}$ in equations (4.6.13). Then it is straightforward to prove by induction statement (A) of the theorem for $p = 4$. The proof of the general case of any $p \geq 3$ is basically the same as for $p = 4$.

It remains to show the monotonicity of μ_p for $p \geq 3$. We consider p as a real variable and differentiate

$$\frac{d}{dp}\mu_p = \frac{d}{dp}\frac{\lambda_p}{p} = \frac{r(p)}{p^2}, \quad r(p) = \left(p\frac{d}{dp} - 1\right)\lambda_p.$$

Hence,

$$r'(p) = \frac{d^2}{d^2p}\lambda_p^2 = -\int_0^1 ds\,\rho(s)\{s^p(\log s)^2 + (1-s)^p\,[\log(1-s)]^2\}$$

and therefore $r(p)$ is a monotone decreasing function. It can have only one zero, say, $r(p_0) = 0$, which corresponds to the maximal value of μ_p. Obviously, $2 < p_0 < 3$; otherwise equality (4.6.16) would be impossible. Hence, μ_p is a monotone decreasing function for $p \geq 3$. This completes the proof. □

The end part of the proof can be modified in order to avoid differentiation under the integral sign. Instead we can consider the difference

$$\Delta(n) = \mu_n - \mu_{n+1} = \frac{1}{n(n+1)} \int_0^1 ds\, \rho(s)\psi_n(s)\,, \quad n = 1, 2, \ldots,$$

where

$$\psi_n(s) = 1 - s^n - (1-s)^n - ns(1-s)\left[s^{n-1} + (1-s)^{n-1}\right].$$

For the proof of Proposition 4.6.1 it is sufficient to show that $\psi_n(s) > 0$ for all $n \geq 3$ and $s \in (0, 1)$. Simple calculations show that

$$\psi_1(s) = -2s(1-s), \quad \psi_2(s) = 0, \quad \psi_n(0) = \psi_n(1) = 0\,.$$

The derivative with respect to s reads

$$\psi_n'(s) = n(n+1)s(1-s)\left[(1-s)^{n-2} - s^{n-2}\right], \quad n \geq 1\,.$$

Hence, $\psi_n(s) > 0$ for $s \in (0, 1)$ and $n \geq 3$. Therefore, $\Delta(n) > 0$ for $n \geq 3$ provided $\rho(s)$ is not concentrated at $s = 0$ or/and $s = 1$. Thus we can always assume that

$$\mu_n - \mu_{n+1} > 0\,, \quad n \geq 3\,. \tag{4.6.17}$$

We also want to discuss the corresponding functions

$$y^{(p)}(x) = 1 + \sum_{m=1}^{\infty} (-1)^{mp} y_{mp}^{(p)} \frac{x^{mp}}{(mp)!}\,, \quad p = 3, 4, \ldots\,, \tag{4.6.18}$$

where coefficients of the series are given in equations (4.6.13), and the function

$$y^{(2)}(x) = 1 + \sum_{n=2}^{\infty} (-1)^n y_n^{(2)} \frac{x^n}{n!} \tag{4.6.19}$$

with coefficients from equations (4.6.14). Our aim is to prove the following statement.

Theorem 4.6.1. *Equation (4.1.4) admits a class of self-similar solutions of the form (4.6.1), where $y(x)$ is represented by Taylor series (4.6.3), only for the following countable set of values of parameter μ in (4.6.1):*

$$\mu_p = \frac{\lambda_p}{p}\,, \quad \lambda_p = \int_0^1 ds\rho(s)\left[1 - s^p - (1-s)^p\right], \quad p = 1, 2, \ldots. \tag{4.6.20}$$

The corresponding solutions are denoted below by $y^{(p)}(x)$ and normalized by condition $y^{(p)}(0) = 1$, $p = 1, 2, \ldots$. These solutions have the following properties:

(a) *If* $p = 1$*, then* $\mu_1 = 0$ *and* $y^{(1)}(x)$ *is a stationary solution of equation (4.1.4). This solution is uniquely defined by the first coefficient* $y_1^{(1)}$ *of the Taylor series (4.6.3) and has the explicit form*

$$y^{(1)}(x) = \exp(-y_1^{(1)}\, x)\,, \tag{4.6.21}$$

where $y_1^{(1)}$ *is a free parameter.*

(b) *If* $p = 3, 4, \dots$*, then* $\mu_p > 0$ *and* $y^{(p)}(x)$ *is given by equations (4.6.13) and (4.6.18). The function* $y^{(p)}(x)$ *behaves like*

$$y^{(p)}(x) = 1 + (-1)^p\, y_p^{(p)}\, x^p \tag{4.6.22}$$

for small $x > 0$*, where* $y_p^{(p)}$ *is a free parameter and dots denote higher-order terms.*

(c) *If* $p = 2$*, then we have a sort of degeneracy because* $\mu_2 = \mu_3$*. This allows to choose freely* <u>*two*</u> *parameters* $y_2^{(2)}$ *and* $y_3^{(2)}$ *and represent* $y^{(2)}(x)$ *by equations (4.6.14) and (4.6.19). The asymptotics of* $y^{(2)}(x)$ *for small* $x > 0$ *is given by the same formula (4.6.22) with* $p = 2$*.*

(d) *The above described Taylor series for* $y^{(p)}(x)$*,* $p \geq 2$*, have infinite radius of convergence. They define entire analytic functions* $y^{(p)}(z)$ *of complex variable* $z \in \mathbb{C}$*, satisfying the estimates*

$$\left|y^{(p)}(z)\right| \leq \exp\{\gamma_p\, |z|\}, \quad p \geq 2\,, \tag{4.6.23}$$

with some positive constant γ_p*.*

Proof. The main part of the proof is already given above. It remains to prove property (d) related to convergence of power series for $y^{(p)}(x)$ and estimates (4.6.23). We begin with the case $p \geq 3$ and consider equations (4.6.13) and (4.6.15). If $y_p^{(p)} = a^p$, then we can denote

$$y_{mp}^{(p)} = a^{mp}\, \tilde{y}_{mp}^{(p)}\,, \quad m \geq 1, \quad p \geq 3,$$

and reduce the problem to $a = 1$. Then we fix some $p \geq 3$ and denote

$$u_m = \tilde{y}_{mp}^{(p)}\,, \quad m \geq 1; \quad u_1 = 1\,.$$

It follows from (4.6.13) that

$$u_1 = 1, \quad |u_m| \leq C(p) \max_{1\leq k\leq m-1} |u_k|\, |u_{m-k}|, \quad m \geq 2\,, \tag{4.6.24}$$

where

$$C(p) = \max_{m\geq 2}\left[|\lambda_{mp} - m\lambda_p|^{-1} \sum_{k=1}^{m-1} \binom{mp}{kp} h_{kp,\,(m-k)p} \right].$$

Let us show that $C(p) < \infty$. Note that

$$\sum_{k=1}^{m-1} \binom{mp}{kp} h_{kp,(m-k)p} \le \sum_{j=1}^{mp-1} \binom{mp}{j} h_{j,mp-j} = \lambda_{mp}, \quad m \ge 2,$$

for any $p \ge 1$ (see also the proof of Lemma 4.5.1). Hence, we obtain

$$C(p) \le \max_{m\ge 2} \left| 1 - \frac{m\lambda_p}{\lambda_{mp}} \right|^{-1} = \max_{m\ge 2} \left| 1 - \frac{\mu_p}{\mu_{mp}} \right|^{-1},$$

in the notation of (4.6.20). It was proved above that μ_n is a decreasing function of n for $n \ge 3$ (see (4.6.17)). Therefore,

$$C(p) = \left(\frac{\mu_p}{\mu_{2p}} - 1 \right)^{-1} = \frac{\lambda_{2p}}{2\lambda_p - \lambda_{2p}} > 1, \quad p \ge 3. \tag{4.6.25}$$

Note that $C(p) > 1$, since

$$\lambda_{2p} - (2\lambda_p - \lambda_{2p}) = 2(\lambda_{2p} - \lambda_p) > 0.$$

Then it is easy to prove that $|u_m| \le C^m(p)$ in (4.6.24) and the estimate (4.6.23) for $p \ge 3$ follows. Indeed, we obtain

$$\left| y^{(p)}(z) \right| \le 1 + \sum_{m=1}^{\infty} \left(C(p)\, a^p \right)^m \frac{x^{mp}}{(mp)!} < e^{\gamma_p x}, \quad \gamma_p = aC^{1/p}(p). \tag{4.6.26}$$

It remains to consider the value $p = 2$ with given

$$y_2^{(2)} = b \ne 0, \quad y_3^{(2)} = d.$$

Then we denote

$$a = \max(|b|^{1/2}, |d|^{1/3}), \quad y_n^{(2)} = a^n \tilde{y}_n,$$

and obtain from equations (4.6.14) the following estimate:

$$|y_2| \le 1, \quad |y_3| \le 1,$$
$$|y_n| \le |\lambda_n - n\mu_2|^{-1} \sum_{k=2}^{n-2} \binom{n}{k} h_{k,n-k} |y_k|\, |y_{n-k}|, \quad n \ge 4.$$

We repeat the same considerations as before and take into account the identity $\mu_2 = \mu_3$. Then we obtain

$$|y_n| \le \frac{\lambda_4}{4\mu_2 - \lambda_4} \max_{2\le k\le n-2} |y_k|\, |y_{n-k}|, \quad n \ge 4.$$

Note that

$$\lambda_4 - (4\mu_2 - \lambda_4) = 2(\lambda_4 - \lambda_2) > 0\,, \quad \text{since} \quad \lambda_{n+1} > \lambda_n, \quad n = 0, 1, \dots$$

Then the estimate

$$|y_n^{(2)}| \le (aC_1)^n\,, \quad C_1 = \frac{\lambda_4}{4\mu_2 - \lambda_4} > 1\,, \quad n \ge 2\,, \tag{4.6.27}$$

can be easily proved by induction. Hence, we obtain

$$|y^{(2)}(z)| \le \exp[aC_1|z|], \quad z \in \mathbb{C}\,.$$

This completes the proof. □

In the course of the proof we have obtained some estimates, which are used in the next section. We collect them in the following lemma.

Lemma 4.6.2.
(a) *If $p \ge 3$, $|y_p^{(p)}| \le a^p$, and $\{y_{mp}^{(p)},\ m = 2, 3, \dots\}$ are defined by recursive formulas* (4.6.13), *then*

$$|y_{mp}^{(p)}| \le \left(\frac{\lambda_{2p} a^p}{2\lambda_p - \lambda_{2p}}\right)^m, \quad m = 1, 2, \dots.$$

(b) *If $|y_2^{(2)}| \le a^2$, $|y_3^{(2)}| \le a^3$, and $\{y_n^{(2)},\ n = 4, \dots\}$ are defined by formulas* (4.6.14), *then*

$$|y_n^{(2)}| \le \left(\frac{\lambda_4 a}{2\lambda_2 - \lambda_4}\right)^n, \quad n = 2, \dots.$$

Proof. Statements (a) and (b) follow from inequalities (4.6.26) and (4.6.27), respectively. This completes the proof. □

All results of this section are obtained for the Fourier transformed Boltzmann equation. The corresponding solutions of the Boltzmann equation are studied in the next section.

4.7 Distribution functions

The solutions obtained in Theorem 4.6.1 do not necessary correspond to any solutions of the Boltzmann equation. For example, in the simplest case (a) of the theorem we can choose $y_1^{(1)} = a > 0$ in (4.6.21). Then the inverse Fourier transform yields

$$\mathfrak{F}^{-1}_{k \to v}(e^{-a|k|^2/2}) = (2\pi a)^{-3/2} \exp\left(-\frac{|v|^2}{2a}\right), \tag{4.7.1}$$

i. e., the Maxwell distribution function, which is the well-known solution of the Boltzmann equation. If, however, we choose $y_1^{(1)} < 0$ in (4.6.21), then this function is obviously not a Fourier transform of any solution of the Boltzmann equation. In cases (b) and (c) of Theorem 4.6.1 we have also another problem with positivity of corresponding distribution functions. Let us consider the asymptotic behavior of solutions (4.6.22) (note that this formula is also valid for $p \geq 2$) for small $x = |k|^2/2$. For all $p \geq 2$, we obtain

$$y^{(p)}\left(\frac{|k|^2}{2}\right) = 1 + O(|k|^4), \quad |k| \to 0. \tag{4.7.2}$$

Hence, $y^{(p)}$ cannot be a Fourier transform of positive function (or measure) $f(|v|)$. Otherwise the comparison with equations (4.1.2) shows that (4.7.2) implies the equality

$$m_2 = \langle f(|v|), |v|^2 \rangle = 0. \tag{4.7.3}$$

Hence, $f(|v|) = \delta(v)$ and therefore $y^{(p)} = 1$. This is an obvious contradiction. We can conclude that the Boltzmann equation (3.3.1) does not have non-negative self-similar solutions of the form

$$f(|v|, t) = e^{3\mu t/2} F(|v| \, e^{\mu t/2}), \tag{4.7.4}$$

which correspond to characteristic functions described in equations (4.6.1), where $x = |k|^2/2$. The same conclusion can be made independently by simple calculation of the second moment of $f(|v|, t)$:

$$m_2(t) = \langle f(|v|, t), |v|^2 \rangle = e^{-\mu t} m_2(0). \tag{4.7.5}$$

This contradicts the energy conservation for the Boltzmann equation. The contradiction vanishes only in two “extreme” cases: (1) $m_2(0) = 0$ or (2) $m_2(0) = \infty$. Case (1) was considered above. It is not related to any non-trivial positive solution of the Boltzmann equation. On the contrary, case (2) of the infinite second moment is related to interesting positive solutions of that equation, which will be discussed in Sections 5.7 and 5.8. In this section we consider only solutions with bounded second moment.

How can we use the results of Section 4.6 for the Boltzmann equation? This is the question we want to examine. It is clear that self-similar solutions (4.6.1) are related to invariance of equation (4.1.4) with respect to the scaling transformation $x \to \lambda x$, $\lambda = \text{const}$. There is also another class of invariant transformations of those equations, and we are going to use it. Indeed, we can multiply the self-similar solutions (4.6.1) by $\exp(-\gamma x)$ and consider a class of *modified* self-similar solutions

$$\varphi_\gamma^{(p)}(x, t) = e^{-\gamma x} y^{(p)}(x e^{-\mu_p t}), \quad p = 2, 3, \dots; \quad \gamma \geq 0, \tag{4.7.6}$$

where $y^{(p)}$ are constructed in Theorem 4.6.1. The estimate (4.6.2) shows that we can always use the standard inverse Fourier transform if the parameter $y \geq 0$ in (4.7.6) is sufficiently large. Hence, we obtain

$$
\begin{aligned}
f_y^{(p)}(|v|,t) &= \frac{1}{(2\pi)^3}\int\limits_{\mathbb{R}^3} dk\, y^{(p)}\Big(\frac{|k|^2}{2}e^{-\mu_p t}\Big)\exp\Big(ik\cdot v - y\frac{|k|^2}{2}\Big)\\
&= e^{\frac{3\mu_p t}{2}} F_y^{(p)}(|v|e^{\frac{\mu_p t}{2}}, t),
\end{aligned}
$$

where

$$
\begin{aligned}
F_y^{(p)}(|v|,t) &= \frac{1}{(2\pi)^3}\int\limits_{\mathbb{R}^3} dk\, y^{(p)}\Big(\frac{|k|^2}{2}\Big)\exp\Big(ik\cdot v - b\frac{|k|^2}{2}\Big),\\
b &= b(t) = ye^{\mu_p t}, \quad p \geq 2,
\end{aligned}
$$

provided y is sufficiently large. We transform this integral to

$$
F_y^{(p)}(|v|,t) = \frac{1}{2\pi^2|v|}\int\limits_0^\infty dr r y^{(p)}\Big(\frac{r^2}{2}\Big)e^{-br^2/2}\sin r|v|\,.
$$

If $p = 2$, we use equation (4.6.19) and obtain

$$
F_y^{(2)}(|v|,t) = \sum_{n=0}^\infty y_n^{(2)} I_n(z,b)\,, \quad z = |v|,
$$

where $y_0^{(2)} = 1$, $y_1^{(2)} = 0$,

$$
I_n(z,b) = \frac{1}{(2n)!!2\pi^2 z}\int\limits_0^\infty dr r^{2n+1}e^{-br^2/2}\sin rz\,, \quad n \geq 0\,. \tag{4.7.7}
$$

Then we use the tables from [94] (pp. 510 and 1051) and obtain

$$
I_n(z,b) = \frac{b^{-n}}{(2\pi b)^{3/2}}e^{-z^2/2b}L_n^{1/2}\Big(\frac{z^2}{2b}\Big), \tag{4.7.8}
$$

where $L_n^{1/2}(\cdot)$ is the nth Laguerre polynomial, $n = 0, 1, \ldots$. Hence,

$$
\begin{aligned}
F_y^{(2)}(|v|,t) &= M_b(|v|)\sum_{n=0}^\infty y_n^{(2)}(-b)^{-n}L_n^{1/2}\Big(\frac{|v|^2}{2}\Big),\\
M_b(|v|) &= (2\pi b)^{-3/2}\exp\Big(-\frac{|v|^2}{2b}\Big), \quad b = ye^{\mu_2 t} > 0\,.
\end{aligned}
$$

Finally, we obtain the following formal solution of the Boltzmann equation (3.3.1):

$$
\begin{aligned}
f_y^{(2)}(|v|,t) &= M_y(|v|)\sum_{n=0}^\infty (-1)^n y_n^{(2)} y^{-n} e^{-n\mu_2 t} L_n^{1/2}\Big(\frac{|v|^2}{2y}\Big),\\
y_0^{(2)} &= 1, \quad y_1^{(2)} = 0\,,
\end{aligned} \tag{4.7.9}
$$

where $\gamma > 0$ and $y_2^{(2)}$ and $y_3^{(2)}$ are free parameters. The coefficients $y_n^{(2)}$ for $n \geq 4$ are given in equations (4.6.14). These equations are invariant under transformation $y_n^{(2)} = \alpha^n \tilde{y}_n$ for any $\alpha = \text{const.}$, $n = 2, 3, \ldots$. On the other hand, if $f(v, t)$ is a solution of the Boltzmann equation (3.3.1), then so is the function

$$f_\beta(v, t) = \beta^3 f(\beta v, t), \quad \beta > 0.$$

Hence, the parameter γ can be removed from equation (4.7.6), i. e., we can assume that $\gamma = 1$ without any loss of generality. Then by taking free parameters $y_{2,3}^{(2)}$ sufficiently small, we can always guarantee that the weighted L^2-norm (4.2.10)–(4.2.11) of $f_1^{(2)}(v, t)$ is bounded. Similar conclusions can be made for other classes ($p = 3, 4, \ldots$) of self-similar solutions from Theorem 4.6.1. The results can be formulated in the following way.

Theorem 4.7.1. *If the kernel $g(\mu)$ of the Boltzmann equation (3.3.1) satisfies the inequality*

$$0 \leq g(\mu)(1 - \mu) \leq c, \quad \mu \in [-1, 1], \tag{4.7.10}$$

then this equation has same classes of particular solutions $f^{(p)}(|v|, t)$, $p = 2, \ldots$, directly related to self-similar solutions of equations (4.1.4) which are described in Theorem 4.6.1. These solutions have the following form:

$$f^{(2)}(|v|, t) = M(|v|)\left[1 + \sum_{n=2}^{\infty} y_n^{(2)} e^{-n\mu_2 t} L_n^{1/2}\left(\frac{|v|^2}{2}\right)\right], \tag{4.7.11}$$

$$f^{(p)}(|v|, t) = M(|v|)\left[1 + \sum_{m=1}^{\infty} y_{mp}^{(p)} e^{-m\lambda_p t} L_{mp}^{1/2}\left(\frac{|v|^2}{2}\right)\right],$$

$$M(|v|) = (2\pi)^{-3/2} \exp(-|v|^2/2), \quad p = 2, 3, \ldots, \tag{4.7.12}$$

in the notation of equations (4.6.2). The coefficients $y_p^{(p)}$ and, in addition, $y_3^{(2)}$ are free parameters for all $p = 2, 3, \ldots$. These parameters can be chosen in such a way that

$$\|f^{(p)}(\cdot, t)\|^2_{L^2_{M^{-1}}} = \int_{\mathbb{R}^3} dv\, M^{-1}(|v|)\, [f^{(p)}(|v|, t)]^2 < \infty, \quad t \geq 0, \quad p \geq 2. \tag{4.7.13}$$

The other coefficients of series (4.7.11) and (4.7.12) are defined by recursive formulas (4.6.14) and (4.6.13), respectively.

Proof. It was shown above how to obtain the solutions (4.7.11)–(4.7.12) of the Boltzmann equation by applying the inverse Fourier transform to modified self-similar solutions (4.7.6) of equation (4.1.4). On the other hand, there exists an independent way to derive the above formulas for $f^{(p)}(|v|, t)$, $p = 2, 3, \ldots$.

We can consider the Boltzmann equation (3.3.1) and look for its solution in the form of Laguerre series (4.2.8) for coefficients $\{u_n(t),\ n = 0, 1, \dots\}$ of the series. Assuming without loss of generality that $u_0(t) = 1$, $u_1(t) = 0$, we can look for special classes of solutions of ODEs (4.2.8) such that

$$u_n(t) = y_n \, e^{-n\mu t}, \quad \mu = \text{const.}, \quad n = 0, 1, \dots .$$

Then we would come to the same solutions $f^{(p)}(|v|, t)$ that are described in the formulation of the theorem. It remains to prove estimates (4.7.13). The proof is based on Lemma 4.6.2 and formula (4.2.11) for norm (4.7.13). Then, under assumption of Lemma 4.6.2, we obtain

$$\begin{aligned}
\|f^{(p)}(|v|, t)\|^2_{L^2_{M^{-1}}} &= C \sum_{m=0}^{\infty} \frac{(2mp+1)!!}{(2mp)!!} [y^{(p)}_{mp} \, e^{-m\lambda_p t}]^2 \\
&\le C \sum_{m=0}^{\infty} \frac{(2mp+1)!!}{(2mp)!!} b^m(p) a^{mp}, \quad p = 3, 4, \dots ; \\
\|f^{(2)}(|v|, t)\|^2_{L^2_{M^{-1}}} &= C \left\{ 1 + \sum_{m=0}^{\infty} \frac{(2n+1)!!}{(2n)!!} [y^{(2)}_n \, e^{-n\mu_2 t}]^2 \right\} \\
&\le C \sum_{m=0}^{\infty} \frac{(2n+1)!!}{(2n)!!} b^n(2) a^n ,
\end{aligned}$$

where C is an irrelevant constant,

$$b(p) = \frac{\lambda_{2p}}{2\lambda_p - \lambda_{2p}}, \quad p = 2, 3 \dots ,$$

in the notation of equations (4.6.2). We note that $a > 0$ is a free parameter. If $a < b^{-1}(2)$ for $p = 2$ or $a < b^{-1/p}(p)$ for $p = 3, 4, \dots$, then inequality (4.7.13) is satisfied. Indeed, the series for the $L^2_{M^{-1}}$-norm are convergent for such values of a because

$$\frac{(2n+1)!!}{(2n)!!} = \frac{2}{\sqrt{\pi}} \frac{\Gamma(n+3/2)}{\Gamma(n+1)} = 2\sqrt{\frac{n}{\pi}} + O(1), \quad n \to \infty .$$

The proof is completed. □

Solutions (4.7.11)–(4.7.12) of the Boltzmann equation (3.3.1) do not look very transparent. In particular, it is not clear if they are non-negative. However, we shall see in the next section that there is at least one family of free parameters $y^{(2)}_2$ and $y^{(3)}_2$ in (4.7.11) for which the solution has a very simple and clear form.

4.8 Exact solutions

The Boltzmann equation was firstly published in his famous paper [57]. The paper was considered by Boltzmann as a further development of ideas of Maxwell from another famous paper [115]. Thus, it so happened that the most important exact solution of the Boltzmann equation, namely, the equilibrium Maxwell distribution, was published several years before the equation itself. By the way, the first achievement of Boltzmann in kinetic theory was the generalization of the Maxwell distribution to the case of a gas in an external field. This is the so-called Maxwell–Boltzmann distribution

$$f_{M-B}(x,v) = \frac{\rho}{(2\pi T)^{3/2}} \exp\left\{-\frac{m|v-u|^2}{2T} - \frac{U(x)}{T}\right\},$$

where $U(x)$ denotes the external potential field. This formula has many applications, but here we mention it just as an exact solution of the Boltzmann equation in the presence of external field $U(x)$. Of course, the equilibrium solutions are valid for any intermolecular forces.

No non-equilibrium solution of the spatially homogeneous Boltzmann equation was known in explicit form for a long time. Then a new and very simple solution was published in 1975–1977 independently in papers [18] and [107, 108]. In the literature, this solution is sometimes called the "BKW-solution," after the names of its authors.[1]

It is surprising that this simple solution remained unknown for such a long time. Still, it is often used for testing numerical methods.

We present this solution here in two different ways. The first way was used in the original publication [18]. It is based on a very simple idea which does not use any information except for the Boltzmann equation (3.3.1) itself. Looking for the radial solution $f(|v|,t)$ of this equation

$$\frac{\partial f}{\partial t} = Q(f,f), \tag{4.8.1}$$

we note that (a) $Q(f,f)$ is a bilinear operator and (b) it maps functions of the form

$$f(|v|) = e^{-\alpha|v|^2} \sum_{n=0}^{N} a_n \, |v|^{2n} \tag{4.8.2}$$

to functions of a similar form

$$Q(f,f) = e^{-\alpha|v|^2} \sum_{m=0}^{M} b_m \, |v|^{2m}, \tag{4.8.3}$$

where $\alpha = \text{const.}$ and $M = 2N$ because of the bilinearity of Q. Hence, we cannot construct a solution of (4.8.1) in the form (4.8.2) where $\{a_n = a_n(t),\ n = 0,1,\dots,N\}$. The obvious reason is that the time derivative f_t would be proportional to polynomial $P_N(|v|^2,t)$ of Nth order in $|v|^2$, whereas the collision integral (4.8.3) would be proportional to polynomial $\tilde{P}_{2N}(|v|^2,t)$ of order $2N$ in $|v|^2$. However, there is a possibility to get

1 This solution was found even earlier by R. S. Krupp in his MS thesis [109], but the thesis was never published and mentioned for the first time in the literature only in 1984 in the paper [81] by M. H. Ernst.

polynomials of the same order on both sides of equation (4.8.1). It is enough to assume that $\alpha = \alpha(t)$ in equation (4.8.2). Then f_t becomes a polynomial of order $N_1 = N + 1$ in variable $|v|^2$. Hence, $N_1 = 2N$ if $N = 1$.

These simple considerations suggest to look for the solution of the Boltzmann equation (4.8.1) (in the notation of equation (3.3.1)) in the form

$$f(|v|, t) = [a_0(t) + a_1(t)|v|^2]e^{-\alpha(t)|v|^2}.$$

Then we assume the same normalization conditions as in (3.3.5) and obtain after straightforward calculations

$$f(|v|, t) = \frac{1}{2}\left(\frac{\beta}{2\pi}\right)^{3/2} e^{-\beta|v|^2/2}\,[5 - 3\beta + \beta(\beta - 1)|v|^2], \tag{4.8.4}$$

where

$$\beta = \beta(t) = 2\alpha(t), \quad \langle f, 1\rangle = 1, \quad \langle f, |v|^2\rangle = 3.$$

Then we try to find $\beta(t)$ from equation (4.8.1). Omitting t, we use the notation of equation (3.3.1) and obtain

$$f(|v'|)f(|w'|) - f(|v|)f(|w|) = \frac{1}{4}\left(\frac{\beta}{2\pi}\right)^3 e^{-\frac{\beta(|v|^2+|w|^2)}{2}}\,\beta^2(\beta - 1)^2\,\Delta(v', w'; v, w),$$
$$\Delta(v', w'; v, w) = |v'|^2\,|w'|^2 - |v|^2\,|w|^2 = |u|^2[(U\cdot\hat{u})^2 - (U\cdot\omega)^2],$$
$$U = \frac{v + w}{2}.$$

Simple calculations lead to

$$\int_{S^2} d\omega\, g(\hat{u}\cdot\omega)\Delta(v', w'; v, w) = \frac{g^{(2)}}{4}\,[|v|^4 + |w|^4 - 4\,|v|^2\,|w|^2 + 2(v\cdot w)^2],$$

$$g^{(2)} = 2\pi \int_{-1}^{1} ds g(s)(1 - s^2),$$

in the notation of equation (3.3.1). Then we evaluate the collision integral with function (4.8.4) and obtain

$$Q(f, f) = \frac{g^{(2)}}{16}(\beta - 1)^2(15 - 10\beta|v|^2 + \beta^2\,|v|^4)M_\beta(|v|),$$
$$M_\beta(|v|) = \left(\frac{\beta}{2\pi}\right)^{3/2}\exp\left(-\frac{\beta|v|^2}{2}\right).$$

Then we obtain for the same function

$$\frac{\partial f}{\partial t} = \beta_t\frac{\partial f}{\partial \beta} = -\frac{\beta_t(\beta - 1)}{4\beta}(15 - 10\beta|v|^2 + \beta^2\,|v|^4)M_\beta(|v|).$$

Hence, equation (4.8.1) is satisfied if

$$\beta_t = -\frac{g^{(2)}}{4}\beta(\beta - 1)\,.$$

The general solution of this equation reads as

$$\beta(t) = \left[1 + \theta e^{-\mu t}\right]^{-1}, \quad \mu = \frac{g^{(2)}}{4} = \frac{\pi}{2}\int_{-1}^{1} ds g(s)(1 - s^2)\,, \tag{4.8.5}$$

where $\theta = \text{const}$.

It is easy to find the Fourier transform of the solution $f(|v|, t)$ given in equations (4.8.4) and (4.8.5). After some calculations, we obtain

$$\varphi(x, t) = \langle f, e^{-ik\cdot v}\rangle = \left[1 + \theta e^{-\mu t}\right] e^{-x(1+\theta e^{-\mu t})}\,, \quad x = \frac{|k|^2}{2}\,. \tag{4.8.6}$$

Note that $\varphi(x, t)$ satisfies equations (4.1.4) and (4.1.5). It can be written as

$$\varphi(x, t) = e^{-x}\, y^{(2)}(x e^{-\mu t})\,, \quad \mu = \mu_2 = \mu_3\,,$$

in the notation of Theorem 4.6.1. This is a particular case of the function $y^{(2)}(x)$ given in equations (4.6.19) and (4.6.14). Note that

$$y^{(2)}(x) = e^{-\theta x}\,(1 + \theta x) = 1 + \sum_{n=1}^{\infty} y_n^{(2)} \frac{x^n}{n!}\,, \tag{4.8.7}$$

where

$$y_n^{(2)} = (1 - n)(-\theta)^n\,, \quad n \geq 2\,. \tag{4.8.8}$$

We remind the reader that $y^{(2)}(x)$ in Theorem 4.6.1 is a solution of equation (4.6.2) in the form of power series (4.6.19), where the first two coefficients $y_{2,3}^{(2)}$ can be chosen arbitrarily. In this particular case we have

$$y_2^{(2)} = -\theta^2\,, \quad y_3^{(2)} = 2\theta^3\,. \tag{4.8.9}$$

It is easy to check that other coefficients $\{y_n^{(2)},\ n \geq 4\}$ satisfy equations (4.6.14), as expected.

It should be stressed that the function $y^{(2)}(x)$ from (4.8.7) is the only known nontrivial, i. e., $y^{(2)}(x) \neq \exp(-\theta x)$, explicit solution of equation (4.6.2). The same is true for the related self-similar solution of (4.1.4). Of course, this solution can also be "guessed" by considerations used for the Boltzmann equation itself at the beginning of Section 4.8.

Until now we did not discuss the important question of positivity of solutions from Sections 4.6 and 4.7. Generally speaking, this question remains open (see, e. g., [13]) with the exception of the explicit solution (4.8.4)–(4.8.5). It is easy to check that this solution is non-negative for all $t \geq 0$ if and only if $-2/5 \leq \theta \leq 0$. Note that the tail temperature $\tau(t)$ (see Definition 4.3.2) of the exact solution $f(|v|, t)$ reads

$$\tau(t) = 1 + \theta e^{-\mu t}.$$

It will be shown in Chapter 5 that $\tau'(t) \geq 0$ for positive ("physical") solutions of the Boltzmann equation. Therefore, it is not surprising that θ is negative for such solutions. We can use $\tau(t)$ as the time-dependent parameter and represent the exact solution of equation (3.3.1) in the form

$$\begin{aligned} f(|v|, t) &= (2\pi\tau)^{-3/2} e^{-|v|^2/2\tau} \frac{1}{2\tau}\left[(5\tau - 3) + (1 - \tau)\frac{|v|^2}{\tau}\right], \\ \tau(t) &= \tau_0 e^{-\mu t} + (1 - e^{-\mu t}); \\ \mu &= \frac{\pi}{2}\int_{-1}^{1} ds g(s)(1 - s^2), \quad 3/5 \leq \tau_0 \leq 1. \end{aligned} \qquad (4.8.10)$$

The initial tail temperature $\tau_0 = \tau(0)$ is a free parameter in this representation of the solution. Its Fourier transform is given in equation (4.8.6) with $\theta = \tau_0 - 1$. We also present below some useful formulas for moments of $f(|v|, t)$, etc.

1. The Laguerre series representation

$$f(|v|, t) = (2\pi)^{-3/2} e^{-|v|^2/2} \sum_{n=0}^{\infty} (1 - n)(1 - \tau_0)^n e^{-n\mu t} L_n^{1/2}(|v|^2/2) \qquad (4.8.11)$$

   follows directly from Theorem 4.7.1 and related series (4.8.7)–(4.8.8) with $\theta = \tau_0 - 1$.

2. Normalized moments of $f(|v|, t)$

$$z_n(t) = \frac{1}{(2n + 1)!!}\langle f, |v|^{2n}\rangle, \quad n = 0, 1, \ldots, \qquad (4.8.12)$$

   can be easily found from the related representation of $\varphi(|k|, t)$ (4.8.6)

$$\varphi(|k|, t) = \sum_{n=0}^{\infty} (-1)^n \frac{z_n(t)}{n!}\left(\frac{|k|^2}{2}\right)^n. \qquad (4.8.13)$$

   Then we obtain

$$z_n(t) = \tau^{n-1}\{1 + (n - 1)[1 - \tau(t)]\}, \quad n = 0, 1, \ldots, \qquad (4.8.14)$$

   in the notation of equations (4.8.10).

Finally, we show how to obtain an exact solution $f(x, v, t)$ of the spatially inhomogeneous Boltzmann equation

$$f_t + v \cdot f_x = Q(f, f),$$

where $x \in \mathbb{R}^3$ denotes the position of the particle. The idea of general transformation of this equation to its spatially homogeneous form is due to A. A. Nikolskii [121–123]. We look for a solution in the form

$$f(x, v, t) = F(vt - x, t). \tag{4.8.15}$$

It is easy to check that $F(v, t)$ satisfies the equation

$$t^3 F_t = Q(F, F).$$

Then we obtain

$$F(v, t) = f_1(v, t_1), \quad t_1 = C - \frac{1}{2t^2},$$

where $f_1(v, t_1)$ is the solution of the spatially homogeneous Boltzmann equation (4.8.1) in variables (v, t_1). Hence, we can take the above discussed exact solution (4.8.10) of that equation as the function $f_1(v, t_1)$ (with obvious change of notation). Then we construct $F(v, t)$ and finally return to initial variables by transformation (4.8.15). Thus we also obtain a related exact solution of the spatially inhomogeneous Boltzmann equation. The resulting formulas can be found in [18]. See also the paper [120] by R.G. Muncaster, who obtained a similar result independently.

Remarks on Chapter 4

1. The presentation in Chapter 4 is partly based on papers [19, 25, 27]. However, in this book we do not consider all questions related to the Poincaré normal form (see, e. g., [6]) of the Fourier transformed Boltzmann equation, resonances, and other questions connected with linearization of the Boltzmann equation. Readers who are interested in that kind of questions can find some information in [22, 25, 27, 146].
2. The review of results on various generalizations of exact BKW-solutions (see Section 4.8) can be found in [21, 80, 81]. New exact solutions obtained in the 2000s are discussed in Section 5.8. An interesting exact solution of the inelastic Maxwell–Boltzmann equation and its connection with BKW-mode is discussed in Section 6.3.
3. The exact solution (4.8.10) was in fact the main result of the unpublished thesis [109]. The Fourier transform was used in [109] only as a tool to get the exact solution in a simpler way. In particular, the identity (3.2.8) was not obtained there.

5 Asymptotic problems

5.1 Formation of Maxwellian tails

The exact solution of the Boltzmann equation studied in Section 4.8 obviously belongs to the class of rapidly decreasing (as $|v| \to \infty$) distribution functions introduced in Section 4.3. The asymptotic behavior of these functions for large velocities can be characterized by the so-called tail temperature (see Definition 4.3.2). Our aim in this section is to study the temporal evolution of this asymptotic characteristics for solutions of the Boltzmann equation. Here and below, we shall mainly consider radial solutions $f(|v|, t)$. For the reader's convenience we recall some results of Chapter 4. Let $f(|v|, t)$ be a solution of the Cauchy problem

$$f_t = Q(f, f), \quad f|_{t=0} = f_0(|v|) \in B_+, \tag{5.1.1}$$

in the notation of equations (3.3.1). We remind the reader that the class B_+ is defined in the following way:

$$B_+ = \{f(|v|) \geq 0 \colon \langle f, 1\rangle = 1, \ \langle f, |v|^2\rangle = 3; \\ \exists\, r > 0 \text{ s.t. } \langle f, e^{r|v|^2/2}\rangle < \infty\}. \tag{5.1.2}$$

The generalized solution $f(|v|, t)$ of the problem (5.1.1) can be defined in a natural way through the solution of the related Fourier transformed problem. The Fourier transformed problem was studied in detail in Section 4.5. The proof of existence and uniqueness for all $t > 0$ of the generalized solution $f(|v|, t) \in B_+$ to the problem (5.1.1) was also given there (see also Section 3.3). In this section we consider the tail temperature $\tau(t)$ of the solution $f(|v|, t) \in B_+$ defined by equality

$$\frac{1}{\tau(t)} = \sup\left\{r > 0 \colon \left\langle f(|v|, t), \exp\left(\frac{r}{2}|v|^2\right)\right\rangle < \infty\right\}. \tag{5.1.3}$$

Note that $\tau(t)$ is a monotone function; it never decreases with time t. Indeed we assume that $g(\mu) \in L^1([-1, 1])$ in equations (5.1.1) and (3.1.1). Then, under standard normalization conditions

$$\langle f, 1\rangle = 1, \quad \langle f, |v|^2\rangle = 3, \tag{5.1.4}$$

we can transform equation (5.1.1) to its integral form

$$f(|v|, t) = f_0(|v|)e^{-g_{\text{tot}}t} + \int_0^t d\tau e^{-g_{\text{tot}}(t-\tau)} Q[f(\tau), f(\tau)], \quad g_{\text{tot}} = 2\pi \int_{-1}^{1} d\mu g(\mu).$$

Obviously,

$$f(|v|, t) \geq f_0(|v|)e^{-g_{\text{tot}}} \tag{5.1.5}$$

https://doi.org/10.1515/9783110550986-006

and therefore

$$\tau(t) \geq \tau(0) \quad \text{if} \quad t \geq 0$$

or, equivalently,

$$\tau(t_1) \geq \tau(t_2) \quad \text{if} \quad t_1 \geq t_2\,.$$

Similarly one can prove that the monotonicity of $\tau(t)$ holds for the Boltzmann equation with any kernel $g(|u|, \mu)$ in (2.1.20) such that [53]

$$g_{\text{tot}}(|u|) = 2\pi \int_{-1}^{1} d\mu g(|u|, \mu) \leq a + b\,|u|^{2-\varepsilon}\,,$$

with some positive constants a, b, and ε.

Our approach to study $\tau(t)$ is based on the formula

$$\tau(t) = \limsup_{n\to\infty} (z_n)^{1/n}, \quad z_n = \frac{1}{(2n+1)!!}\langle f(|v|, t), |v|\rangle^{2n}\,, \tag{5.1.6}$$

derived in Section 4.4, Lemma 4.4.1. Another tool that we use is the set of ordinary differential equations (ODEs) (4.5.8)

$$\dot z_0 = \dot z_1 = 0, \quad \dot z_n + \lambda_n z_n = \sum_{k=1}^{n-1} \binom{n}{k} h_{k,n-k} z_k z_{n-k}\,, \quad n = 2, 3, \ldots, \tag{5.1.7}$$

in the notation of equations (4.2.3) and (4.2.14). The problem (5.1.1) implies the following initial data for these equations:

$$z_n(0) = z_n^{(0)} = \frac{1}{(2n+1)!!}\langle f_0, |v|\rangle^{2n}\,, \quad n = 0, 1, \ldots. \tag{5.1.8}$$

Note that $z_0^{(0)} = z_1^{(0)} = 1$ since $f_0 \in B_+$.

It is important that coefficients λ_n, $h_{k,l}$ and initial data $z_n^{(0)}$ in equations (5.1.7) and (5.1.8) are non-negative for all $n, k, l \geq 0$. We can also prove that $z_n(t)$ depends monotonically on these parameters.

Lemma 5.1.1. *Let $\tilde z_n(t)$ be a solution of the set of equations obtained from* (5.1.7) *and* (5.1.8) *by replacing λ_n, $h_{k,l}$, and $z_n^{(0)}$ by $\tilde\lambda_n$, $\tilde h_{k,l}$, and $\tilde z_n^{(0)}$ for all $n = 0, 1, \ldots; k, l = 1, \ldots$. Then if $\tilde\lambda_n \geq \lambda_n$, $0 \leq \tilde h_{k,l} \leq h_{k,l}$, $0 \leq \tilde z_n^{(0)} \leq z_n^{(0)}$, we have $\tilde z_n(t) \leq z_n(t)$ for all $t > 0$. Conversely, if $\tilde\lambda_n \leq \lambda_n$, $\tilde h_{k,l} \geq h_{k,l}$, $\tilde z_n^{(0)} \geq z_n^{(0)}$, we have $\tilde z_n(t) \geq z_n(t)$ for all $t > 0$.*

Proof. It is sufficient to consider recurrence formulas of the form (4.5.9) for solution of the problem (5.1.7)–(5.1.8) and to use induction over n. This completes the proof. □

The lemma enables us to construct minorants and majorants not only for the tail temperature $\tau(t)$, but also for the whole set of normalized moments $z_n(t)$, $n = 0, 1, \ldots$. To use the lemma it is necessary to know some exact solutions of system (5.1.7). These are:

1. the steady-state solution

$$z_n = c^n, \quad n = 0, 1, \ldots, \tag{5.1.9}$$

   where c is an arbitrary positive number, and
2. the simplest non-stationary solution (4.8.14) in slightly modified form

$$z_n(t) = \left(1 - \theta e^{-\lambda t}\right)^{n-1} \left[1 + (n-1)\theta e^{-\lambda t}\right], \quad n = 0, 1, \ldots, \tag{5.1.10}$$

   where $0 \le \theta \le 1$, $\lambda = \lambda_2/2 = \lambda_3/3$.

In addition, we will further consider different versions of the "faulty" system (5.1.7). The simplest version consists of neglecting the non-linear terms of this system, and gives the following obvious estimate, which holds for all $t \ge 0$:

$$z_n(t) \ge z_n(0) \exp[-\lambda_n t], \quad n = 2, 3, \ldots. \tag{5.1.11}$$

Using the lemma and equations (5.1.9) and (5.1.10), we can now establish the general properties of the tail temperature $\tau(t)$ of the solution $f(|v|, t)$ of the Cauchy problem (5.1.1).

Proposition 5.1.1. *The function $\tau(t)$ does not decrease as t decreases, and for all $t \ge 0$ we have the inequality*

$$1 - e^{-\lambda t} \le \tau(t) \le a, \tag{5.1.12}$$

where λ is the same as in (5.1.10), *while the constant a depends on the initial condition*

$$a = \sup_{n=0,1,\ldots} \sqrt[n]{z_n(0)}. \tag{5.1.13}$$

Proof. The upper estimate in (5.1.12) follows by comparing the solution of problem (5.1.7)–(5.1.8) with accurate steady-state solutions of $\tilde{z}_n = a^n$, $n = 0, 1, \ldots$, and was essentially obtained in Section 4.5. The lower estimate in (5.1.12) is obtained by comparing the solution of this problem with the exact solution (5.1.10) for $\theta = 1$, since the latter corresponds to the initial conditions

$$\tilde{z}_0^{(0)} = \tilde{z}_1^{(0)} = 1, \quad \tilde{z}_n^{(0)} = 0, \quad n = 0, 1, \ldots, \tag{5.1.14}$$

which sets a lower limit to the initial condition (5.1.8) for any distribution function $f_0(|v|) \in B_+$. It is clear from these discussions that we have the stronger inequalities

$$\left(1 - \theta e^{-\lambda t}\right)^{n-1} \left[1 + (n-1)\theta e^{-\lambda t}\right] \le z_n(t) \le a^n, \quad n = 0, 1, \ldots, \tag{5.1.15}$$

a consequence of which is inequality (5.1.12).

It remains to establish the monotonic dependence of $\tau(t)$ on time. It follows from (5.1.11) that for any $t > 0$

$$\tau(t) \geq \limsup_{n\to\infty} \left\{ [z_n(0)]^{1/n} \exp\left(-\frac{\lambda_n t}{n}\right)\right\},$$

and to prove the inequality $\tau(t) \geq \tau(0)$ it is sufficient to prove that

$$\lim_{n\to\infty} \frac{\lambda_n}{n} = 0. \tag{5.1.16}$$

On the other hand, the inequality $\tau(t) \geq \tau(0)$ for $t \geq 0$ is equivalent to the inequality $\tau(t_1) \geq \tau(t_2)$ for all $t_1 \geq t_2$, since equations (5.1.7) are invariant under a time shift. Consequently, to conclude the proof of the proposition it is sufficient to prove equation (5.1.16), which is obvious for pseudo-Maxwell molecules. In general, we will use condition (4.5.6), and we obtain the estimate

$$\lambda_n < c \int_0^1 ds s^{-2+\varepsilon} [1 - s^n - (1-s)^n], \quad n = 2, 3, \dots,$$

where c and ε are fixed positive numbers (without loss of generality we can assume that $\varepsilon < 1$). Expressing the integral on the right hand side in terms of the gamma function, we obtain

$$\lambda_n < \frac{c}{1-\varepsilon}\left\{\frac{-n}{n-1+\varepsilon} + \Gamma(\varepsilon)\frac{n!}{\Gamma(n+\varepsilon)}\right\} \underset{n\to\infty}{\approx} c\frac{\Gamma(\varepsilon)}{1-\varepsilon} n^{1-\varepsilon},$$

whence equation (5.1.16) follows. The proposition is proved. □

Remark. The monotonic dependence of $\tau(t)$ on time was already proved for pseudo-Maxwell molecules on the basis of estimate (5.1.5). The proposition just extends this proof to the case on non-integrable kernels, satisfying condition (4.5.6). In particular, this condition is satisfied for true Maxwell molecules.

The interest from the point of view of physics of the problem is the fact that the Maxwellian tail is formed during relaxation of the initial condition with compact support, i. e., in the case when $f_0(|v|) \equiv 0$ when $|v| > v_0$, where v_0 is a certain limit velocity. For such initial conditions, obviously, $\tau(0) = 0$. Estimate (5.1.12) shows that the equation $\tau(t) = 0$ is only possible when $t = 0$, and when $t > 0$ the solution $f(|v|, t)$ of problem (5.1.1), roughly speaking, decreases as $|v| \to \infty$ no more rapidly than $\exp[-|v|^2/2(1-e^{-\lambda t})]$, but also no more slowly that $\exp(-|v|^2/2a)$. The fact that any continuous non-negative solution of the Boltzmann equation when $t > 0$ is bounded below by a function of the form $A\exp[-B|v|^{2+\varepsilon}]$ for any $\varepsilon > 0$ was discovered by Carleman [64] (see also [129]), but for Maxwell molecules we can obtain a much more accurate integral estimate of the decrease in the solution as $v \to \infty$. In this case, since we are considering generalized solutions, we estimate not the distribution function itself, but a certain integral characteristic of it, namely, the tail temperature $\tau(t)$.

In the next section we will extend our study of the process by which Maxwellian tails are formed, by giving particular attention to the case of initial conditions with compact support.

5.2 Refined estimates of the normalized moments and the tail temperature

We will begin with an example of the construction of more accurate estimates of the law of time evolution of the tail temperature (and normalized moments) for a certain class of initial conditions with compact support. A typical function of this case is the monoenergetic distribution

$$f_0(|v|) = \frac{1}{2\pi\sqrt{3}}\,\delta(|v|^2 - 3)\,, \tag{5.2.1}$$

where the constants are chosen so as to satisfy the normalization conditions (5.1.4). Assuming the possibility of "spreading" of the delta function over a certain finite interval (not too great, see below), we obtain the class of initial distributions for which the following proposition applies.

Proposition 5.2.1. *Suppose that in (5.1.1) $f_0(|v|) \equiv 0$ when $|v| > v_0$, where $3 \le v_0^2 \le 5$. We then have the inequalities*

$$\begin{aligned} (1 - e^{-\lambda t})^{n-1}[1 + (n-1)e^{-\lambda t}] &\le z_n(t) \\ &\le (1 - \theta e^{-\lambda t})^{n-1}[1 + (n-1)\theta e^{-\lambda t}], \quad n = 0, 1, \ldots; \end{aligned} \tag{5.2.2}$$

$$1 - e^{-\lambda t} \le \tau(t) \le 1 - \theta e^{-\lambda t}\,, \tag{5.2.3}$$

where λ is the same as in (5.1.10) and $\theta = (1 - v_0^2/5)^{1/2}$.

Proof. It is sufficient to establish that inequalities (5.2.2) hold when $t = 0$ and then use Lemma 5.1.1. It follows from the condition of the proposition that

$$z_0^{(0)} = z_1^{(0)} = 1, \quad z_n^{(0)} \le \frac{3v_0^{2(n-1)}}{(2n+1)!!}, \quad n = 2, 3, \ldots\,,$$

and, consequently, it remains to prove the elementary inequalities

$$\frac{3v_0^{2(n-1)}}{(2n+1)!!} \le (1-\theta)^{(n-1)}[1 + (n-1)\theta], \quad \theta = \sqrt{1 - \frac{v_0^2}{5}}, \quad n = 2, \ldots.$$

Assuming here that

$$\alpha = 1 - \theta, \quad v_0^2 = 5\alpha(2 - \alpha),$$

we reduce the problem to proving the inequalities (for $0 < \alpha < 1$)

$$[n - \alpha(n-1)]\,(2-\alpha)^{1-n}\,(2n+1)!! \geq 3 \cdot 5^{n-1}, \quad n = 2, \ldots .$$

By differentiating the left hand side with respect to α it can be shown that it is sufficient to consider the case when $\alpha = 0$, and proof of the corresponding numerical inequalities requires no explanation. The proposition is proved, since (5.2.3) follows immediately from (5.2.2), in accordance with equation (5.1.6). □

Hence, for the initial conditions considered above the moments of the solution of the Cauchy problem (5.1.1) for all $t \geq 0$ lie between the corresponding moments of the two exact solutions from Section 4.8, which differ solely in the choice of the time origin. Relaxation of the moments is here almost monotonic functions of time t. The tail temperature $\tau(t)$ also approaches the equilibrium value $\tau(\infty) = 1$ as t increases.

We will now consider a somewhat different type of relaxation which shows that for initial conditions with compact support ($\tau(0) = 0$) the tail temperature $\tau(t)$ may increase as much as desired in a specified time interval $\Delta(t)$. To do this we will use Lemma 5.5.1 to construct a minorant of the solution of the Cauchy problem (5.1.7)–(5.1.8) as follows:

1. fixing the natural number $m \geq 2$, we replace the initial conditions (5.1.8) by

$$\tilde{z}_0 = 1, \quad \tilde{z}_n^{(0)} = \tilde{z}_m^{(0)}\,\delta_{nm}, \quad n = 1, 2, \ldots ; \tag{5.2.4}$$

2. using the inequality

$$\lambda_n = \int_0^1 ds\rho(s)[1 - s^n - (1-s)^n] \leq n \int_0^1 ds\rho(s)s\,, \quad n = 2, 3, \ldots ,$$

   based on the elementary estimate

$$ns - [1 - s^n - (1-s)^n] \geq 0\,, \quad 0 \leq s \leq 1, \quad n \geq 2\,,$$

   we replace the eigenvalues λ_n, $n = 2, \ldots$, of the linear part of system (5.1.7) by

$$\tilde{\lambda}_n = n\delta, \quad \delta = \int_0^1 ds\rho(s)s, \quad n = 2, \ldots ; \tag{5.2.5}$$

3. we keep the right hand sides of equations (5.1.7) unchanged.

The solution $\{\tilde{z}_n(t),\ n = 0, 1, \ldots\}$ of the “faulty” Cauchy problem thus obtained, as can be easily seen, will be expressed by the equations

$$\tilde{z}_n(t) = \begin{cases} 0, & \text{for } n \neq pm,\ p = 0, 1, \ldots , \\ y_p(t), & \text{for } n = np,\ p = 0, 1, \ldots , \end{cases}$$

$$y_0(t) = 1, \quad y_p(t) = u_p\big[z_m(0)\big]^p t^{p-1} e^{-pm\delta t}, \quad p = 1, \dots ,$$

where δ is the same as in (5.2.5), while the numerical sequence $\{u_p,\ p = 1, \dots\}$ is defined recurrently:

$$u_1 = 1, \quad u_p = \frac{1}{p-1} \sum_{k=1}^{p-1} \binom{pm}{km} h_{km,(p-k)m}\, u_k\, u_{p-k}, \quad p = 2, \dots . \tag{5.2.6}$$

According to Lemma 5.1.1, for any $m = 2, 3, \dots$, we have the inequality

$$z_n(t) \geq \tilde{z}_n(t), \quad n = 0, 1, \dots ,$$

where $\{z_n(t),\ n = 0, 1, \dots\}$ is the solution of problem (5.1.7)–(5.1.8). Hence, we obtain immediately an estimate for the tail temperature

$$\tau(t) = \limsup_{n\to\infty} \sqrt[n]{z_n(t)} \;\geq\; t^{1/m} e^{-\delta t} \sqrt[m]{z_m(0)}\; \limsup_{p\to\infty} \sqrt[pm]{u_p}\,. \tag{5.2.7}$$

It only remains to establish the existence of a non-zero limit on the right hand side. To do this we can assume that the function $\rho(s) \geq 0$ in (4.2.3) and (4.2.14) satisfies the stronger inequality

$$\rho(s) \geq \varepsilon > 0, \quad 0 < s < 1\,.$$

Then

$$h_{kl} = \int_0^1 ds \rho(s) s^k (1-s)^l \geq \varepsilon \frac{k!l!}{(k+l+1)!}, \tag{5.2.8}$$

and the recurrent sequence (5.2.6) is estimated from below by a recurrent sequence of the form

$$\tilde{u}_1 = 1, \quad \tilde{u}_p = \frac{\varepsilon}{(p-1)(pm+1)} \sum_{k=1}^{p-1} \tilde{u}_k\, \tilde{u}_{p-k}, \quad p = 2, \dots .$$

Hence, after using induction and carrying out elementary calculations, based on the identity

$$\sum_{k=1}^{k} (p-k) = \frac{1}{6} p\,(p^2 - 1)\,,$$

we obtain the following final estimate for the sequence (5.2.6):

$$u_p \geq \tilde{u}_p \geq p \left(\frac{\varepsilon}{6m}\right)^{p-1}, \quad p = 1, 2 \dots .$$

Substituting these inequalities into (5.2.7) and combining the result with estimate (5.1.12), we arrive at the conclusion which proves the following proposition.

Proposition 5.2.2. *The function $\tau(t)$ for all $m = 2, 3, \dots$ satisfies the following inequalities:*

$$\left(\frac{\varepsilon t}{6m}\right)^{1/m} e^{-\delta t} \sqrt[m]{z_m(0)} \le \tau(t) \le \sup_{n=0,1,\dots} \sqrt[n]{z_n(0)}, \tag{5.2.9}$$

where

$$\delta = \int_0^1 ds \rho(s) s, \quad \varepsilon = \inf \rho(s), \quad s \in (0,1).$$

It is convenient to use a simplified notation $|v| = v \ge 0$ in the rest of this section. Then it is easy to see that even in the class of initial conditions $f_0(v) \in B_+$ with compact support the values of the normalized moments

$$\sqrt[m]{z_m(0)} = \frac{4\pi}{(2m+1)!!} \int_0^\infty dv f_0(v) v^{2(m+1)}, \quad m = 0, 1, \dots,$$

for $m \ge 2$ can reach as high a value as desired for fixed values of $z_0(0) = z_1(0) = 1$. In fact, we can specify the function $f_0(v)$ to be a decreasing one, for example, proportional to v^{-6}, as $v \to \infty$, and then “cut off” its tail at a fairly large distance $R \gg 1$ from the origin of coordinates. Then the normalized moments z_0 and z_1 will be practically independent of the number R, while the moments z_m will increase without limit as R increases for $m \ge 2$.

To clarify the situation we will give the following elementary example. We specify the arbitrary number $M \gg 1$, and also numbers $A(M)$ and $v_0(M)$, which depend on M, such that

$$\frac{1}{A(M)} = \int_0^M \frac{dx x^2}{1+x^6}, \quad \frac{3}{A(M) v_0^2(M)} = \int_0^M \frac{dx\, x^4}{1+x^6}.$$

We will consider as the initial conditions for the Boltzmann equation the distribution function $f_0(v)$ with compact support, defined by the equation

$$f_0(v) = \frac{A(M)}{4\pi v_0^3(M)} \left[1 + \left(\frac{v}{v_0}\right)^6\right]^{-1} \eta[M v_0 - v], \tag{5.2.10}$$

where $\eta(x)$ is a unit function. By calculating the first moments of the function $f_0(v)$, we can show that $z_0(0) = z_1(0) = 1$, in agreement with (5.2.3), and

$$z_2(0) = \frac{A(M)\, v_0^4(M)}{15} \left[M - \int_0^M \frac{dx}{1+x^6}\right].$$

After simple calculations, we obtain the following asymptotic formulas as $M \to \infty$:

$$A(M) \cong A(\infty) = \frac{6}{\pi}, \quad v_0^2(M) \cong v_0^2(\infty) = \frac{3}{2}, \quad z_2(0) \cong \frac{9M}{10\pi},$$

which fairly accurately estimate the increase in $z_2(0)$ as the number M increases.

The lower estimate from (5.2.9) can now be rewritten for $m = 2$ and $M \gg 1$ in the following approximate form:

$$\tau(t) \geq \sqrt{\frac{\varepsilon z_2(0)}{12} t\, e^{-\delta t}} = \sqrt{\frac{3\varepsilon M}{40\pi} t\, e^{-\delta t}},$$

whence it is clear that by fixing the arbitrary (as small as desired) initial time interval we can obtain as large an increase in the value of $\tau(t)$ as we want in this time interval by a corresponding increase in the number M, which occurs in the definition of the initial condition (5.2.10). Here, at the instant of time t the equation $\tau(0) = 0$ will always be satisfied since the function (5.2.10) has a compact support for any value of M. Noting, finally, that according to the results in Section 5.1 the function $\tau(t)$ cannot decrease, we can formulate the proved property of the solutions $f(|v|, t) \in B_+$ of the Cauchy problem (5.1.7)–(5.1.8) in the form of the following theorem.

Theorem 5.2.1. *Let the scattering kernel $g(\mu)$ on the right hand side of* (5.1.1) *be such that*

$$\inf_{-1 \leq \mu \leq 1} g(\mu) > 0. \tag{5.2.11}$$

Then for any numbers $N > 0$ and $\Delta t > 0$ we can specify a certain initial condition $f_0(|v|) \in B_+$ in (5.1.1) *such that (a) $f_0(|v|)$ has a compact support and (b) for all $t \geq \Delta t$ the inequality $\tau(t) > N$ is satisfied.*

We note that the example of the initial condition (5.2.10), constructed to prove the theorem, obviously belongs to the Hilbert space $\hat{L}^2_{M^{-1}}$ with norm (4.2.10). Hence, the following corollary holds.

Corollary 1. *If inequality* (5.2.11) *is satisfied, the Cauchy problem* (5.1.1) *does not have, generally speaking, a global solution with respect to time t in the Hilbert space $L^2_{M^{-1}}$ with norm* (4.2.10) *or at least in the subset of non-negative functions of this space.*

This corollary is also a rigorous justification of the insufficiency of the space $L^2_{M^{-1}}$ for constructing the global in time existence theory. Note that inequality (5.2.11) may be considerably weakened when the correctness of the theorem and the corollary is preserved. However, this would require some extension of the proofs whereas inequality (5.2.11) is obviously satisfied for the models often used (isotropic scattering $\rho(s) \equiv 1$ and true Maxwell molecules $\rho(s) \geq \rho(0) > 0$). That is why we have chosen this approach.

To conclude this section we note that to construct an asymptotic theory of relaxation (as $|v| \to \infty$), we can also use a more general method based on the Fourier transformation in the complex domain (see the next section).

5.3 Application of complex Fourier transform

In Sections 5.1 and 5.2 we studied asymptotic properties of radial solutions $f(|v|, t) \in B_+$. In this section we want to extend our study to the more general class of anisotropic distribution functions $f(v, t)$, $v \in \mathbb{R}^3$. We keep the same notation B_+ for the extended class of functions

$$B_+ = \{f(v) \geq 0 \colon \langle f, 1\rangle = 1, \langle f, v\rangle = 0, \langle f, |v|^2\rangle = 3; \\ \exists r > 0 \text{ s.t. } \langle f, e^{r|v|^2/2}\rangle < \infty\}. \tag{5.3.1}$$

Then we consider the Cauchy problem

$$f_t = Q(f, f), \quad f|_{t=0} = f_0(v) \in B_+, \tag{5.3.2}$$

in the notation of equation (5.3.1). Moreover we consider the case of pseudo-Maxwell molecules. Then we can assume that

$$g_{\text{tot}} = 2\pi \int_{-1}^{1} d\mu g(\mu) = 1 \tag{5.3.3}$$

in the collisional integral $Q(f, f)$. Note that the solution of the problem (5.3.2) was constructed in Section 3.3 in the form of the Wild sum (3.3.10). This solution $f(v, t)$ is a probability density (perhaps in the generalized sense); it preserves mass, momentum, and energy. However, we would like to prove that $f(v, t) \in B_+$ for all $t > 0$. It is also desirable to prove that the tail temperature $\tau(t)$ of $f(v, t)$ (defined in the same way as in equation (5.1.3)) is uniformly bounded for all $t > 0$. These problems are solved in Sections 5.1 and 5.2 for radial solutions by using equations for moments. In principle, a similar approach can be applied to anisotropic solutions. However, the anisotropic moment equations are more complicated and we prefer another approach. This approach is partly based on the paper [49].

It is not difficult to show that for some short time $t \in [0, T(f_0)]$ the solution $f(v, t)$ does remain in the class B_+. It can be shown by straightforward integration of the Wild sum (3.3.10). Hence, at least for $0 < t \leq T(f_0)$ we can consider the complex Fourier transform

$$\psi(k, t) = \varphi(ik, t) = \langle f, e^{k \cdot v}\rangle, \quad k \in \mathbb{R}^3, \tag{5.3.4}$$

where $\varphi(k,t)$, $k \in \mathbb{R}^3$, stands for the usual Fourier transform (3.3.14). Repeating almost without changes the derivation of the transformed equation from Sections 3.2 and 3.3, we obtain exactly the same equation as equation (3.3.15) for $\psi(k,t)$, i. e.,

$$\psi_t + \psi = \hat{Q}_+(\psi,\psi) = \int_{S^2} d\omega\, g(\hat{k}\cdot\omega)\psi(k_+)\,\psi(k_-)\,,$$
$$k_\pm = (k \pm |k|\omega), \quad \hat{k} = k/|k|, \quad k \in \mathbb{R}^3. \tag{5.3.5}$$

It is not surprising, since equation (3.3.15) is invariant under scaling transformations $k \to \lambda k$, $\lambda = \text{const}$. We assume that $f_0 \in B_+$. Therefore,

$$C^2(a_0) = \langle f_0, e^{a_0|v|^2}\rangle < \infty \tag{5.3.6}$$

for some $a_0 > 0$. Hence,

$$\begin{aligned}\psi_0 = \langle f_0, e^{k\cdot v}\rangle &= \Big\langle \sqrt{f_0}\, e^{\frac{a_0}{2}|v|^2}, \sqrt{f_0}\, e^{-\frac{a_0}{2}|v|^2 + k\cdot v}\Big\rangle \\ &\le C(a_0)\left\{\left\langle f_0, \exp\left[-a_0\left(v - \frac{k}{a_0}\right)^2\right]\right\rangle e^{\frac{|k|^2}{a_0}}\right\}^{1/2} \\ &\le C(a_0)\,\langle f, 1\rangle\, e^{\frac{|k|^2}{2a_0}},\end{aligned}$$

in the notation of equation (5.3.6). Therefore, we obtain the estimate

$$\psi|_{t=0} = \psi_0(k) \le C(a_0)\exp\left(\frac{|k|^2}{2a_0}\right) \tag{5.3.7}$$

for the initial data to equation (5.3.5). Then the following statement is valid.

Lemma 5.3.1. *If $f(v) \in B_+$, then there exists such $y > 0$ that*

$$\psi(k) = \langle f, e^{k\cdot v}\rangle \le e^{y|k|^2}, \quad k \in \mathbb{R}^3. \tag{5.3.8}$$

Proof. It is clear that we have an inequality like (5.3.7) with some $a > 0$ and $C > 1$ for the function $\psi(v)$ from (5.3.8), i. e.,

$$\psi(k) \le C\exp(a|k|^2), \quad k \in \mathbb{R}^3.$$

Then, for any (small) $r_0 > 0$ we can find such $b = b(r_0, C)$ that

$$\psi(k) \le \exp(b|k|^2) \quad \text{if} \quad |k| \ge r_0\,.$$

It is enough to take $b = a + (\log C)/r_0^2$. Hence, it remains to consider the values $|k| < r_0$. Since $f \in B_+$, we obtain

$$\psi(k) = 1 + \langle f, e^{k\cdot v} - 1 - k\cdot v\rangle$$

and use elementary inequalities

$$e^x - 1 - x \le x^2 e^x, \quad k \cdot v \le |k|\,|v|.$$

Then we obtain

$$\psi(k) \le 1 + g|k|^2, \quad g = \langle f|v|^2, \exp(r_0|v|)\rangle, \quad |k| \le r_0.$$

Hence, inequality (5.3.8) holds for all $k \in \mathbb{R}^3$ with $\gamma = \max(b, g)$. This completes the proof. □

The next step is to prove the following lemma.

Lemma 5.3.2. *If $f_0(v) \in B_+$, then there exists $\gamma > 0$ such that the solution $\psi(k,t)$ of equation* (5.3.5) *with initial data*

$$\psi|_{t=0} = \psi_0(k) = \langle f_0, e^{k\cdot v}\rangle, \quad k \in \mathbb{R}^3, \tag{5.3.9}$$

satisfies the estimate

$$\psi(k,t) \le \exp\left(\frac{\gamma|k|^2}{2}\right), \quad k \in \mathbb{R}^3, \quad t \ge 0. \tag{5.3.10}$$

Proof. We can construct the solution $\psi(k,t)$ in the form of the Wild sum like it was done in Section 3.3 for the Boltzmann equation (see equations (3.3.8)–(3.3.10)). Then we obtain

$$\psi(k,t) = e^{-t} \sum_{n=0}^{\infty} (1 - e^{-t})^n \psi_n(k), \tag{5.3.11}$$

where $\psi_0(k)$ is given in equation (5.3.9) and

$$\psi_{n+1}(k) = \frac{1}{n+1} \sum_{k=0}^{n} \hat{Q}(\psi_k, \psi_{n-k}), \quad n = 0, 1, \dots,$$

in the notation of equation (5.3.5). The obvious consideration by induction leads to the conclusion that

$$\psi_n(k) \le \exp(\gamma|k|^2), \quad n = 1, 2, \dots, \tag{5.3.12}$$

provided that this inequality is fulfilled for $\psi_0(k)$. This assumption for $\psi_0(k)$ follows directly from assumption (5.3.9) and Lemma 5.3.1. Hence, the estimates (5.3.12) hold for all $n = 0, 1, \dots$. Then we substitute this estimate in the Wild sum (5.3.11) and obtain the estimate (5.3.10) for $\psi(k,t)$. The proof is completed. □

Now we can prove the main result of this section.

Theorem 5.3.1. *Let the distribution function $f(v,t)$ be a solution of the problem* (5.3.2)–(5.3.3) *and let $\gamma > 0$ be a number such that*

$$\psi_0(k) = \langle f_0, e^{k\cdot v}\rangle \le \exp\left(\frac{\gamma|k|^2}{2}\right), \quad k \in \mathbb{R}^3. \tag{5.3.13}$$

Then $f(v,t) \in B_+$ for all $t \ge 0$ and the tail temperature $\tau(t)$ of the distribution $f(v,t)$ satisfies the inequality $\tau(t) \le \gamma$ for any $t \ge 0$.

Proof. Firstly we note that the existence of γ from assumption (5.3.13) of the theorem follows from Lemma 5.3.1. Then we consider the integral

$$I(\lambda,t) = \int_{\mathbb{R}^3} dk\psi(k,t)e^{-\lambda|k|^2/2}, \quad \lambda > \gamma.$$

It follows from Lemma 5.3.2 that the integral converges for all $\lambda > \gamma$ and

$$0 < I(\lambda,t) < C(\lambda-\gamma)^{-3/2}, \quad C = \text{const.}$$

On the other hand, we can represent $I(\lambda,t)$ as a double integral and change the order of integration. Then we obtain

$$I(\lambda,t) = \int_{\mathbb{R}^3} dke^{-\lambda|k|^2/2}\langle f(v,t), e^{k\cdot v}\rangle = \int_{\mathbb{R}^3} dvf(v,t)\Phi_\lambda(v),$$

where

$$\Phi_\lambda(v) = \int_{\mathbb{R}^3} dke^{-\frac{\lambda}{2}(k-v/\lambda)^2+\frac{|v|^2}{2\lambda}} = \left(\frac{2\pi}{\lambda}\right)^{3/2} e^{\frac{|v|^2}{2\lambda}}.$$

Hence, the integral

$$I\left(\frac{1}{r},t\right) = (2\pi r)^{3/2}\langle f(v,t), e^{r|v|^2/2}\rangle \le \frac{C_1 r^3}{(1-r\gamma)^{3/2}}, \quad C_1 = \text{const.},$$

converges for any $r > \gamma^{-1}$. Then the estimate $\tau(t) \le \gamma$ follows from the definition (5.1.3) of the tail temperature $\tau(t)$ extended to anisotropic solutions $f(v,t)$, $v \in \mathbb{R}^3$. This completes the proof. □

With this we conclude our investigation of asymptotic forms as $|v| \to \infty$; in the next section we consider the asymptotic form as $t \to \infty$.

5.4 The general approach to the large time asymptotic problem

We come back in this section to the same Cauchy problem (5.3.2) and consider the Boltzmann equation

$$f_t = Q(f,f), \quad f|_{t=0} = f_0(v), \tag{5.4.1}$$

with the same normalization (5.3.3) of the kernel $g(\mu) \in L^1_+(-1,1)$. We always assume in this section that the solution $f(v,t)$ of (5.4.1) is the generalized density of a probability measure satisfying the conservation laws

$$\langle f, 1\rangle = 1, \quad \langle f, v\rangle = 0, \quad \langle f, |v|^2\rangle = 3. \tag{5.4.2}$$

The characteristic function $\varphi(k,t)$ defined as

$$\varphi(k,t) = \langle f,\ e^{-ik\cdot v}\rangle, \quad k \in \mathbb{R}^3, \tag{5.4.3}$$

satisfies the equation

$$\varphi_t + \varphi = \hat{Q}_+(\varphi,\varphi) = \int_{S^2} d\omega\, g(\hat{k}\cdot\omega)\varphi(k_+)\varphi(k_-), \tag{5.4.4}$$

in the notation of equations (5.3.5). The initial conditions for this equation are the following:

$$\begin{aligned} &\varphi|_{t=0} = \varphi_0(k) = \langle f_0,\ e^{-ik\cdot v}\rangle;\\ &\varphi_0(0) = 1, \quad \nabla\varphi_0|_{k=0} = 0, \quad \Delta\varphi_0|_{k=0} = -3. \end{aligned} \tag{5.4.5}$$

Due to conservation laws (5.4.2), these conditions hold not only for φ_0 at initial moment $t = 0$, but also for the solution $\varphi(k,t)$ at any $t \geq 0$. The only Maxwellian distribution that satisfies the initial conditions (5.4.5) is

$$\hat{M}(k) = \exp\left(-\frac{|k|^2}{2}\right), \quad k \in \mathbb{R}^3. \tag{5.4.6}$$

Consequently, we expect that at any $k \in \mathbb{R}^3$

$$\lim_{t\to\infty} \varphi(k,t) = \hat{M}(k). \tag{5.4.7}$$

Note that the solution $\varphi(k,t)$ of that problem can be constructed in the form of the Wild sum (like in equation (5.3.11)). Hence, $\varphi(k,t)$ is the characteristic function at any $t \geq 0$. Thereby the pointwise convergence of $\varphi(k,t)$ implies the convergence of the function $f(v,t)$ to the Maxwellian

$$M(v) = (2\pi)^{-3/2}\exp\left(-\frac{|v|^2}{2}\right), \quad v \in \mathbb{R}^3, \tag{5.4.8}$$

in the sense of convergence of probability measures [85].

Our aim is to prove the limiting equality (5.4.7) and to try to estimate the rate of convergence. We introduce the set Φ of characteristic functions as a subset of the space of complex-valued continuous functions $C(\mathbb{R}^3;\mathbb{C})$. Then

$$\Phi = \{\varphi \in C(\mathbb{R}^3;\mathbb{C}):\ \varphi(k) = \langle f,\ e^{-ik\cdot v}\rangle;\, f \in F_+(\mathbb{R}^3)\}, \tag{5.4.9}$$

where $F_+(\mathbb{R}^3)$ is the set of generalized densities $f(v)$ of probability measures in $\mathbb{R}^3$. Roughly speaking, this means that $f(v) \geq 0$ and $\langle f, 1\rangle = 1$. We also introduce the norm

$$\|\varphi\| = \sup_{k\in\mathbb{R}^3} |\varphi(k)| \tag{5.4.10}$$

and note that for any $\varphi \in \Phi$ we have

$$\|\varphi\| = \varphi(0) = 1. \tag{5.4.11}$$

Then we consider the bilinear operator

$$\hat{Q}_+(\varphi, \psi)(k) = \int_{S^2} d\omega\, g(\hat{k}\cdot\omega)\varphi(k_+)\,\psi(k_-),$$
$$k_\pm = (k \pm |k|\omega)/2, \quad \hat{k} = k/|k|, \quad k \in \mathbb{R}^3, \tag{5.4.12}$$

and introduce the linear operator

$$\hat{L}^+\varphi(k) = Q(1,\varphi) + Q(\varphi, 1) = \int_{S^2} d\omega\, g(\hat{k}\cdot\omega)\big[\varphi(k_+) + \psi(k_-)\big]. \tag{5.4.13}$$

Then we can formulate the following lemma.

Lemma 5.4.1. *The following estimate is valid for any $k \in \mathbb{R}^3$ and for any pair of functions such that $\|\varphi_{1,2}\| \leq 1$:*

$$\big|\hat{Q}_+(\varphi_1, \varphi_1) - \hat{Q}_+(\varphi_2, \varphi_2)\big|(k) \leq \hat{L}^+|\varphi_1 - \varphi_2|(k). \tag{5.4.14}$$

Proof. We note that the left hand side can be controlled by the integral

$$\int_{S^2} d\omega\, g(\hat{k}\cdot\omega)\,|\varphi_1(k_+)\varphi_1(k_-) - \varphi_2(k_+)\varphi_2(k_-)|$$

and apply the elementary inequality

$$|ab - cd| = \frac{1}{2}\big|(a+c)(b-d) + (b+d)(a-c)\big|$$
$$\leq |a-c| + |b-d| \quad \text{if} \quad \max(|a|, |b|, |c|, |d|) \leq 1,$$

where $a = \varphi_1(k_+)$, $b = \varphi_1(k_-)$, $c = \varphi_2(k_+)$, $d = \varphi_2(k_-)$. The estimate (5.4.14) follows and this completes the proof. □

This simple property of a class of bilinear operators was discussed in detail in [44] (see also Section 6.8 below). The operators satisfying (5.4.14) were termed in [44] as "$\hat{L}^+$-Lipschitz operators" due to the fact that they satisfy the Lipschitz condition with respect to the linear operator $\hat{L}^+$ (not to some constant!) pointwise in $\mathbb{R}^3$. This property is stronger than the classical Lipschitz condition

$$\|\hat{Q}_+(\varphi_1,\varphi_1)-\hat{Q}_+(\varphi_2,\varphi_2)\|\le 2\|\varphi_1-\varphi_2\|\,,$$

which obviously follows from the estimate (5.4.14).

Let us compare two solutions $\varphi_{1,2}\in\Phi$ of equation (5.4.12) satisfying the initial conditions

$$\varphi_i(k,0)=\varphi_i^{(0)}(k),\quad i=1,2. \tag{5.4.15}$$

Then we denote

$$u(k,t)=\varphi_1(k,t)-\varphi_2(k,t),\quad u_0(k)=\varphi_1^{(0)}(k)-\varphi_2^{(0)}(k), \tag{5.4.16}$$

and obtain

$$\begin{aligned} u(k,t)&=u_0(k)\,e^{-t}+\int_0^t d\tau\, e^{-(t-\tau)}\Delta(k,t),\\ \Delta(k,t)&=\left[\hat{Q}_+(\varphi_1,\varphi_1)-\hat{Q}_+(\varphi_2,\varphi_2)\right](k,t). \end{aligned} \tag{5.4.17}$$

We recall that $\varphi_{1,2}\in\Phi$ and therefore at any $t\ge 0$ the functions $\varphi_{1,2}$ satisfy conditions of Lemma 5.4.1. Hence, we obtain

$$\left|u(k,t)\right|\le\left|u_0(k)\right|e^{-t}+\int_0^t d\tau\, e^{-(t-\tau)}\hat{L}^+\left|u(k,\tau)\right|,\quad k\in\mathbb{R}^3. \tag{5.4.18}$$

The next step is to prove the following lemma.

Lemma 5.4.2. *Let $\varphi_{1,2}\in\Phi$ be two solutions of equation (5.4.4) satisfying initial conditions (5.4.15). Then the following estimate is valid for all $k\in\mathbb{R}^3$ and $t\ge 0$:*

$$\left|\varphi_1(k,t)-\varphi_2(k,t)\right|\le y(k,t), \tag{5.4.19}$$

where $y(k,t)$ is the solution of the Cauchy problem for the linear equation

$$y_t+y=\hat{L}^+y,\quad y|_{t=0}=\left|u_0(k)\right|, \tag{5.4.20}$$

in the notation of equations (5.4.13) and (5.4.16).

Proof. We transform equation (5.4.20) to integral form for $y(k,t)$ and get

$$y(k,t)=\left|u_0(k)\right|e^{-t}+\int_0^t d\tau\, e^{-(t-\tau)}\hat{L}^+y(k,\tau).$$

We also have the exponential form of $y(k,t)$,

$$y(k,t) = e^{-t}\exp(t\hat{L}^+)\,|u_0(k)|,$$

where $y(k,t) \geq 0$ and $\hat{L}^+ y(k,t) \geq 0$ for all $k \in \mathbb{R}^3$ and $t \geq 0$. Then the standard comparison argument applied to $|u(k,t)|$ from equation (5.4.18) and $y(k,t)$ shows that the estimate (5.4.19) holds. This completes the proof. □

Hence, the absolute value of the difference between two solutions $\varphi_{1,2} \in \Phi$ of the non-linear equation (5.4.4) can be controlled by a solution of the linear equation (5.4.20). In fact, the linear operator $\hat{L}^+$ was studied in detail in Sections 3.5–3.7 as a part of the linearized Boltzmann collision operator $\hat{L} = \hat{L}^+ - I$, where I is the identity. Note that $u_0(0) = 0$ and hence $y(0,t) = 0$ for all $t \geq 0$ in equations (5.4.20). The general solution of the linear problem was constructed in Section 3.7. It is rather complicated, but we can simplify the estimate (5.4.19) in the following way.

Theorem 5.4.1. *Let functions $\varphi_{1,2}(k,t)$ satisfy assumptions of Lemma 5.4.2 and the corresponding initial conditions (5.4.15) satisfy inequality*

$$|\varphi_1^{(0)}(k) - \varphi_2^{(0)}(k)| \leq C\,|k|^{2p}, \quad k \in \mathbb{R}^3, \tag{5.4.21}$$

with some constants $C > 0$ and $p > 0$. Then the following estimate holds for all $k \in \mathbb{R}^3$ and $t \geq 0$:

$$|\varphi_1(k,t) - \varphi_2(k,t)| \leq C\,|k|^{2p}\exp[-\lambda_p t], \tag{5.4.22}$$

where

$$\lambda_p = 2\pi \int_{-1}^{1} d\mu\, g(\mu)\left[1 - \left(\frac{1+\mu}{2}\right)^p - \left(\frac{1-\mu}{2}\right)^p\right], \tag{5.4.23}$$

with the same constants C and p as in equation (5.4.21).

Proof. We note that the operator $\hat{L}^+$ (5.4.13) is obviously monotone in the following sense:

$$0 \leq \hat{L}^+|u_1(k)| \leq \hat{L}^+|u_2(k)| \quad \text{if} \quad 0 \leq |u_1(k)| \leq |u_2(k)|,$$

for any $u_{1,2} \in C(\mathbb{R}^3;\mathbb{C})$ and all $k \in \mathbb{R}^3$. Note that $\hat{L}^+$ is bounded in $C(\mathbb{R}^3;\mathbb{C})$. Hence, the exponential operator

$$\exp(t\hat{L}^+) = \sum_{n=0}^{\infty} \frac{t^n}{n!}(\hat{L}^+)^n, \quad t \geq 0, \tag{5.4.24}$$

is also a monotone bounded operator in $C(\mathbb{R}^3;\mathbb{C})$. Under assumption of the theorem it follows from Lemma 5.4.2 that

$$|\varphi_1(k,t) - \varphi_2(k,t)| \leq e^{-t}\exp(t\hat{L}^+)\,|u_0(k)|, \tag{5.4.25}$$

in the notation of equation (5.4.17). On the other hand, the assumption (5.4.21) reads as

$$|u_0(k)| \le C|k|^{2p}, \quad p > 0, \tag{5.4.26}$$

in the same notation. Since $|k|^{2p}$ is the eigenfunction of $\hat{L}^+$ satisfying the equation

$$\hat{L}^+|k|^{2p} = \lambda_p^+|k|^{2p}, \quad \lambda_p^+ = 2\pi \int_{-1}^{1} d\mu g(\mu) \left[\left(\frac{1+\mu}{2}\right)^p + \left(\frac{1-\mu}{2}\right)^p \right],$$

we obtain from (5.4.24)–(5.4.26)

$$|\varphi_1(k,t) - \varphi_2(k,t)| \le C\,e^{-t} \exp(t\hat{L}^+)\,|k|^{2p} = C\,|k|^{2p} \exp[t(\lambda_p^+ - 1)]\,.$$

Finally, we note that $\lambda_p^+ - 1 = \lambda_p$ in the notation of equation (5.4.23) because the condition (5.3.3) is always assumed in this section. Therefore, we obtain the estimate (5.4.23) and this completes the proof. □

Corollary 2. *Lemma* 3.3.2 *follows from Theorem* 5.4.1 *for* $p = 1$.

Proof. The assumption (5.4.21) for $p = 1$ obviously holds for

$$C = \sup_{k\in\mathbb{R}^3} \left| \frac{\varphi_1^{(0)}(k) - \varphi_2^{(0)}(k)}{|k|^2} \right|.$$

Since $\lambda_1 = 0$, it follows from (5.4.22) that

$$\left| \frac{\varphi_1(k,t) - \varphi_2(k,t)}{|k|^2} \right| \le C\,, \quad t > 0, \quad k \in \mathbb{R}^3\,,$$

with the same constant C. Taking the supremum of the left hand side, we obtain the result of Lemma 3.3.2. This completes the proof. □

Remark. Note that in the proofs of Lemmas 5.4.1 and 5.4.2 and Theorem 5.4.1 we did not use assumptions (5.4.2) for $f(v,t)$ and assumption (5.4.5) for $\varphi(k,t)$. The only needed assumption of that kind was the usual equality $\langle f, 1\rangle = \varphi|_{k=0} = 1$ for the probability measure and its characteristic function.

Let us now use the assumptions (5.4.5) that correspond to the unique characteristic function $\hat{M}(k)$ (5.4.6). Obviously, this function is a stationary solution of equation (5.4.5). Then we can prove the following statement.

Corollary 3. *Let* $\varphi(k,t) \in \Phi$ *be a solution of equation* (5.4.4) *with initial condition* $\varphi_0(k)$, *satisfying the inequality*

$$|\varphi_0(k) - \hat{M}(k)| \le C|k|^{2p}, \quad k \in \mathbb{R}^3, \tag{5.4.27}$$

with some constant $C > 0$ and $p > 1$. Then

$$\left|\varphi(k,t) - \hat{M}(k)\right| \le \min\left(C|k|^{2p}e^{-\lambda_p t}, 2\right) \tag{5.4.28}$$

for all $t > 0$ and $k \in \mathbb{R}^3$. The limiting equality (5.4.7) is valid for any $p > 1$ since $\lambda_p > 0$ for $p > 1$.

Proof. It is sufficient to substitute $\varphi_1(k,t) = \varphi(k,t)$, $\varphi_2(k,t) = \hat{M}(k)$ into inequality (5.4.22). Evidently all conditions of Theorem 5.4.1 are satisfied. This proves the corollary. □

In the next two sections we discuss the rate of convergence to equilibrium for different classes of initial data.

5.5 Convergence to equilibrium

The convergence to equilibrium in the Fourier representation, i. e., the limiting equality (5.4.7), obviously follows from inequality (5.4.28) provided the condition (5.4.27) is satisfied for some $p > 1$. For example, all radial solutions of the Boltzmann equation satisfy this condition. However, the condition (5.4.27) is not satisfied for an important class of initial data such that

$$\left\langle f_0, v_\alpha v_\beta - \frac{1}{3}\delta_{\alpha\beta}|v|^2\right\rangle \neq 0, \quad \alpha, \beta = 1, 2, 3.$$

We recall the expansion of the corresponding characteristic functions:

$$\varphi_0(k) = \langle f_0, e^{ik\cdot v}\rangle = 1 - \frac{m_{\alpha\beta}^{(2)}(0)}{2}k_\alpha k_\beta + o(|k|^2),$$

for $k \to 0$. This form of expansion follows from normalization conditions (5.4.5), which also imply that (see equation (3.8.10))

$$m_{\alpha\beta}^{(2)}(0) = \delta_{\alpha\beta} + g_{\alpha\beta}^{(0)}, \quad g_{\alpha\alpha}^{(0)} = 0.$$

The connection of the symmetric tensor $g_{\alpha\beta}^{(0)}$ with $f_0(v)$ is given by the formula

$$g_{\alpha\beta}^{(0)} = \left\langle f_0, v_\alpha v_\beta - \frac{1}{3}\delta_{\alpha\beta}|v|^2\right\rangle, \quad \alpha, \beta = 1, 2, 3.$$

The time evolution of the second moment $m_{\alpha\beta}^{(2)}(t)$ was studied in Section 3.8 (see equations (3.8.10) and (3.8.11)). The result reads as

$$m_{\alpha\beta}^{(2)}(t) = \delta_{\alpha\beta} + g_{\alpha\beta}(t), \quad g_{\alpha\beta}(t) = g_{\alpha\beta}^{(0)}\, e^{-\lambda t},$$

$$\lambda = \frac{3}{2}\lambda_2 = \frac{3\pi}{2}\int_{-1}^{1} d\mu g(\mu)(1-\mu^2) < 1, \tag{5.5.1}$$

in the notation of equation (5.4.23). Now we can prove the following statement.

Lemma 5.5.1. *Let $\varphi(k,t) \in \Phi$ be a solution of equation* (5.4.4) *with initial condition $\varphi_0(k) \in \Phi$ such that*

$$\varphi_0(k) = 1 - \frac{1}{2}|k|^2 - \frac{1}{2}g^{(0)}_{\alpha\beta}\, T_{\alpha\beta}(k) + \psi_0(k), \tag{5.5.2}$$

where

$$\begin{aligned} |\psi_0(k)| &\le C|k|^{2p}, \quad p > 1; \quad T_{\alpha\beta}(k) = k_\alpha k_\beta - \frac{1}{3}\delta_{\alpha\beta}|k|^2, \\ g^{(0)}_{\alpha\alpha} &= 0; \quad \alpha,\beta = 1,2,3. \end{aligned} \tag{5.5.3}$$

Then for all $k \in \mathbb{R}^3$ and $t > 0$

$$|\varphi(k,t) - e^{-|k|^2/2}| \le C_1[|k|^{2q}\, e^{-\lambda_q t} + |k|^2 e^{-\lambda t}], \quad q = \min(p,2), \tag{5.5.4}$$

in the notation of equations (5.4.23) *and* (5.5.1).

Proof. We note that the representation of $\varphi(k,t)$ in the form (5.5.2) for $t > 0$ reads as

$$\begin{aligned} \varphi(k,t) &= 1 - \frac{1}{2}|k|^2 + w(k,t) + O(|k|^{2p}), \\ w(k,t) &= -\frac{1}{2}g^{(0)}_{\alpha\beta}\, e^{-\lambda t}\, T_{\alpha\beta}(k), \quad \alpha,\beta = 1,2,3, \end{aligned} \tag{5.5.5}$$

in the notation of equation (5.5.3). We denote

$$\varphi - e^{-|k|^2/2} = u_1, \quad \varphi + e^{-|k|^2/2} = 2u_2$$

and obtain

$$u_{1_t} + u_1 = \hat{Q}_+(u_2,u_1) + \hat{Q}_+(u_1,u_2),$$

in the notation of equation (5.4.12). Then we split $u_1(k,t)$ into two parts,

$$u_1(k,t) = u(k,t) + w(k,t),$$

and note that $w(k,t)$ satisfies the equation

$$w_t + w = \hat{L}^+ w = \hat{Q}_+(1,w) + \hat{Q}_+(w,1),$$

in the same notation of equation (5.4.12). Hence, the equation for $u = u_1 - w$ reads

$$u_t + u = \hat{Q}_+(u_2,u) + \hat{Q}_+(u,u_2) + S, \tag{5.5.6}$$

and

$$S = \hat{Q}_+(u_2 - 1, w) + \hat{Q}_+(w, u_2 - 1). \tag{5.5.7}$$

Note that $|u_2| \leq 1$ and therefore $|u_2 - 1| \leq 2$. Hence, we use equation (5.5.5) and obtain

$$\big|\hat{Q}_+(u_2, u) + \hat{Q}_+(u, u_2)\big|(k,t) \leq \hat{L}^+|u|(k,t),$$
$$\big|S(k,t)\big| \leq \Gamma_0 |k|^2 e^{-\lambda t}, \quad \Gamma_0 = 2\sum_{\alpha,\beta=1}^{3} |g_{\alpha\beta}^{(0)}|. \tag{5.5.8}$$

The substitution of these inequalities into the integral form of equation (5.5.6) yields

$$\big|u(k,t)\big| = \big|u(k,0)\big|e^{-t} + \int_0^t d\tau e^{-(t-\tau)}\big[\hat{L}^+\big|u(k,\tau)\big| + \Gamma_0|k|^2 e^{-\lambda t}\big]. \tag{5.5.9}$$

It follows from equations (5.5.2) and (5.5.3) that

$$\begin{aligned} u(k,0) &= \varphi_0(k) + \frac{1}{2} g_{\alpha\beta}^{(0)} T_{\alpha\beta}(k) - e^{-|k|^2/2} \\ &= 1 - \frac{1}{2}|k|^2 - e^{-|k|^2/2} + \psi_0(k) = O(|k|^{2q}), \end{aligned} \tag{5.5.10}$$

where

$$q = \min(p, 2), \quad p > 1. \tag{5.5.11}$$

Then it is easy to see that inequality (5.5.9) implies that

$$\big|u(k,t)\big| \leq y(k,t), \tag{5.5.12}$$

where $y(k,t)$ is the solution of the problem

$$y_t + y = \hat{L}^+ y + \Gamma_0|k|^2 e^{-\lambda t}, \quad y|_{t=0} = C|k|^{2q},$$

where the constant C is the same as in the O-symbol in equation (5.5.10). The explicit solution of this problem reads

$$y(k,t) = C|k|^{2q} e^{-\lambda_q t} + \Gamma_0|k|^2 \frac{e^{-\lambda t} - e^{-t}}{1-\lambda}, \tag{5.5.13}$$

in the notation of equations (5.4.23), (5.5.1), and (5.5.11). We use the definition of $u(k,t)$ and obtain

$$\varphi(k,t) - e^{-|k|^2/2} = u(k,t) + O(|k|^2 e^{-\lambda t}).$$

Then the resulting estimate (5.5.4) follows from equations (5.5.12) and (5.5.13). This completes the proof. □

Now we can prove the convergence to equilibrium for solutions of the Boltzmann equation.

Theorem 5.5.1. *Let $f(v,t)$ be a solution of the Cauchy problem* (5.4.1), *satisfying conservation laws* (5.4.2) *and such that*

$$\langle f_0, |v|^{2+\varepsilon}\rangle < \infty \tag{5.5.14}$$

for some $\varepsilon > 0$. Then

$$f(v,t) \underset{t\to\infty}{\longrightarrow} M(v) = (2\pi)^{-3/2}\exp(-|v|^2/2), \tag{5.5.15}$$

where the convergence is understood in the sense of convergence of probability measures.

Proof. It is sufficient to prove the pointwise convergence of $\varphi(k,t)$ for every $k \in \mathbb{R}^3$ as $t \to \infty$. In other words, it is enough to check that all conditions of Lemma 5.5.1 are satisfied. Then the convergence of $\varphi(k,t)$ to $\exp(-|k|^2/2)$ obviously follows from estimate (5.5.4). Thus we need to show that

$$\varphi_0(k) = \langle f_0, e^{-ik\cdot v}\rangle = 1 - \frac{m^{(2)}_{\alpha\beta}(0)}{2}k_\alpha k_\beta + O(|k|^{2+\varepsilon}),$$
$$m^{(2)}_{\alpha\beta}(0) = \delta_{\alpha\beta} + g^{(0)}_{\alpha\beta}, \quad g^{(0)}_{\alpha\alpha} = 0,$$

where ε is the same as in equation (5.5.14). The structure of terms of orders $0 \le n \le 2$ evidently follows from conservation laws (see equations (5.4.5)). Hence, we need to estimate the function

$$\psi_0(k) = \left\langle f_0, e^{-ik\cdot v} - \left[1 - ik\cdot v - \frac{(k\cdot v)^2}{2}\right]\right\rangle. \tag{5.5.16}$$

We consider the function

$$y(x) = e^{-ix} - 1 + ix + \frac{x^2}{2}, \quad x \in \mathbb{R},$$

and note that

$$|y(x)| \le 2 + |x| + \frac{x^2}{2}.$$

For any $x_0 > 0$ and $\delta > 0$ we can easily find such $C(x_0,\delta)$ that

$$|y(x)| \le C|x|^{2+\delta} \quad \text{if} \quad |x| \ge x_0. \tag{5.5.17}$$

On the other hand,

$$y(x) = i\frac{x^3}{3!} + O(x^4)$$

for small $|x|$. Hence, for any $\delta \in (0,1)$ we can find such C_δ that inequality (5.5.17) holds for all $x \in \mathbb{R}$. Taking $\delta = \varepsilon$, we obtain

$$|y(x)| \le C_\varepsilon |x|^{2+\varepsilon}, \quad x \in \mathbb{R},$$

provided $0 < \varepsilon < 1$ in (5.5.14). Then we estimate $\psi_0(k)$ from (5.5.16) and obtain

$$|\psi_0(k)| \le \langle f_0, |y(k \cdot v)| \rangle \le C_\varepsilon |k|^{2+\varepsilon} \langle f_0, |v|^{2+\varepsilon} \rangle.$$

Hence, the condition (5.5.3) of Lemma 5.5.1 is fulfilled. The case $\varepsilon \ge 1$ is trivial because in that case the condition (5.5.14) is automatically satisfied for any $0 < \varepsilon < 1$. Therefore, the theorem is also valid for all $\varepsilon \ge 1$. This completes the proof. □

It is clear that the estimate (5.5.4) from Lemma 5.5.1 defines the rate of pointwise convergence of characteristic function $\varphi(k,t)$ to equilibrium. We recall that

$$\lambda_p = 2\pi \int_{-1}^{1} d\mu g(\mu) \left[1 - \left(\frac{1+\mu}{2} \right)^p - \left(\frac{1-\mu}{2} \right)^p \right], \quad \lambda = \frac{3}{2}\lambda_2 = \frac{3\pi}{2}. \tag{5.5.18}$$

Since λ_p is an increasing function of $p \ge 1$ and $\lambda_1 = 0$, we obtain from the estimate (5.5.4) the following asymptotic formula:

$$\varphi(k,t) = e^{-|k|^2/2} + O(|k|^{2q} e^{-\lambda_q t}), \quad t \to \infty, \tag{5.5.19}$$

where $q = \min(p, 2)$, $p > 1$.

Of course, we are interested in the rate of convergence to equilibrium of the corresponding solution $f(v,t)$ of the Boltzmann equation (5.4.1). This question is not trivial because the correction term in equation (5.5.19) tends to infinity for large $|k|$. The question about using different metrics for estimates of the distance between $f(v,t)$ and the Maxwellian $M(v)$ is very interesting. For the sake of brevity we omit discussion of that question. Interested readers can find related information in papers [87, 142, 143, 147]. Some simple estimates for radial solutions can be found in [27].

Another question that arises in connection with the asymptotic formula (5.5.19) is the possible existence of solutions to the Boltzmann equation, whose relaxation to equilibrium takes a very long time. It happens in the case when $q \to 1$ in equation (5.5.19). Such solution is studied in the next section.

5.6 Slow relaxation of solutions with power-like tails

Looking at eigenvalues λ_p (5.5.8) of the linearized collision operator (see also equations (3.6.5) in slightly different notation), one can see that $\lambda_p \to 0$ as $p \to 1$. The asymptotic formula (5.5.4) also suggests the existence of solutions with very slow rates of relaxation. Our aim in this section is to construct an example of such (positive!) solution of the Boltzmann equation.

We shall need below a simple inequality for a series with positive terms.

Lemma 5.6.1. *The inequality,*

$$S_\delta = \sum_{n=1}^{\infty} f(n + [n\delta]) \le \sum_{n=1}^{\infty} f(n) < \infty,$$

where $f(n) \ge 0$ for all $n = 1, 2, \dots$ and $[x]$ denotes the integer part of $x > 0$, holds for any convergent series and any $\delta \in [0, 1]$.

Proof. The half-infinite interval $[1, \infty)$ can be represented as a union of half-open intervals

$$\Delta_0 = \left[1, \frac{1}{\delta}\right); \quad \Delta_k = \left[\frac{k}{\delta}, \frac{k+1}{\delta}\right), \quad k = 1, 2, \dots,$$

such that $[n\delta] = k$ for all $n \in \Delta_k$. It is obvious that $\Delta_i \cap \Delta_k = \emptyset$ if $i \ne k$. Hence,

$$S_\delta = \sum_{k=0}^{\infty} \sum_{n\in\Delta_k} f(n+k) = \sum_{k=0}^{\infty} \sum_{n\in\Delta_k + k} f(n),$$

where

$$\Delta_k + k = \tilde{\Delta}_k = \left[\frac{k}{\delta} + k, \frac{k+1}{\delta} + k\right), \quad k = 1, 2, \dots.$$

It is easy to verify that the distance between two intervals $\tilde{\Delta}_k$ and $\tilde{\Delta}_{k+1}$ is equal to one for all $k \ge 0$. This completes the proof. □

Now we can prove the main result of this section.

Theorem 5.6.1. *We consider the problem (5.4.1) for the Boltzmann equation with bounded kernel*

$$0 \le g(\mu) \le g_1, \quad \mu \in [-1, 1], \tag{5.6.1}$$

of the collision integral. For any $\varepsilon > 0$ there exists a positive solution $f_\varepsilon(v, t)$ of that problem, which satisfies conservation laws (5.4.2) and inequalities

$$\begin{aligned} \|f - M\|_{L^1} &= \int_{\mathbb{R}^3} dv\, |f(v,t) - M(v)| \ge c_1 e^{-\varepsilon t}, \\ \|f - M\|_{L^\infty} &= \sup_{v\in\mathbb{R}^3} |f(v,t) - M(v)| \le c_2 e^{-\varepsilon t}, \quad c_{1,2} > 0, \end{aligned} \tag{5.6.2}$$

in the notation of equation (5.4.7). The corresponding characteristic function

$$\varphi(k, t) = \langle f, e^{-ik\cdot v} \rangle$$

satisfies the inequality

$$c_3 e^{-\varepsilon t} \le \sup_{k\in\mathbb{R}^d} |\varphi(k,t) - M(k)| \le c_4 e^{-\varepsilon t}, \quad c_{3,4} > 0, \tag{5.6.3}$$

in the notation of equation (5.4.1).

Proof. The idea of the proof is rather transparent. For simplicity we consider radial solutions $f(|v|,t)$ and use the same notations as in Section 4.1. Then we obtain (see equations (4.1.3) and (4.1.4))

$$\varphi(k,t) = \tilde{\varphi}(x,t) = \langle f(|v|,t), e^{-ik\cdot v}\rangle, \quad x = |k|^2/2, \tag{5.6.4}$$

$$\tilde{\varphi}_t = \int_0^1 ds\rho(s)\{\tilde{\varphi}(sx)\tilde{\varphi}[(1-s)x] - \tilde{\varphi}(0)\tilde{\varphi}(x)\},$$
$$x \ge 0, \quad \rho(s) = 4\pi g(1-2s). \tag{5.6.5}$$

Below we omit tildes and consider the following initial data:

$$\varphi_{t=0} = \varphi_0(x) = e^{-x}(1+\theta x^\alpha), \quad \theta > 0,\ \alpha > 1. \tag{5.6.6}$$

The motivation for such choice of $\varphi_0(x)$ is very clear: x^α is the eigenfunction of the linearized operator. It is logical to look for the solution in the form

$$\varphi(x,t) = e^{-x}\, u(z,t), \quad z = \theta x^\alpha. \tag{5.6.7}$$

Then we obtain

$$u_t = \int_0^1 ds\rho(s)\{u(s^\alpha z)\, u[(1-s)^\alpha z] - u(0)u(z)\}, \quad u|_{t=0} = 1 + z\,. \tag{5.6.8}$$

Assuming that

$$u(z,t) = \sum_{n=0}^{\infty} u_n(t) z^n, \quad u_0(t) = 1, \tag{5.6.9}$$

we obtain

$$\dot{u}_1 + \lambda_\alpha u_1 = 0, \quad u_1(0) = 1;$$
$$\dot{u}_n + \lambda_{n\alpha} u_n = \sum_{k=1}^{n-1} h_{k\alpha,(n-k)\alpha}\, u_k\, u_{n-k}, \quad u_n(0) = 0; \quad n = 2,3,\dots,$$

where $\lambda_{n\alpha}$ are given in equations (5.5.8),

$$h_{k\alpha,\, l\alpha} = \int_0^1 ds\rho(s) s^{k\alpha}(1-s)^{l\alpha}, \quad k,l = 0,1,\dots.$$

Hence,

$$u_1(t) = e^{-\lambda_\alpha t};$$
$$u_n(t) = \sum_{k=1}^{n-1} h_{k\alpha,\,(n-k)\alpha} \int_0^t d\tau e^{-\lambda_{n\alpha}(t-\tau)} u_k(\tau) u_{n-k}(\tau), \quad n = 2,3,\dots. \tag{5.6.10}$$

Obviously, we have the estimate

$$0 \le \rho(s) \le \rho_M = 4\pi g_1, \tag{5.6.11}$$

in the notation of equation (5.6.1). Therefore,

$$h_{k\alpha,\,(n-k)\alpha} \le \rho_M B(k\alpha + 1, l\alpha + 1) = \rho_M \frac{\Gamma(k\alpha + 1)\Gamma(l\alpha + 1)}{\Gamma[(k + l)\alpha + 2]},$$

where $B(x, y)$ and $\Gamma(x)$ denote beta and gamma functions, respectively [94]. Our aim is to show that

$$u_n(t) \le \left(\frac{\rho_M}{\lambda_\alpha}\right)^{n-1} \frac{\Gamma(\alpha + 1)]^n}{\Gamma(n\alpha + 1)} e^{-\lambda_\alpha t}, \quad n = 1, 2, \dots . \tag{5.6.12}$$

Obviously, it is correct for $n = 1$. Then we assume that it is correct for $k = 1, \dots, n$. We denote

$$u_{n+1}(t) = y_{n+1}\left(\frac{\rho_M}{\lambda_\alpha}\right)^{n} \frac{[\Gamma(\alpha + 1)]^{n+1}}{\Gamma[(n + 1)\alpha + 1]} e^{-\lambda_\alpha t}$$

and obtain from inequalities (5.6.11)

$$y_{n+1} \le \lambda_\alpha I_{n+1}(t) \sum_{k=1}^{n} \frac{\Gamma[(n + 1)\alpha + 1)}{\Gamma[(n + 1)\alpha + 2]}, \quad n = 1, 2, \dots, \tag{5.6.13}$$

where

$$I_{n+1}(t) = e^{\lambda_\alpha t} \int_0^t d\tau e^{-\lambda_{(n+1)\alpha}(t-\tau) - 2\lambda_\alpha \tau}$$
$$= F(t;\, \lambda_\alpha,\, \lambda_{(n+1)\alpha} - \lambda_\alpha), \quad F(t;\, \lambda, \mu) = \frac{e^{-\lambda t} - e^{-\mu t}}{\mu - \lambda}.$$

One can easily check that

$$0 \le F(t;\, \lambda, \mu) \le \frac{1}{\lambda}, \quad \lambda > 0, \quad \mu \ge 0, \quad t \ge 0.$$

Then we use the identity $\Gamma(z + 1) = z\Gamma(z)$ and reduce (5.6.11) to

$$u_{n+1} \le \frac{n}{(n + 1)\alpha + 1}.$$

The right hand side is obviously less than unity for any $\alpha > 1$ and $n = 1, 2, \dots$. Hence, the estimate (5.6.12) holds for all coefficients $\{u_n(t),\ n = 1, 2, \dots\}$ of series (5.6.9). Note that all $u_n(t)$ are positive. Thus we obtain the following estimate of the solution $u(z,t)$ of the problem (5.6.8):

$$1 + ze^{-\lambda_\alpha t} \le u(z,t) \le 1 + \frac{\Gamma(\alpha+1)}{D} w(Dz)e^{-\lambda_\alpha t},$$
$$D = \frac{\rho_M}{\lambda_\alpha}\Gamma(\alpha+1), \quad w(z) = \sum_{n=1}^{\infty} \frac{z^n}{\Gamma(n\alpha+1)}, \quad z \ge 0, \quad \alpha > 1. \tag{5.6.14}$$

We use the transformation (5.6.7) and obtain

$$e^{-x}[1 + \theta x^\alpha e^{-\lambda_\alpha t}] \le \varphi(x,t)$$
$$\le e^{-x}\left[1 + \frac{\Gamma(\alpha+1)}{D} w(\theta Dz)\, e^{-\lambda_\alpha t}\right], \quad x \ge 0. \tag{5.6.15}$$

Let us consider the function $w(Dz)$ from equations (5.6.14),

$$w(\theta D x^\alpha) = S(y) = \sum_{n=1}^{\infty} \frac{z^{\alpha n}}{\Gamma(n\alpha+1)}, \quad y = x(\theta D),$$

assuming that $\alpha = 1 + \delta$, $0 < \delta < 1$. If $0 \le y \le 1$, then obviously

$$S(y) \le \sum_{n=1}^{\infty} \frac{y^n}{\Gamma(n+1)} = e^y - 1 \le ye^y.$$

A similar estimate holds for $y > 1$ as we shall see further. Note that the following inequalities are valid for any real $a \ge 0$: $[a] \le a \le [a] + 1$, where $[a]$ is the integer part of a. Thereby

$$\Gamma(n\alpha+1) = \Gamma(n + n\delta + 1) \le (n + [n\delta])!\,,$$
$$y^{n\alpha} = y^{n(1+\delta)} \le y^{n+[n\delta]+1}, \quad y \ge 1.$$

Therefore,

$$S(y) \le y \sum_{n=1}^{\infty} f(n + [n\delta]), \quad f(m) = \frac{y^m}{m!}, \quad m = 1, 2, \dots.$$

Then it follows from Lemma 5.6.1 that

$$S(y) \le y \sum_{n=1}^{\infty} f(n) = y\,(e^y - 1) \le y\,e^y.$$

Hence, we obtain from (5.6.15) a simple upper bound for $\varphi(k,t)$:

$$\varphi(k,t) \le e^{-x}(1 + Cye^{y-\lambda_\alpha t},), \quad 1 < \alpha < 2,$$
$$C = \frac{\Gamma(\alpha+1)}{D}, \quad y = (\theta D)^{1/\alpha} x, \quad D = \frac{\rho_M}{\lambda_\alpha}\Gamma(\alpha+1)\,. \tag{5.6.16}$$

The lower bound from (5.6.15) reads

$$\varphi(k,t) \geq e^{-x}(1+\theta x^{\alpha} e^{-\lambda_\alpha t}), \quad 1<\alpha<2,$$
$$\lambda_\alpha = \int_0^1 ds\rho(s)[1-s^{\alpha}-(1-s)^{\alpha}], \quad \rho(s)=4\pi g(1-2s), \quad 0<s\leq 1. \tag{5.6.17}$$

Note that $\alpha > 1$ and $\theta > 0$ are free parameters. It is clear that for any sufficiently small $\varepsilon > 0$ we can choose $\alpha(\varepsilon)$ such that $\lambda_{\alpha(\varepsilon)} = \varepsilon$. This follows from the fact that λ_α is a continuous function of α and $\lambda_1 = 0$. Then we need to choose $\theta(\varepsilon)$ assuming that $\lambda_\alpha = \varepsilon$. The choice of θ must also guarantee the positivity of the solution $f(v,t)$. It is enough to guarantee the positivity of the initial condition (5.6.6). Making use of the inverse Fourier transform and tables from [94], we obtain

$$f(v,0) = \frac{1}{(2\pi)^3}\int_{\mathbb{R}^3} dk\, e^{ik\cdot v-|k|^2/2}[1+\theta\,(|k|^2/2)^{\alpha}]$$
$$= \frac{1}{(2\pi)^{3/2}} e^{-|v|^2/2}\left[1+\frac{2}{\sqrt{\pi}}\,\theta\,\Gamma(\alpha+3/2)\,{}_1F_1(-\alpha,3/2;\,|v|^2/2)\right], \tag{5.6.18}$$

where ${}_1F_1(a,b;z)$ is the degenerate hypergeometric function (see Section 3.6). This function has no singularities for $0 \leq z \leq R$ with any $R > 0$. Its behavior for large $z > 0$ is described by the formula

$${}_1F_1(a,b;z) = \frac{\Gamma(a)}{\Gamma(b)} e^{z} z^{a-b}\left[1+O\left(\frac{1}{z}\right)\right], \quad z \geq 1,$$

provided $a \neq 0,-1,\dots$. In our case $a = -\alpha$, $b = 3/2$, and therefore this formula is valid. Hence, the function $f(v,0)$ (5.6.18) is non-negative for sufficiently small $\theta > 0$ that $f(v,0) \geq 0$ for any θ satisfying the condition

$$0 \leq \theta \leq \theta_0(\alpha)\,. \tag{5.6.19}$$

Then a simple calculation shows that the choice of such $\alpha > 1$ in the initial condition (5.6.6) that $\lambda_\alpha = \varepsilon$ and the choice of θ satisfying the condition (see equations (5.6.11) and (5.6.12))

$$0 < \theta \leq \min\left(\frac{\varepsilon}{8\rho_M},\ \theta_0(\alpha)\right)$$

leads to the solution $f(|v|,t)$ of the Boltzmann equation that satisfies the theorem. Note that for sufficiently small ε we can always choose $\theta(\varepsilon) = \varepsilon/8\rho_M$. Then for such parameters $\alpha(\varepsilon)$ and $\theta(\varepsilon)$, we obtain from (5.6.15) the following estimate:

$$a(x)e^{-\varepsilon t} \leq \varphi(x,t) - e^{-x} \leq b(x)e^{-\varepsilon t},$$
$$a(x) = \theta x^{\alpha} e^{-x}, \quad b(x) = 4\theta x e^{-x/2}, \quad \varepsilon = \lambda_\alpha, \quad \theta = \varepsilon/8\rho_M. \tag{5.6.20}$$

We recall the notation $x = |k|^2/2$ from equations (5.6.4). Then the estimate (5.6.3) holds with

$$c_3 = \max_{x\geq 0} a(x), \quad c_4 = \max_{x\geq 0} b(x). \tag{5.6.21}$$

It remains to prove estimates (5.6.2) for $f(v,t)$. It follows from (5.6.20) that

$$\int_{\mathbb{R}^3} dk|\varphi(|k|^2/2,t) - e^{-|k|^2/2}|\,|k|^{2N} < \infty \tag{5.6.22}$$

for any $N > 0$ and all $t \geq 0$. In particular, this means that we can use the standard inverse Fourier transform and obtain

$$f(v,t) - M(v) = \frac{1}{(2\pi)^3}\int_{\mathbb{R}^3} dk\,[\varphi(|k|^2/2,\,t) - \hat{M}(k)]\,e^{ik\cdot v}, \tag{5.6.23}$$

in the notation of equations (5.4.1) and (5.4.7). On the other hand, we have

$$\varphi(|k|^2/2,t) - \hat{M}(k) = \int_{\mathbb{R}^3} dv[f(v,t) - M(v)]\,e^{-ik\cdot v}.$$

Hence,

$$\|f(v,t) - M(v)\|_{L^\infty} \leq \frac{e^{-\varepsilon t}}{(2\pi)^3}\int_{\mathbb{R}^3} dk\,b(|k|^2/2),$$

$$\|f(v,t) - M(v)\|_{L^1} \geq c_3\,e^{-\varepsilon t},$$

in the notation of equations (5.6.20) and (5.6.21). This completes the proof. □

Remark. The solution $f(v,t)$ of the Boltzmann equation given in equation (5.6.23) has bounded derivatives of all orders with respect to components of $v \in \mathbb{R}^3$ as it follows from equation (5.6.22). It is possible to weaken the condition (5.6.1) of Theorem 5.6.1 and to extend the proof to the case of true Maxwell molecules. However, this needs relatively long calculations and therefore we omit it. Some details can be found in [27].

In the next two sections we consider some classes of solutions to the Boltzmann equation with infinite second moment. Of course, in that case we cannot expect that the solution relaxes to any Maxwellian distribution.

5.7 A class of solutions with infinite second moment

We continue to study the Cauchy problem (5.4.1) for different classes of initial data $f_0(v)$. In this section we consider only radial solutions $f(|v|,t)$ such that

$$E(f) = \langle f(|v|,t),\,|v|^2\rangle = \infty, \quad t \geq 0.$$

Note that the second moment $E(f)$ has the physical meaning of average kinetic energy of gas molecules with accuracy to an irrelevant constant factor. There are, however, some physical problems which lead to solutions with infinite energy (see, e. g., a discussion in [40] related to papers [71, 93, 126, 140] on infinitely strong shock waves). It is also interesting from a mathematical point of view to understand what happens with the solution $f(|v|, t)$ of the problem (5.4.1) if $E(f_0) = \infty$. On the one hand, the solution $f(|v|, t)$ in the form of a Wild sum (see Section 3.3, equations (3.3.8)–(3.3.10)) for the cut-off case does exist for any $t > 0$. On the other hand, the asymptotic behavior of $f(|v|, t)$ for large $t > 0$ is unclear.

The aim of this section is to clarify this question for some classes of initial data $f_0(|v|)$ in equations (5.4.1). Thus we consider such non-negative initial distribution $f_0(|v|)$ that

$$\langle f_0(|v|), 1\rangle = 1, \quad \langle f_0(|v|), |v|^2\rangle = \infty \tag{5.7.1}$$

and pass to the Fourier representation. The resulting Cauchy problem for the characteristic function

$$\varphi(x, t) = \langle f_0(|v|), e^{ik\cdot v}\rangle, \quad x = |k|^2/2, \quad k \in \mathbb{R}^3, \tag{5.7.2}$$

reads as

$$\begin{gathered} \varphi_t = \int_0^1 ds\rho(s)\{\varphi(sx)\varphi[(1-s)x] - \varphi(0)\varphi(x)\}, \quad x \geq 0, \\ \rho(s) = 4\pi g(1-2s); \quad \varphi|_{t=0} = \varphi_0(x), \quad \varphi(0, t) = 1. \end{gathered} \tag{5.7.3}$$

We introduce a class of functions having the form

$$\varphi(x, t) = \sum_{n=0}^{\infty} \varphi_n(t) \frac{x^{n\alpha}}{\Gamma(n\alpha + 1)} \tag{5.7.4}$$

and assume that

$$0 < \alpha < 1, \quad |\varphi_n(0)| \leq a^n, \quad n = 0, 1, \dots, \tag{5.7.5}$$

for some $a > 0$. We recall that similar solutions with $\alpha = 1$ in (5.7.4) were considered in detail in Chapter 4. Note that the value $\alpha = 1$ corresponds to distribution functions with *finite* energy. The solution of equation (5.7.3) represented by series (5.7.4) with $1 < \alpha < 2$ was also studied in Section 5.6.

The assumption (5.7.4) with any fixed $\alpha > 0$ allows to reduce, at least formally, the equation (5.7.3) to a set of ODEs for coefficients $\{\varphi_n, n = 0, 1, \dots\}$. Substituting the series (5.7.4) into this equation and collecting terms of the same degree in x, we finally obtain

$$\dot{\varphi}_0 = 0, \quad \dot{\varphi}_1 = \lambda(\alpha)\varphi_1;$$

$$\dot{\varphi}_n - \lambda(n\alpha)\varphi_n = \sum_{k=1}^{n-1} B_\alpha(k, n-k)\, \varphi_k\, \varphi_{n-k}\,, \quad n = 2, 3, \dots\,, \tag{5.7.6}$$

where

$$\lambda(n\alpha) = \int_0^1 ds\rho(s)\left[s^{n\alpha} + (1-s)^{n\alpha} - 1\right], \quad n = 1, 2, \dots\,;$$

$$B_\alpha(k,l) = \frac{\Gamma[(k+l)\alpha + 1]}{\Gamma(k\alpha+1)\,\Gamma(l\alpha+1)} \int_0^1 ds\rho(s) s^{k\alpha}(1-s)^{l\alpha}; \quad k, l = 1, 2, \dots. \tag{5.7.7}$$

Note that we use a simplified notation for eigenvalues $\lambda(p)$ of linearized collision operator: $\lambda(p) = \lambda_0(2p)$, in the "old" notation of equations (3.6.5). We assume in this section that

$$0 \le \rho(s) \le Cs^{-(1+\gamma)}, \quad 0 < \gamma < 1. \tag{5.7.8}$$

This inequality is satisfied for true Maxwell molecules with $\gamma = 1/4$. Then the function $\lambda(p)$ has the following properties:

$$\begin{aligned} &\text{(a)} \quad \lambda(p) \le \lambda(p_0), \quad \lambda'(p) \le 0, && \text{if} \quad p \ge p_0 > \gamma; \\ &\text{(b)} \quad \lambda(1) = 0; \quad \lambda(p) > 0, && \text{if} \quad 1 > p \ge p_0 > \gamma; \\ &\text{(c)} \quad \lambda(p) < 0, && \text{if} \quad p > 1. \end{aligned}$$

We can also prove the following lemma.

Lemma 5.7.1. *If $\rho(s)$ satisfies inequality (5.7.8) with some $0 < \gamma < 1$, then there exists a number $R = R(\alpha, \gamma)$ such that for any $\alpha > \gamma$*

$$\frac{1}{n-1} \sum_{k=1}^{n-1} B_\alpha(k, n-k) \le C\, R(\alpha, \gamma), \quad n = 2, 3, \dots, \tag{5.7.9}$$

in the notation of equations (5.7.7) and (5.7.8).

Proof. It follows from (5.7.8) that

$$\int_0^1 ds\rho(s)\, s^{k\alpha}\,(1-s)^{(n-k)\alpha} \le C\,\frac{\Gamma(k\alpha - \gamma)\,\Gamma[(n-k)\alpha + 1]}{\Gamma(n\alpha + 1 - \gamma)}, \quad k = 1, \dots, n-1,$$

where $\Gamma(z)$ denotes the gamma function.

Hence,

$$B_\alpha(k, n-k) \le C\frac{\Gamma(n\alpha+1)\Gamma(k\alpha-\gamma)}{\Gamma(n\alpha+1-\gamma)\,\Gamma(k\alpha+1)}, \quad n \ge 2.$$

It is known that [94]

$$\lim_{r\to\infty} \frac{\Gamma(z)\, z^r}{\Gamma(z+r)} = 1.$$

Therefore,

$$\frac{\Gamma(k\alpha-\gamma)}{\Gamma(k\alpha+1)} \underset{k\to\infty}{\approx} (k\alpha)^{-(1+\gamma)}, \qquad \frac{\Gamma(n\alpha+1)}{\Gamma(n\alpha+1-\gamma)} \underset{n\to\infty}{\approx} (n\alpha)^{\gamma}.$$

Thus we obtain

$$\frac{1}{n-1}\sum_{k=1}^{n-1} B_\alpha(k, n-k) \le C\, \frac{\Gamma(n\alpha+1)}{\Gamma(n\alpha+1-\gamma)} \sum_{k=1}^{n-1} \frac{\Gamma(k\alpha-\gamma)}{\Gamma(k\alpha+1)} \le C\, R(\alpha,\gamma).$$

This completes the proof. □

Below we will also need a simple inequality for eigenvalues

$$n\lambda(\alpha) - \lambda(n\alpha) \ge (n-1)\lambda(\alpha), \quad \alpha > \gamma, \quad n \ge 1,$$

which follows from inequality $\lambda'(p) < 0$ if $p > \gamma$.

Let us consider ODEs (5.7.6) with initial data

$$\varphi_n|_{t=0} = \varphi_n^{(0)}, \quad n = 0, 1, \dots; \quad \varphi_0^{(0)} = 1.$$

Then the first coefficients can be found immediately:

$$\varphi_0(t) = 1; \quad \varphi_1 = \varphi_1^{(0)} e^{\lambda(\alpha)t}, \quad \lambda(\alpha) > 0.$$

In other words, the solution $\varphi(x,t)$ (5.7.4) for $0 < \alpha < 1$ behaves for small x like

$$\varphi(x,t) = 1 + \varphi_1^{(0)} \left(x\, e^{\mu_\alpha(t)}\right)^\alpha + \cdots, \quad \mu_\alpha = \lambda_\alpha/\alpha.$$

It is convenient to fix a value $\gamma < \alpha < 1$ and to pass to new variables by the transformation

$$\varphi(x,t) = \psi(\tilde{x},t), \quad \tilde{x} = x\, e^{\mu_\alpha(t)}.$$

Then the equation for $\psi(x,t)$ reads

$$\psi_t + \mu_\alpha x\, \psi_x = \int_0^1 ds\, \rho(s) \big[\psi(sx)\psi[(1-s)x] - \psi(0)\psi(x)\big],$$
$$\mu_\alpha = \lambda_\alpha/\alpha, \quad \psi|_{t=0} = \varphi_0(x). \tag{5.7.10}$$

Substituting the series

$$\psi = \sum_{n=0}^{\infty} \psi_n(t) \frac{x^{n\alpha}}{\Gamma(n\alpha + 1)}, \tag{5.7.11}$$

we obtain ODEs for $\psi_n(t)$:

$$\begin{aligned} \dot{\psi}_0 &= \dot{\psi}_1 = 0, \\ \dot{\psi}_n(t) + r_n(\alpha)\psi_n &= \sum_{k=1}^{n-1} B_\alpha(k, n-k)\psi_k\psi_{n-k}, \\ r_n(\alpha) &= n\lambda(\alpha) - \lambda(n\alpha), \quad n \geq 2; \\ \psi|_{t=0} &= \varphi_n^{(0)}, \quad n = 0, 1, \ldots. \end{aligned} \tag{5.7.12}$$

The recurrence relations for $\psi_n(t)$ follow from these equations:

$$\begin{aligned} &\psi_0(t) = 1, \quad \psi_1(t) = \psi_1^{(0)}; \\ &\psi_n(t) = \varphi_n(0)\, e^{-r_n(\alpha)t} \\ &\quad + \sum_{k=1}^{n-1} B_\alpha(k, n-k) \int_0^t d\tau \psi_n(\tau)\psi_{n-k}(\tau) e^{-r_n(\alpha)(t-\tau)}, \quad n \geq 2. \end{aligned} \tag{5.7.13}$$

It is clear from equations (5.7.13) that

$$\lim_{t\to\infty} \psi_n(t) = u_n, \quad n = 0, 1, \ldots,$$

where $\{u_n\}$ is the unique steady-state solution of (5.7.12) given by the recurrence formulas

$$u_0 = 1, \quad u_1 = \varphi_1^{(0)}; \quad u_n = \sum_{k=1}^{n-1} B_\alpha(k, n-k)\, u_k\, u_{n-k}, \quad n \geq 2. \tag{5.7.14}$$

It is not difficult to prove estimates of the form

$$|\psi_n(t)| \leq C_1 A_1^n, \quad |u_n| \leq C_2 A_2^n, \quad n = 1, 2, \ldots,$$

for solutions of equations (5.7.13) and (5.7.14), respectively. The analysis of these equation is very similar to the analysis performed in Sections 4.5 and 4.6 for analytic radial solutions. Therefore, below we formulate some results without proofs. The detailed proofs can be found in [40].

Lemma 5.7.2. *If $\rho(s)$ satisfies conditions* (5.7.8) *with $0 < \gamma < 1$ and inequalities* (5.7.5) *hold for some $a > 0$, then, for any $\gamma < \alpha < 1$, the following estimates are valid:*

$$|\psi_n(t)| \leq a^n \left[1 + \frac{C}{\lambda(\alpha)} R(\alpha, \gamma)\right]^{n-1}, \quad n \geq 1, \quad t \geq 0,$$

for solutions of equations (5.7.13) *and*

$$|u_n| \le |\varphi_1^{(0)}| \left[\frac{C}{\lambda(\alpha)} R(\alpha, \gamma) \right]^{n-1}, \quad n \ge 1,$$

for solutions of equations (5.7.14). *The parameter* $R(\alpha, \gamma)$ *is defined in Lemma* 5.7.1.

The main result of this section can be formulated in the following way. We consider equation (5.7.3) and assume that:

(A) There exist two numbers $C > 0$ and $0 < \gamma < 1$ such that

$$0 \le p(s) \le C s^{-(1+\gamma)}, \quad 0 < s \le 1 . \tag{5.7.15}$$

(B) The initial condition has the following form:

$$\varphi(x, 0) = \varphi_0(x) = \sum_{n=0}^{\infty} \varphi_n^{(0)} \frac{x^{n\alpha}}{\Gamma(n\alpha + 1)}, \quad \varphi_0^{(0)} = 1, \quad \varphi_1^{(0)} \ne 0, \tag{5.7.16}$$

with some $\gamma < \alpha < 1$ and such that

$$\sup_{n=1,2,\dots} |\varphi_n^{(0)}|^{1/n} < \infty.$$

Remark. Condition (A) is equivalent to condition (4.5.6) of Theorem 4.5.1 provided $\gamma = 1 - \varepsilon$. This condition is satisfied for true Maxwell molecules with $\gamma = 1/4$.

Theorem 5.7.1. *Under assumption* (A) *and* (B) *there exists a unique solution of equation* (5.7.3) *satisfying the initial condition* (5.7.16) *and represented in the form*

$$\varphi(x, t) = \psi(x\, e^{\mu_\alpha t}), \quad \mu_\alpha = \frac{1}{\alpha} \int_0^1 ds p(s) [s^\alpha + (1-s)^\alpha - 1] > 0 , \tag{5.7.17}$$

where $\psi(x, t)$ *is given by formulas* (5.7.11) *and* (5.7.13). *The series* (5.7.11) *converges for all* $x > 0$. *Moreover,*

$$\sup_{n=1,2,\dots} |\psi_n(t)|^{1/n} \le b$$

for all $t > 0$ *and some constant* b *depending only on* $p(s)$ *and on the initial condition.*

Proof. The proof follows from the above described construction of solution and from Lemma 5.7.2. The solution is unique by the standard uniqueness theorem on ODEs and is given, by construction, by (5.7.11), (5.7.13). This completes the proof. □

The second result concerns self-similar solutions and their connection with solutions described by Theorem 5.7.1.

Theorem 5.7.2. *If $\rho(s)$ satisfies assumption* (A) *with some $\gamma < 1$, then for any number μ_α given by* (5.7.17) *with $0 < \gamma < 1$, there exists a self-similar solution $\varphi(x,t) = u^{(\alpha)}(xe^{\mu_\alpha t})$ given by the following series:*

$$u^{(\alpha)}(x) = \sum_{n=0}^{\infty} u_n^{(\alpha)} \frac{x^{n\alpha}}{\Gamma(n\alpha+1)}, \qquad \sup_{n=1,2,\dots} \left|u_n^{(\alpha)}\right|^{1/n} \le \infty, \tag{5.7.18}$$

where $u_0^{(\alpha)} = 1$, $u_1^{(\alpha)} \neq 0$ can be chosen arbitrarily, and $u_n^{(\alpha)}$ $(n \ge 2)$ are given by the recurrence formulas in (5.7.14)*. If $\psi(x,t)$ is the function defined in Theorem* 5.7.1 *(for a given value of α), then*

$$\lim_{t\to\infty} \psi(x,t) = u^{(\alpha)}(x) \tag{5.7.19}$$

for any $x \ge 0$, provided $u_1^{(\alpha)} = \psi_1(0)$.

Proof. The proof is based on the construction described in the first part of this section (the function $u^{(\alpha)}(x)$ is obviously a steady solution of equation (5.7.10)) and on Lemma 5.7.2. It was already shown that $u^{(\alpha)}(x)$ is formally the limit of the series (5.7.11) as $t \to \infty$. On the other hand, for any $x > 0$ the series converges uniformly on $t \in [0,\infty)$ (Lemma 5.7.2). Therefore, the limit is rigorously justified. Inequality (5.7.18) also follows from Lemma 5.7.2. Thus the theorem is proved. □

Theorem 5.7.2 explains the exact meaning of the asymptotic equality

$$\varphi(x,t) = \psi(xe^{\mu_\alpha t}, t) \underset{t\to\infty}{\simeq} u^{(\alpha)}(xe^{\mu_\alpha t}) \tag{5.7.20}$$

and the corresponding relation

$$\begin{aligned} f(|v|,t) &= e^{-3\mu_\alpha t/2} F(|v|e^{-\mu_\alpha t/2}, t) \underset{t\to\infty}{\simeq} e^{-3\mu_\alpha t/2}\, \Phi_\alpha(|v|e^{-\mu_\alpha t/2}), \end{aligned} \tag{5.7.21}$$

where

$$\int_{\mathbb{R}^3} dv \Phi_\alpha(|v|,t) e^{-ik\cdot v} = u^{(\alpha)}\left(\frac{|k|^2}{2}\right)$$

for the corresponding solutions of the Boltzmann equation (5.4.1). We stress that the transition from (5.7.19) to (5.7.20) is not justified as yet, since we still must prove that $u^{(\alpha)}(|k|^2/2)$ is a characteristic function (Fourier transform of a probability measure).

Theorem 5.7.3. *All self-similar solutions described in Theorem* 5.7.1 *are characteristic functions.*

Proof. The idea of the proof is rather simple if we postulate that the Boltzmann equation preserves (in time) the positivity of its solution. Then we need to find at least one example of a non-negative function $f_0(|v|)$, whose Fourier transform $\varphi_0(|k|^2/2)$ can be represented in the form of series (5.7.16). After that we solve equation (5.7.3) with this particular initial condition. The solution $\varphi(x,t)$, $x = |k|^2/2$, will be a characteristic function at any $t > 0$ in accordance with our postulate. On the other hand,

$$\varphi(x,t) = \varphi(xe^{-\mu_a t}, t)$$

is also a characteristic function and so is its limit (5.7.19). In other words, the corresponding distribution function $\Phi_0(|v|)$ from equations (5.7.21) is non-negative, as expected. □

The above postulate (self-evident for physicists) follows from a uniqueness lemma proved in Section 3.3. The more important assumption related to examples of characteristic functions in the form of series (5.7.16) will be justified in the next section.

5.8 Exact self-similar solutions

In this section we consider the simplest case of the Boltzmann equation for pseudo-Maxwell molecules (3.3.1) with constant kernel $g(\mu) = (4\pi)^{-1}$, satisfying condition (3.3.6). Then the Fourier transform of the isotropic solution $f(|v|,t)$ leads to the equation

$$\varphi_t(x,t) = \int_0^1 ds\{\varphi(sx)\varphi[(1-s)x] - \varphi(0)\varphi(x)\}, \quad x \geq 0, \tag{5.8.1}$$

in the notation of equations (5.7.2) and (5.7.3). We look for self-similar solution in the form $\varphi(x,t) = \psi(x\, e^{\mu t})$, $\psi(0) = 1$, and obtain

$$\mu x \psi_x = \int_0^1 ds \psi(sx)\psi[(1-s)x] - \psi(x). \tag{5.8.2}$$

It is easy to notice that this equation can be simplified by Laplace transform. Let

$$y(p) = p\mathcal{L}[\psi] = p\int_0^\infty dx \psi(x) e^{-px}. \tag{5.8.3}$$

Then we obtain a non-linear ODE

$$\mu p^2 y'' - py' + y(1-y) = 0, \quad y(p) \underset{p\to\infty}{\to} 1. \tag{5.8.4}$$

The case $\mu = 0$ is trivial; thus we assume $\mu \neq 0$. Then equation (5.7.4) can be reduced to a standard form by the following substitutions:

$$\text{(a)}\quad \mu \neq -1: \quad y(p) = \frac{1}{2} - \frac{1}{A}\left[\frac{1}{2} - p^{2\beta} w(p^{\beta})\right],$$

$$\beta = \frac{1+\mu}{5\mu}, \quad A = \frac{1}{6\mu\beta^2} = \frac{25\mu}{6(1+\mu)^2}; \tag{5.8.5}$$

$$\text{(b)}\quad \mu = -1: \quad y(p) = \frac{1}{2} - 6w(\log p). \tag{5.8.6}$$

These substitutions lead to two equations for $w(\tau)$ (we omit the intermediate calculations):

$$\text{(a)}\quad \mu \neq -1: \quad w'' = 6w^2 + 3\frac{1-A^2}{2\tau^4}; \tag{5.8.7}$$

$$\text{(b)}\quad \mu = -1: \quad w'' = 6w^2 - \frac{1}{24}. \tag{5.8.8}$$

Here and below, primes denote derivatives with respect to τ. Each of these equations has two particular solutions easy to find:

$$\text{(a)}\quad w = \frac{1 \pm A}{2\tau^2}; \quad \text{(b)}\quad w = \pm\frac{1}{12}.$$

These solutions correspond to the trivial ones ($y = 0$, $y = 1$) of the original equation.

The case $\mu = -1$ and the special case $A = \pm 1$ for $\mu \neq 1$ lead to the following first-order ODEs:

$$\text{(a)}\quad (w')^2 = 4w^3 + \text{const.}; \quad \text{(b)}\quad (w')^2 = 4w^3 - \frac{w}{12} + \text{const.}$$

These equations can be solved in terms of the Weierstrass elliptic function $\mathcal{P}(\tau; g_2, g_3)$, satisfying the equation [94]

$$(\mathcal{P}')^2 = \mathcal{P}^3 - g_2\mathcal{P} - g_3.$$

If $|A| \neq 1$ in equation (5.8.5), then the equation is not of the Painlevé type (see [125]). It has moving logarithmic critical points and therefore does not have any "simple" analytic solution. Hence, relatively simple solutions can exist only for values of μ satisfying the equality

$$A^2 = \frac{625\mu^2}{36(1+\mu)^4} = 1$$

and for $\mu = -1$. Thus there are five possible values of μ:

$$\mu = -1, \quad \mu = 2/3, \quad \mu = 3/2, \quad \mu = -1/6, \quad \mu = -6.$$

Not all these values of μ correspond to true solutions of the transformed Boltzmann equation (5.8.2), since we also need the correct asymptotics (5.8.4) for $p \to \infty$. The qualitative behavior of the solutions of equations (5.8.7) with $A^2 = 1$ and (5.8.8) is best understood by an analogy with classical mechanics. We re-write these equations as

$$\text{(a)}\quad w'' = -\frac{\partial U_1}{\partial w}, \quad U_1 = -2w^3; \tag{5.8.9}$$

$$\text{(b)}\quad w'' = -\frac{\partial U_2}{\partial w}, \quad U_2 = -w^3 + \frac{w}{24}, \tag{5.8.10}$$

and consider them as the equation of motion of a mass point on a line under an external force with potential energy $U_{1,2}(w)$. Then it becomes clear that the condition $w(\tau) \underset{\tau\to\infty}{\longrightarrow} -1/12$, where $\tau = \log p$, cannot be satisfied by a solution of equation (5.8.10). Thus the value $\mu = -1$ does not lead to correct asymptotics. Similarly, in the case $\mu = -6$ we need a solution $w(\tau)$ of equation (5.8.9) such that $\tau^2 w(\tau) \underset{\tau\to\infty}{\longrightarrow} 0$, where $\tau = p^{1/6}$, and this is impossible.

Hence, there are only three relevant values of μ:

$$\mu = 2/3, \quad \mu = 3/2, \quad \mu = -1/6.$$

In the first two cases the solution of equation (5.8.9) must be chosen in such a way that $\tau^2 w(\tau) \underset{\tau\to\infty}{\longrightarrow} 1$. General properties of the Weierstrass function $\mathcal{P}(\tau; 0, g_3)$ show that the only possible choice is $g_3 = 0$. We obtain

$$w(\tau) = \mathcal{P}(\tau; 0, 0) = \frac{1}{\tau^2}, \quad \mu = 2/3, \quad \mu = 3/2, \tag{5.8.11}$$

where, of course, we can change τ to $\tau_0 + \tau$ with arbitrary $\tau_0 > 0$.

In the case $\mu = -1/6$ we also need to satisfy the condition $\tau^2 w(\tau) \underset{\tau\to\infty}{\longrightarrow} 1$. This condition is satisfied by a wider class of solutions $w(\tau) = \mathcal{P}(\tau_0 + \tau;\, 0, g_3)$ with arbitrary values of $\tau_0 > 0$ and g_3. The simplest solution of this type is

$$w(\tau) = \frac{1}{(\tau_0 + \tau)^2}, \quad \tau_0 > 0, \quad \mu = -1/6. \tag{5.8.12}$$

The corresponding solution of equation (5.8.4) reads

$$y(p) = 1 - \frac{1}{(1 + \tau_0 p)^2}, \quad \mu = -1/6. \tag{5.8.13}$$

The previous two exact solutions (see (5.8.11)) provide two solutions of equation (5.8.4) in the following form:

$$y(p) = 1 - \frac{1}{(1 + \tau_0 p^{-\beta})^2}; \quad \beta = 1/2, \quad \mu = 2/3; \quad \beta = 1/3, \quad \mu = 3/2. \tag{5.8.14}$$

The first solution (5.8.13) corresponds to the well-known BKW-mode for the Boltzmann equation which was discussed in detail in Section 4.8. The other two cases of the closed form solution (5.8.14) lead to the similarity solutions with infinite second moment found by Bobylev and Cercignani in 2002 [39, 40] (see also [41]). Our investigation in this section shows that there is almost no hope to find any other closed form solution of equation (5.8.2).

Let us now consider the two solutions (5.8.14) and invert the Laplace transform (5.8.3). Without loss of generality, we consider the case $\tau_0 = 1$ and obtain

$$\psi(x) = \mathcal{L}^{-1}\left[\frac{1}{p(1+p^{-\beta})^2}\right] = \sum_{n=0}^{\infty}(-1)^n\frac{(n+1)x^{n\beta}}{\Gamma(n\beta+1)},$$

$$\beta = 1/2 \quad (\mu = 2/3); \quad \beta = 1/3 \quad (\mu = 3/2). \tag{5.8.15}$$

This can be seen by taking the Laplace transform, term by term, of the power series in $p^{-\beta}$. The process is easily justified for $|p| > 1$, but the result is valid for any complex p by analytic continuation. We can see now that these solutions belong to the general class of self-similar solutions discussed in detail in Section 5.7.

It remains to show that $\psi(|k|^2/2)$ is the characteristic function. We note that the standard inversion formula

$$\psi(x) = \frac{1}{2\pi i}\int_{a+i\infty}^{a-i\infty} e^{px}\Psi(p), \quad a > 0,$$

where

$$\Psi(p) = \frac{1}{p(1+p^{-\beta})^2}; \quad \beta = 1/2, \quad \beta = 1/3, \tag{5.8.16}$$

is not very useful in that case. However, it is not difficult to show that for a certain class of functions (see [41] for details), which includes $\Psi(p)$ from (5.8.16), we can use another formula:

$$\psi(x) = \frac{1}{\pi}\int_0^{\infty} ds e^{-sx}\Im\Psi(s\,e^{-i\pi}), \tag{5.8.17}$$

where $\Im z$ denotes the imaginary part of z, $z \in \mathbb{C}$. We substitute here functions (5.8.16) and obtain after straightforward calculations

$$\psi(x) = \tilde{\psi}_\beta(x) = 2\frac{\sin\beta\pi}{\beta\pi}\int_0^{\infty}\frac{ds\,(1+s\cos\beta\pi)}{(1+s^2+2s\,\cos\beta\pi)^2}\,e^{-xs^{-1/\beta}},$$

where $\beta = 1/2$, $\beta = 1/3$. Then we remark that

$$x = \frac{|k|^2}{2}, \quad e^{-\theta|k|^2/2} = \mathcal{F}[M_\theta(|v|)] = \int_{\mathbb{R}^3} dv M_\theta(|v|)\,e^{-ik\cdot v},$$

where $M_\theta(|v|)$ denotes the Maxwellian distribution

$$M_\theta(|v|) = (2\pi\theta)^{-3/2} e^{-|v|^2/2\theta}$$

with "temperature" $\theta > 0$. Hence, the corresponding solution of the Boltzmann equation (3.3.1) with constant kernel $g(\mu) = (4\pi)^{-1}$ reads

$$\begin{aligned} f_\beta(|v|, t) &= F_\beta(|v| e^{-\frac{\mu_\beta t}{2}}) e^{-\frac{3}{2}\mu_\beta t}, \\ F_\beta(|v|) &= 2\frac{\sin\beta\pi}{\beta\pi} \int_0^\infty \frac{ds\,(1 + s\cos\beta\pi)\, M_{\theta_\beta(s)}(|v|)}{(1 + s^2 + 2s\,\cos\beta\pi)^2}, \end{aligned} \tag{5.8.18}$$

where $\theta_\beta(s) = s^{-1/\beta}$, $\beta = 1/2$, $\mu_{1/2} = 2/3$; $\beta = 1/3$, $\mu_{1/3} = 3/2$.

A byproduct result of this section is obviously an explicit example of a characteristic function in the form of series (5.7.14). This example is needed for completing the proof of Theorem 5.7.3. Note that the function $F_\beta(|v|)$ given in (5.7.18) is positive for $0 \le \beta \le 1/2$. Its Fourier transform is given in equation (5.7.15), which coincides with (5.7.14) for $\alpha = \beta$ and

$$\varphi_n^{(0)} = (-1)^n (n + 1), \quad n = 0, 1, \dots .$$

To complete the proof of Theorem 5.7.3 we need to present a similar example of a characteristic function of the form (5.7.6) for $1/2 \le \alpha \le 1$. To this goal, let us consider the so-called Mittag-Leffler function

$$\varphi_0(x) = \sum_{n=0}^\infty (-1)^n \frac{x^{n\alpha}}{\Gamma(n\alpha + 1)}, \quad 0 < \alpha < 1 .$$

It has a relatively simple Laplace transform,

$$\mathcal{L}(\varphi_0) = \int_0^\infty dx\, e^{-px} \varphi_0(x) = \frac{p^{\alpha-1}}{1 + p^\alpha} .$$

Then we can use the same inversion formula (5.8.17) because this function also belongs to the class considered in [41]. Then we obtain

$$\varphi_0(x) = \frac{\sin\alpha\pi}{\alpha\pi} \int_0^\infty \frac{ds e^{-xs^{-1/\alpha}}}{1 + s^2 + 2s\cos\alpha\pi}, \quad 0 < \alpha < 1,$$

and repeat the above considerations to show that $\varphi_0(|k|^2/2)$ is a characteristic function for any $\alpha \in (0, 1)$. This completes the proof of Theorem 5.7.3.

Remarks on Chapter 5

1. Chapter 5 contains three parts related to asymptotic properties of solutions $f(v, t)$ to the spatially homogeneous Boltzmann equation for Maxwell molecules:
   (1) asymptotic behavior for large velocities as $|v| \to \infty$ (Sections 5.1–5.3);
   (2) asymptotic behavior for $t \to \infty$ (Sections 5.4–5.6);
   (3) asymptotic properties of solutions with infinite second moment (Sections 5.7 and 5.8).

   Most of the material is based on papers [21, 27, 40, 41]. We briefly mention below some related publications of different authors.
2. Concerning the large time asymptotics of solutions for the Fourier transformed Boltzmann equation, an important first step was made in the paper [87]. A lot of work was done in the 1990s and 2000s by different authors on the so-called "Cercignani's conjecture" [69]. It became clear after papers [21] and [37] that the conjecture cannot be true in its original form. Finally, the conjecture was proved in a weaker form. The complete information and all related references can be found in the papers [147, 148] by C. Villani. We also mention related works [65–67] on Kac and Boltzmann equations by E. Carlen and his co-authors.
3. The work on solutions with infinite second moment discussed in Sections 5.7 and 5.8 was continued by two different groups of authors (see, e. g., [63, 119] and references therein). Similar questions for non-Maxwell molecules remain open.
4. The solutions from Sections 5.7 and 5.8 are positive and eternal, i. e., they are bounded for all (positive and negative) values of time. Some related questions were discussed in [60, 61]. The absence of positive eternal solutions with finite moments of all orders was proved in [40] in connection with Villani's conjecture from [147].

6 Generalized Maxwell models

6.1 The Boltzmann equation for inelastic interactions

In this chapter we consider various physical systems which differ from classical rarefied gas. It is convenient to begin with the simplest generalization of the Boltzmann equation for hard spheres of diameter d. We will use the notations of Sections 2.1 and 2.2 and consider equation (2.1.2) for the distribution function $f(x, v, t)$:

$$f_t + v \cdot f_x = Q(f,f)\,. \tag{6.1.1}$$

Then we multiply this equation by the test function $\psi(v)$ and integrate over the velocity space $\mathbb{R}^3$. The result reads

$$\partial_t \langle f\,, \psi\rangle + \partial_x \cdot \langle f,\, v\psi\rangle = \big\langle \psi,\, Q(f,f)\big\rangle\,, \tag{6.1.2}$$

where

$$\big\langle f,\, \psi(v)\big\rangle = \int\limits_{\mathbb{R}^3} dv f(x, v, t)\psi(v)\,. \tag{6.1.3}$$

It follows from equation (2.1.2) that

$$\begin{aligned}\big\langle \psi,\, Q(f,f)\big\rangle &= \frac{d^2}{2} \int\limits_{\mathbb{R}^3\times\mathbb{R}^3\times S^2} dv dw dn\, \psi(v)\, |u \cdot n|\, \big[f(v')f(w') - f(v)f(w)\big] \\ &= \frac{d^2}{2} \int\limits_{\mathbb{R}^3\times\mathbb{R}^3} dv dw f(v)f(w)\, |u|\, [R\psi](v, w),\end{aligned} \tag{6.1.4}$$

where

$$\begin{gathered}[R\psi](v, w) = \int\limits_{S^2} dn|\hat{u} \cdot n|\big[\psi(v') - \psi(v)\big], \\ u = v - w, \quad \hat{u} = u/|u|, \quad v' = v - (u \cdot n)n, \quad w' = w + (u \cdot n).\end{gathered} \tag{6.1.5}$$

Equations (6.1.2)–(6.1.5) can be considered as a weak form of the Boltzmann equation (6.1.1) for hard spheres. Note that the post-collisional relative velocity reads

$$u' = v' - w' = u - 2(u \cdot n)n, \quad n \in S^2, \tag{6.1.6}$$

where the vector $r = dn$ connects the centers of the colliding spheres (see Section 1.6) at the moment of collision. If

$$u = u_{\|} n + u_{\perp}, \quad u_{\|} = u \cdot n,$$

https://doi.org/10.1515/9783110550986-007

then

$$u' = u'_{\|} n + u'_{\perp}, \quad u'_{\|} = (u' \cdot n) = -u_{\|}, \quad u'_{\perp} = u_{\perp}. \tag{6.1.7}$$

This is the law of specular reflection for u if n is considered as an outer unit normal vector to the reflecting surface. This is a standard interpretation of the collision of two perfectly elastic spheres. However, there were many works in the last three decades devoted to applications of kinetic equations to so-called granular gases (see, e. g., the book [127] and references therein). In that case collisions are not elastic. The simplest way to introduce inelasticity in the rules of collision is to postulate that equations (6.1.7) change to

$$u' = u'_{\|} n + u'_{\perp}, \quad u'_{\|} = u' \cdot n = -e u_{\|}, \quad u'_{\perp} = u_{\perp}, \tag{6.1.8}$$

where $0 \le e \le 1$ is called the restitution coefficient.

Obviously, the value $e = 1$ corresponds to elastic collision. The transformation (6.1.8) is equivalent to transformation

$$u' = u - (1 + e)(u \cdot n)n, \quad 0 \le e \le 1, \quad u = v - w, \tag{6.1.9}$$

which coincides with equation (6.1.6) in the elastic limit $e = 1$. We also assume that the velocity of the center of mass of colliding particles remains unchanged after their collision, i. e.,

$$v' + w' = v + w. \tag{6.1.10}$$

Combining equations (6.1.9) and (6.1.10), we obtain the following rules of inelastic collision:

$$v' = v - \frac{1+e}{2}(u \cdot n)n, \quad w' = w + \frac{1+e}{2}(u \cdot n)n, \quad u = v - w, \; n \in S^2, \tag{6.1.11}$$

which coincide with corresponding formulas from (6.1.5) in the elastic case $e = 1$.

In this chapter we consider only spatially homogeneous systems. Then the inelastic Boltzmann equation for the distribution function $f(v,t)$ is defined in the following way:

$$f_t = Q_i(f,f), \tag{6.1.12}$$

where the weak form of the operator $Q_i(f,f)$

$$\begin{aligned} \langle \psi(v), Q_i(f,f) \rangle &= \int_{\mathbb{R}^3} dv \psi(v) [Q_i(f,f)](v) \\ &= \frac{d^2}{2} \int_{\mathbb{R}^3 \times \mathbb{R}^3 \times S^2} dv dw dn \, |u \cdot n| f(v) f(w) [\psi(v') - \psi(v)] \\ &= \frac{d^2}{4} \int_{\mathbb{R}^3 \times \mathbb{R}^3 \times S^2} dv dw dn \, |u \cdot n| f(v) f(w) [\psi(v') + \psi(w') - \psi(v) - \psi(w)], \end{aligned} \tag{6.1.13}$$

in the notation of (6.1.11).

It is easy to verify that the strong form of the collision integral in equation (6.1.12) reads

$$Q_i(f,f) = \frac{d^2}{2} \int\limits_{\mathbb{R}^3 \times S^2} dwdn\, |u \cdot n| \left[\frac{1}{e^2} f(v_*) f(w_*) - f(v) f(w) \right], \tag{6.1.14}$$

where $u = v - w$, $n \in S^2$, and v_*, w_* are pre-collisional velocities given by equalities

$$v_* = v - \frac{1+e}{2}(u \cdot n)n, \quad w_* = w + \frac{1+e}{2}(u \cdot n)n. \tag{6.1.15}$$

The restitution coefficient $0 \le e \le 1$ is considered as a given constant parameter. Usually we are interested in the solution $f(v,t)$ of the initial value for equation (6.1.12). It is clear from equation (6.1.13) that the equation preserves total mass and momentum. Without loss of generality we can assume that

$$\langle f, 1 \rangle = 1, \quad \langle f, v \rangle = 0. \tag{6.1.16}$$

We also introduce the usual notation for granular temperature

$$T(t) = \frac{1}{3} \langle f, |v|^2 \rangle . \tag{6.1.17}$$

The equation for $T(t)$ follows from equations (6.1.12) and (6.1.13). We obtain after simple calculations

$$T'(t) = -(1 - e^2) \frac{\pi d^2}{24} \int\limits_{\mathbb{R}^3 \times \mathbb{R}^3} dvdwf(v,t) f(w,t) |v - w|^3 . \tag{6.1.18}$$

Hence, if $e < 1$, then $T(t)$ decreases monotonically in time and we expect that

$$f(v,t) \underset{t \to \infty}{\to} \delta(v) \tag{6.1.19}$$

in contrast with the elastic case $e = 1$. Some details of the cooling process can be understood from the simplified model considered in the next section.

6.2 Inelastic Maxwell model

The idea of the simplification is the following. We consider the weak form (6.1.13) of the collision integral and change the factor $|u \cdot n|$ under the integral sign to $\langle |u| \rangle |u \cdot n| / |u|$, where the average $\langle |u| \rangle$ is a function of time t only. For example, we note that

$$\langle |u|^2 \rangle = \int\limits_{\mathbb{R}^3 \times \mathbb{R}^3} dvdwf(v,t) f(w,t) |v - w|^2 = 6\, T(t) ,$$

in accordance with equations (6.1.16) and (6.1.17). Therefore, a reasonable approximation for $\langle|u|\rangle$ is

$$\langle|u|\rangle \cong \gamma\sqrt{T(t)}, \quad \gamma = \text{const.}, \tag{6.2.1}$$

where γ can be different for different problems. Then we replace equation (6.1.12) in the weak form (6.1.2) by the model equation

$$\partial_t\langle f,\psi\rangle = d^2\gamma\sqrt{T(t)}\,\langle\psi, Q_i^{(M)}(f,f)\rangle\,, \tag{6.2.2}$$

where

$$\begin{aligned}\langle\psi, Q_i^{(M)}(f,f)\rangle &= \frac{1}{2}\int_{\mathbb{R}^3\times\mathbb{R}^3} dv dw\left|\frac{u\cdot n}{|u|}\right| f(v,t)f(w,t)\,[\psi(v')-\psi(v)]\\ &= \frac{1}{4}\int_{\mathbb{R}^3\times\mathbb{R}^3\times S^2} dv dw dn\left|\frac{u\cdot n}{|u|}\right| f(v,t)f(w,t)[\psi(v')+\psi(w')-\psi(v)-\psi(w)],\end{aligned} \tag{6.2.3}$$

in the notation of equations (6.1.11). Equation (6.2.2) is the Maxwell model of the spatially homogeneous Boltzmann equation (6.1.12) for inelastic hard spheres. Its generalization to the spatially inhomogeneous case is obvious from equation (6.1.2). We will consider below only spatially homogeneous systems. Then equation (6.2.2) can be simplified by substitution,

$$f(v,t) = \tilde{f}(v,\tilde{t}), \quad \tilde{t} = \gamma d^2\int_0^t d\tau\sqrt{T(\tau)}\,. \tag{6.2.4}$$

Finally, we omit tildes and obtain

$$\partial_t\langle f,\psi\rangle = \langle\psi, Q_i^{(M)}(f,f)\rangle, \tag{6.2.5}$$

where the right hand side is given in equations (6.2.3) for any "good" test function $\psi(v)$. This is the inelastic Maxwell–Boltzmann equation introduced in [36] (see also [14] for one-dimensional models).

The obvious advantage of this model is a possibility to further simplify it by Fourier transform. Indeed, we can introduce the characteristic function

$$\varphi(k,t) = \langle f, e^{-ik\cdot v}\rangle, \quad k\in\mathbb{R}^3\,, \tag{6.2.6}$$

and obtain from equations (6.2.5) and (6.2.3)

$$\varphi_t = Q_i^{(M)}(\varphi,\varphi) = \int_{\mathbb{R}^3\times\mathbb{R}^3} dv dw f(v) f(w)\, I_k(v,w)\,,$$

where

$$I_k(v,w) = \frac{1}{2}\int_{S^2} dn\left|\frac{u\cdot n}{|u|}\right|(e^{-ik\cdot v'} - e^{-ik\cdot v}),$$

in the notation of equations (6.1.11). Then we transform the integral to

$$I_k(v,w) = \frac{1}{2}e^{-ik\cdot v}J(k,u),$$

where

$$J(k,u) = \int_{S^2} dn\left|\frac{u\cdot n}{|u|}\right|\left\{\exp\left[i\frac{(1+e)}{2}(u\cdot n)(k\cdot n)\right]-1\right\},\quad u = v-w.$$

Then we use the same considerations as in Section 3.2 and obtain

$$J(k,u) = J(u,k) = \int_{S^2} dn\left|\frac{k\cdot n}{|k|}\right|\left\{\exp\left[i\frac{(1+e)}{2}(k\cdot n)(u\cdot n)\right]-1\right\}.$$

Hence,

$$Q_i^{(M)}(\varphi,\varphi) = \frac{1}{2}\int_{S^2} dn\left|\frac{k\cdot n}{|k|}\right|[\varphi(k_+)\varphi(k_-) - \varphi(0)\varphi(k)],$$

$$k_+ = z(k\cdot n),\quad k_- = k-k_+,\quad z = \frac{1+e}{2}, \tag{6.2.7}$$

where $0 \le e \le 1$ is the restitution coefficient. The Fourier transformed inelastic Maxwell–Boltzmann equation reads

$$\varphi_t = \hat{Q}_i^{(M)}(\varphi,\varphi),\quad \varphi|_{t=0} = \varphi_0(k), \tag{6.2.8}$$

where the operator $\hat{Q}_i^{(M)}(\varphi,\varphi)$ is given in (6.2.7).

The operators $\hat{Q}_i^{(M)}(\varphi,\varphi)$ and $\hat{Q}_i^{(M)}(f,f)$ from equations (6.2.3) admit some obvious generalizations. In particular, we can replace the factor $|u\cdot n|/|u|$ by an arbitrary "good" non-negative function $G(|u\cdot n|/|u|)$ under the integral sign in (6.2.3). Then we repeat the above considerations and obtain the same equation (6.2.8) for the characteristic function $\varphi(k,t)$, where

$$Q_i^{(M)}(\varphi,\varphi) = \frac{1}{2}\int_{S^2} dn\, G(|k\cdot n|)\,[\varphi(k_+)\varphi(k-k_+) - \varphi(0)\varphi(k)],$$

$$k_+ = z(k\cdot n)n,\quad \hat{k} = k/|k|,\quad z = \frac{1+e}{2}. \tag{6.2.9}$$

The original operator (6.2.7) can be considered as a particular case of operator (6.2.9) with

$$G(r) = r,\quad r\in[0,1]. \tag{6.2.10}$$

Another possible generalization of operators $\hat{Q}_i^{(M)}(\varphi,\varphi)$ and $\hat{Q}_i^{(M)}(f,f)$ can be obtained by considering integrals over $\mathbb{R}^n$ with any $n \geq 1$ in equations (6.2.3). Then also all our considerations for $n \geq 2$ remain valid. In particular, the Fourier transform of $\hat{Q}_i^{(M)}(f,f)$ is based on the fact that any isotropic function of two vectors $u \in \mathbb{R}^n$ and $k \in \mathbb{R}^n$ depends only on three scalar arguments, namely, $|u|$, $|k|$, and the scalar product $u \cdot k$. This is true for any dimension $n \geq 2$. Therefore, we obtain for any $n \geq 2$ the same equation (6.2.8), where $k \in \mathbb{R}^n$ and the integral (6.2.9) over S^2 is replaced by a similar integral over S^{n-1}, where S^{n-1} denotes the unit sphere in $\mathbb{R}^n$.

The equation (6.2.8) can be also extended to the case $n = 1$. We formally extend the identity

$$\int_{S^{n-1}} dn\, F(n) = 2 \int_{\mathbb{R}^n} dx\, \delta(|x|^2 - 1)\, F(x)\,, \quad n \geq 2,$$

to the case $n = 1$ and obtain from (6.2.9)

$$\hat{Q}_i^{(M)}(\varphi,\varphi) = G(1)\,\{\varphi(zk)\varphi[(1-z)k] - \varphi(0)\varphi(k)\}, \quad k \in \mathbb{R}\,. \tag{6.2.11}$$

Thus, equations (6.2.8) and (6.2.11) describe the one-dimensional inelastic Maxwell model in Fourier representation. Note that $z = (1+e)/2 \to 1$ in the elastic limit $e \to 1$. Then $\hat{Q}_i^{(M)} = 0$ in equation (6.2.11). This is correct, since the elastic collisions are trivial (just an exchange of velocities) in the one-dimensional case. On the contrary, the inelastic collisions remain relevant for the evolution of particles even in the case $n = 1$. We shall see below that the large time asymptotic behavior of solutions to spatially inhomogeneous inelastic Maxwell models looks very similar for dimensions $n = 1$ and $n = 3$.

Our goal is to study the solution of the problem (6.2.8)–(6.2.9). We are interested mainly in the realistic three-dimensional case. However, it is convenient to begin with the one-dimensional model in order to understand the qualitative behavior of solutions. This is done in the next section.

6.3 One-dimensional model and its exact self-similar solution

We consider the problem (6.2.8), (6.2.11) for the one-dimensional characteristic function

$$\varphi(k,t) = \int_{\mathbb{R}} dv\, f(v,t)\, e^{ikv}, \quad k \in \mathbb{R}\,, \tag{6.3.1}$$

where $f(v,t)$ is the corresponding distribution function. Without loss of generality we assume that

$$G(1) = 1, \quad \int_{\mathbb{R}} dv\, f(v,0) \begin{Bmatrix} 1 \\ v \\ |v|^2 \end{Bmatrix} = \begin{pmatrix} 1 \\ 0 \\ 1 \end{pmatrix}. \tag{6.3.2}$$

Then the reduced problem for $\varphi(k,t)$ reads

$$\varphi_t = \varphi(zk)\varphi[(1-z)k] - \varphi(0)\varphi(k), \quad k \in \mathbb{R}, \quad t \in \mathbb{R}_+; \quad \varphi|_{t=0} = \varphi_0(k),$$
$$z = \frac{1+e}{2} \in (1/2, 1]; \quad \varphi(0) = -\varphi''(0) = 1, \quad \varphi'(0) = 0, \tag{6.3.3}$$

where primes denote derivatives with respect to k. This equation looks very similar to equation (4.1.4) for radial solutions to the classical Fourier transformed Boltzmann equation. This similarity becomes exact if we assume that $\varphi_0(k)$ in equations (6.3.3) is an even function of k for any $t > 0$. We denote

$$x = |k|, \quad \varphi(k,t) = \tilde{\varphi}(x,t) \tag{6.3.4}$$

and obtain, omitting tildes, the equation

$$\varphi_t = \varphi(zx)\varphi[(1-z)x] - \varphi(0)\varphi(x), \quad x \geq 0, \tag{6.3.5}$$

which coincides with equation (4.1.4), where

$$\rho(s) = \delta(s-z). \tag{6.3.6}$$

Note that $z = (1+e)/2$ can be any fixed number between $z_{\min} = 1/2$ and $z_{\max} = 1$. The initial condition reads

$$\varphi|_{t=0} = \varphi_0(x), \quad \varphi_0(0) = 1, \quad \varphi'(0) = 0, \quad \varphi''(0) = -1. \tag{6.3.7}$$

We assume that

$$\varphi_0(|k|) = \int_{\mathbb{R}} dv\, f_0(|v|)\, e^{-ikv}, \tag{6.3.8}$$

where $f_0(|v|) = f(v,0)$ satisfies conditions (6.3.2) and has power moments of all orders,

$$m_{2n}^{(0)} = \int_{\mathbb{R}} dv f_0(|v|)\, v^{2n} < \infty, \quad n = 0, 1, \ldots. \tag{6.3.9}$$

Then we formally obtain from equations (6.3.7) and (6.3.8)

$$\varphi_0(x) = \sum_{n=0}^{\infty} \frac{(-1)^n m_{2n}^{(0)}}{(2n)!} x^{2n}. \tag{6.3.10}$$

Hence, we can look for solution of the problem (6.3.5), (6.3.7) in the form of the Taylor series

$$\varphi(x,t) = \sum_{n=0}^{\infty} \frac{(-1)^n m_{2n}(t)}{(2n)!} x^{2n}, \quad m_0 = 1. \tag{6.3.11}$$

In fact this series is a particular case of the more general series

$$\varphi(x,t) = \sum_{n=0}^{\infty} \frac{(-1)^n z_n(t)}{n!} x^n, \quad z_0 = 1, \tag{6.3.12}$$

which was studied in detail in Chapter 4. In particular, it follows from the proof of Theorem 4.5.1 that

$$m_{2n}(t) \le a^{2n} \quad \text{if} \quad a = \sup_{n\ge 0} \left[m_{2n}^{(0)}\right]^{1/2n} < \infty. \tag{6.3.13}$$

We can also prove it independently. The equation for moments $\{m_{2n}(t),\ n = 0,1,\dots\}$ have the following form:

$$\begin{gathered} \dot{m}_0 = 0; \quad \dot{m}_2 + \lambda_2 m_2 = 0; \\ \dot{m}_{2n} + \lambda_{2n} m_{2n} = \sum_{k=1}^{n-1} \binom{2n}{2k} b_{k,n-k} m_{2k} m_{2(n-k)}, \quad n \ge 2, \end{gathered} \tag{6.3.14}$$

where

$$\lambda_{2n} = 1 - z^{2n} - (1-z)^{2n}, \quad n \ge 1; \quad b_{k,l} = z^{2k}(1-z)^{2l}, \quad k \ge 1, \quad l \ge 1, \tag{6.3.15}$$

and dots denote time derivatives. This is a particular case of equations (4.2.2). The solution of these ordinary differential equations with initial data (6.3.9) reads

$$\begin{aligned} m_0 &= 1, \quad m_2(t) = m_2^{(0)}\, e^{-\lambda_2 t}, \\ m_{2n}(t) &= m_{2n}(0)\, e^{-\lambda_{2n} t} \\ &\quad + \sum_{k=1}^{n-1} \binom{2n}{2k} b_{k,n-k} \int_0^1 d\tau\, e^{-\lambda_{2n}(t-\tau)}\, m_{2k}(\tau) m_{2(n-k)}(\tau), \quad n = 2,3,\dots. \end{aligned} \tag{6.3.16}$$

The identity

$$\frac{1}{2}\left[(a_1+a_2)^{2n} + (a_1-a_2)^{2n}\right] = \sum_{k=0}^{n} \binom{2n}{2k} a_1^{2k} a_2^{2(n-k)}, \quad n \ge 0,$$

with $a_1 = z$, $a_2 = 1 - z$ leads to the estimate

$$\sum_{k=1}^{n-1} \binom{2n}{2k} b_{k,n-k} \le \lambda_{2n}, \quad n \ge 2,$$

in the notation of equations (6.3.15). Then we can apply this estimate to equations (6.3.16) and prove by induction inequality (6.3.13). Hence, the series (6.3.11) for $\varphi(x,t)$ represents the entire analytic function of x at any $t > 0$ provided that $a < \infty$ in equations (6.3.13). It is also easy to prove that the corresponding distribution function $f(v,t)$ defined by equation (6.3.1) is non-negative for all $t > 0$ if $f(v,0) \ge 0$.

Thus everything looks formally very similar to radial solutions of the classical Fourier transformed Boltzmann equation (4.1.4). Of course, there is an important difference in large time asymptotic behavior. It is easy to check by using estimates (4.5.14) that

$$m_0 = 1; \quad 0 < m_{2n}(t) \le \alpha\, b^{2n}\, e^{-\lambda_2 t}\,, \quad n \ge 1;$$

$$\alpha = 1 - \frac{\lambda_2}{\lambda_4}, \quad b = \sup\left\{\left(\frac{m_{2n}^{(0)}}{\alpha}\right)^{1/2n}, n = 1, 2, \dots\right\}.$$

This estimate shows that $\varphi(x,t) \underset{t\to\infty}{\to} 1$, as expected (see equation (6.1.19)). There is, however, an interesting peculiarity in the asymptotic behavior of $\varphi(x,t)$, that is, it is difficult to guess from moment equations (6.3.14).

Following considerations from Section 4.6 we can construct self-similar solutions of equation (6.3.5) in the form

$$\varphi(x,t) = \psi(x\, e^{-\mu t}), \quad \mu = \text{const.} \tag{6.3.17}$$

Then the equation for $\psi(x)$ reads

$$-\mu x \psi'(x) = \psi(zs)\, \psi[(1-z)x] - \psi(0)\psi(x), \quad x \ge 0\,. \tag{6.3.18}$$

Without loss of generality we assume that

$$\psi(0) = 1, \quad \psi'(0) = 0, \quad \psi''(0) = -1\,. \tag{6.3.19}$$

Note that the assumption $\psi''(0) = 0$ cannot be fulfilled for the characteristic function

$$\psi(|k|) = \int_{-\infty}^{\infty} dv f(|v|) e^{-ikv}, \quad f(|v|) \ge 0, \tag{6.3.20}$$

because

$$\psi''(0) = -\int_{-\infty}^{\infty} dv f(|v|)|v|^2 < 0. \tag{6.3.21}$$

Of course, we do not consider the trivial solution $\psi(x) = 1$ of equation (6.3.18), which corresponds to $f(v) = \delta(v)$. Thus we assume that

$$\psi(x) = 1 - \frac{x^2}{2} + \cdots, \quad x \to 0, \tag{6.3.22}$$

where dots stand for higher-order terms. We substitute this expansion into equation (6.3.18) and collect all terms proportional to x^2. This gives the following equality for μ:

$$\mu = \frac{\lambda_2}{2} = z(1-z), \tag{6.3.23}$$

in the notation of equation (6.3.15). This value is in complete agreement with Theorem 4.6.1, where $\rho(s) = \delta(s - z)$. The theorem describes, in particular (the case $p = 2$), all solutions of equations (6.3.18), (6.3.19), and (6.3.23) in the form of power series

$$\psi(x) = \sum_{n=0}^{\infty} \psi_n \frac{x^n}{n!}, \quad \psi_0 = -\psi_2 = 1, \quad \psi_1 = 0. \tag{6.3.24}$$

It is easy to verify directly by substitution of the series into equation (6.3.18) that the coefficient ψ_3 can be chosen arbitrarily. Then other coefficients are uniquely defined by the recursive relations

$$\psi_n = -\left[nz(1-z) - \lambda_n(z)\right]^{-1} \sum_{k=1}^{n-1} \binom{n}{k} z^k (1-z)^{n-k} \psi_k \psi_{n-k}, \quad n \geq 4,$$

where

$$\lambda_n(z) = 1 - z^n - (1-z)^n, \quad z = \frac{1+e}{2} \in (1/2, 1].$$

The proof of inequality (4.6.17) shows that $nz(1-z) - \lambda_n(z) > 0$, $n = 4, 5, \dots$, whereas $3z(1-z) - \lambda_3(z) = 0$. This identity explains why ψ_3 is a free parameter. Thus, the solution of equation (6.3.18) in the form of series (6.3.24) is not unique. At the same time we know a particular value of ψ_3, for which this solution has a very simple form (see Section 4.8). It is the value $\psi_3 = 2$, and the solution reads

$$\psi(x) = e^{-x}(1+x). \tag{6.3.25}$$

It is easy to check directly that

$$\varphi(|k|, t) = (1 + |k| e^{-\mu t})\, e^{-|k|\, e^{-\mu t}}, \quad \mu = z = \frac{1+e}{2},$$

is a self-similar solution of equation (6.3.3). The corresponding distribution function reads

$$f(|v|, t) = \frac{1}{2\pi} \int_{\mathbb{R}} dk \varphi(|k|, t)\, e^{ikv} = e^{\mu t} f(|v|\, e^{\mu t}), \tag{6.3.26}$$

where (see also (6.3.20))

$$f(|v|) = \frac{1}{2\pi} \int_{\mathbb{R}} dk \psi |k|\, e^{-ikv},$$

in the notation of equation (6.3.25). Evaluating the integral, we obtain

$$f(|v|) = \frac{2}{\pi} \frac{1}{(1+|v|^2)^2}. \tag{6.3.27}$$

Note that the exact solution $y^{(2)}(x)$ of equations (4.6.2) and (6.3.6) given in equation (4.8.7) is the same as the function $\psi(x)$ from (6.3.25). However, the corresponding distribution functions (see equations (4.8.10), (6.3.26), and (6.3.27)) are completely different.

In particular, the function $f(|v|)$ (6.3.27) has a power-like tail. This was a result obtained in the beginning of the 2000s by Baldassarri et al. [9]. Soon after that, Ernst and Brito suggested that the self-similar solution (6.3.26)–(6.3.27) and its generalization to higher dimensions describes large time asymptotic behavior for a large class of solutions [82]. The close connection with the BKW-solution of the classical Boltzmann equation was discussed in this section above in the spirit of the paper [42]. The results of the papers [42, 43] on the self-similar asymptotics for the three-dimensional Maxwell model of the inelastic Boltzmann equation are discussed in Sections 6.4–6.6.

6.4 Self-similar solutions to the three-dimensional inelastic Maxwell model

We return to the Cauchy problem (6.2.8)–(6.2.9) in the realistic three-dimensional case and consider radial solutions of equation (6.2.8). Note that

$$|k_+|^2 = z^2\,(k\cdot n)^2, \quad |k-k_+|^2 = |k|^2 - z(2-z)\,(k\cdot n)^2\,,$$

in the notation of equations (6.2.9). Similarly to Section 6.3, we denote

$$\varphi(k,t) = \tilde{\varphi}(x,t), \quad x = |k|; \quad G(r) = 2r\rho(r^2), \quad r \in [0,1]\,, \tag{6.4.1}$$

in the notation of equations (6.2.9). Then equation (6.2.8) for $\varphi(x,t)$ reads

$$\varphi_t = \int_0^1 ds\rho(s)\,\{\varphi(szx)\,\varphi[\sqrt{1-\beta s}\,x] - \varphi(0)\varphi(x)\}, \quad x \ge 0\,, \tag{6.4.2}$$

where tildes are omitted,

$$z = \frac{1+e}{2}, \quad \beta = z(2-z), \quad 0 \le e \le 1\,. \tag{6.4.3}$$

It can be shown that the equation for radial solution has the same form (6.4.2) for any dimension $n \ge 2$ [40]. We will consider in this section a class of self-similar solutions

$$\varphi(x,t) = \psi(x\,e^{-\mu t})\,, \tag{6.4.4}$$

assuming that

$$0 \le \rho(s) \le \rho_{\max} < \infty, \quad 0 \le s \le 1\,. \tag{6.4.5}$$

In particular, it follows from equations (6.4.1) and (6.2.10) that $\rho = 1/2$ for our original three-dimensional model. Without loss of generality we assume that

$$\psi(0) = 1, \quad \int_0^1 ds\rho(s) = 1. \tag{6.4.6}$$

We can also introduce a more general class of bilinear operators,

$$[K(\psi_1, \psi_2)] = \int_0^1 ds\rho(s)\,\psi_1[a(s)x]\,\psi_2[b(s)x], \quad x \geq 0, \tag{6.4.7}$$

where

$$0 \leq a(s) \leq 1, \quad 0 \leq b(s) \leq 1, \quad a^2(s) + b^2(s) \leq 1, \quad s \in [0,1]. \tag{6.4.8}$$

Then equation (6.4.2) can be written as

$$\varphi_t = K(\varphi, \varphi) - \varphi, \tag{6.4.9}$$

with

$$a(s) = z\sqrt{s}, \quad b(s) = \sqrt{1 - \beta s}. \tag{6.4.10}$$

The conditions (6.4.8) are satisfied. Then equation (6.4.2) can be written as

$$\beta - z^2 = 2z(1 - z) \geq 0.$$

The equation for $\psi(x)$ from (6.4.4) reads

$$-\mu x\psi'(x) + \psi = K(\psi, \psi). \tag{6.4.11}$$

We assume that $\psi(|k|)$ is a characteristic function in $\mathbb{R}$ having the following asymptotic form for small $|k| = x$:

$$\psi(x) = 1 - \frac{x^2}{2} + \cdots, \tag{6.4.12}$$

where dots denote higher-order terms. Hence, we obtain from equation (6.4.9)

$$\mu = \frac{1}{2}\int_0^1 ds\rho(s)[1 - a^2(s) - b^2(s)] \geq 0. \tag{6.4.13}$$

We assume below that $\mu > 0$ and transform equation (6.4.11) to

$$\frac{d}{dx}[x^{-r}\psi] = -rx^{-(1+r)}\,[K(\psi, \psi)](x), \quad r = \mu^{-1}. \tag{6.4.14}$$

By assumption we have

$$|\psi(|k|)| \le \psi(0) = 1\,. \tag{6.4.15}$$

Therefore, $\psi(x)$ satisfies the integral equation

$$\psi(x) = R(\psi) = r\,x^r \int_x^\infty dy\, y^{-(r+1)}\,[K(\psi,\psi)](y)\,. \tag{6.4.16}$$

We intend to solve this equation by a standard iteration scheme,

$$\psi_{n+1} = R(\psi_n), \quad n = 0, 1, \ldots\,. \tag{6.4.17}$$

Obviously, the operator R has the following properties: (1) $0 \le R(\psi) \le 1$ if $0 \le \psi(x) \le 1$; (2) $R(\psi) \le R(\varphi)$ if $0 \le \psi \le \varphi$. Hence, the iteration process converges pointwise if we choose the initial approximation $0 \le \psi_0(x) \le 1$ in such a way that

$$\psi_0(x) \le R(\psi_0) \quad \text{or} \quad \psi_0(x) \ge R(\psi_0)\,.$$

It is natural to choose $\psi(x)$ in the form (6.3.25) considered in Section 6.3. We denote

$$\psi^{(1)}(x) = e^{-x}\,(1+x)\,, \quad \psi^{(2)}(x) = e^{-x^2/2}\,, \quad x \ge 0\,, \tag{6.4.18}$$

and formulate the following elementary lemma.

Lemma 6.4.1. *The following inequalities are valid for all $x \ge 0$:*

$$\psi^{(1)}(x) \ge R(\psi^{(1)})\,, \quad \psi^{(2)}(x) \le R(\psi^{(2)}), \tag{6.4.19}$$

in the notation of equations (6.4.16) *and* (6.4.18), *provided that*

$$a(s) + b(s) \ge 1, \quad a^2(s) + b^2(s) \le 1, \quad 0 \le s \le 1, \tag{6.4.20}$$

in equation (6.4.7).

Proof. We denote

$$\Delta(x) = \mu\,x\,\psi'(x) + K(\psi,\psi) - \psi, \quad \mu = \frac{1}{r} = \frac{1}{2}\int_0^1 ds p(s)[1 - a^2(s) - b^2(s)],$$

for any function $\psi(x)$. Then it is easy to check the following identity:

$$R(\psi) - \psi = r \int_1^\infty \frac{\tau}{\tau^{r+1}}\Delta(\tau x),$$

in the notation of equation (6.4.16). On the other hand, it follows from equation (6.4.7) that

$$\Delta(x) = \int_0^\infty ds\rho(s)\, g_\psi[a(s),\, b(s);\, x]\,,$$

where

$$g_\psi[a,\, b;\, x] = cx\psi'(x) - \psi(x) + \psi(ax)\psi(bx), \quad c = \frac{1 - a^2 - b^2}{2}.$$

Hence, it is enough for the proof to show that

$$g_{\psi^{(1)}}(a,\, b;\, x) \le 0, \quad g_{\psi^{(2)}}(a,\, b;\, x) \ge 0, \quad x \ge 0, \tag{6.4.21}$$

where $a \ge 0$ and $b \ge 0$ satisfy conditions (6.4.20). We note that

$$\psi^{(1)\prime}(x) = -x\, e^{-x}, \quad \psi^{(2)\prime}(x) = -x\, e^{-x^2/2}\,.$$

Then we obtain

$$g_{\psi^{(1)}}(a,\, b;\, x) = e^{-(a+b)x} A(a, b; x)\,,$$

where

$$A(a, b; x) = e^{(a+b-1)x}\,(1 + x + cx^2) - (1 + ax)(1 + bx) = \sum_{n=0}^{\infty} a_n x^n\,.$$

Since $a + b \ge 1$ and $c \ge 0$, it is obvious that $a_n > 0$ for all $n \ge 3$. Hence, it is enough to consider the first coefficients a_0, a_1, and a_2 in order to prove positivity of $A(a, b; x)$. Straightforward calculations show that $a_0 = a_1 = a_2 = 0$. Hence, $A(a, b; x) \ge 0$ and therefore $g_{\psi^{(1)}}(a,\, b;\, x) \le 0$.

It remains to consider the second function from equations (6.4.21). We obtain

$$\begin{aligned} g_{\psi^{(2)}}(a,\, b;\, x) &= e^{-x^2}\,[e^{(1-a^2-b^2)x^2/2} - (1 + cx^2)] \\ &= e^{-x^2}\,[e^{cx^2} - (1 + cx^2)] \ge 0\,. \end{aligned}$$

This completes the proof. □

Our goal in this section is to solve equation (6.4.16), where $K(\varphi, \varphi)$ is given by equations (6.4.7) and (6.4.10). We consider the iteration scheme (6.4.17) and choose the initial approximation in the form

$$\psi_0(x) = \psi^{(2)}(x) = e^{-x^2/2}\,. \tag{6.4.22}$$

Then we want to apply Lemma 6.4.1 in order to show that

$$\psi_0(x) \le \psi_1(x) = R(\psi_0) . \tag{6.4.23}$$

It is clear that we can do it only if the conditions (4.6.20) are fulfilled for $a(s)$ and $b(s)$ from equations (6.4.10). One of the conditions was already proved in Section 6.3. The second condition can be written as

$$w(s) = a(s) + b(s) - 1 = z\sqrt{s} + \sqrt{1 - \beta s} - 1 \ge 0, \quad s \in [0,1], \quad z \in (1/2, 1], \quad \beta z(2-z).$$

We note that

$$w(0) = w(1) = 0, \quad w'(s) = \frac{1}{2}[zs^{-1/2} - \beta(1-\beta s)^{-1/2}],$$
$$w''(s) = -\frac{1}{4}[zs^{-3/2} + \beta^2(1-\beta s)^{-1/2}] .$$

Then it is easy to see that $w'(s) > 0$ if $s \in (0, s_0)$ and $w'(s) < 0$ if $s \in (s_0, 1)$, where

$$s_0 = [2(2-z)]^{-1} \in \left(\frac{1}{3}, \frac{1}{2}\right] .$$

Hence, $w(s) \ge 1$, and therefore all conditions of Lemma 6.4.1 are satisfied for $a(s)$ and $b(s)$ from equations (6.4.10). Then inequality (6.4.23) is fulfilled and we obtain from equations (4.6.22) and (4.6.17) the monotone increasing sequence

$$\psi_0(x) \le \psi_1(x) \le \psi_2(x) \le \dots, \quad x \ge 0 . \tag{6.4.24}$$

Since $0 \le \psi_n(x) \le 1$ for all $n \ge 0$, the sequence is convergent at any $x \ge 0$. However, we need to show that

$$\psi(x) = \lim_{n\to\infty} \psi_n(x)$$

is not a trivial solution of equation (6.4.16)

$$\psi(x) = 1. \tag{6.4.25}$$

Lemma 6.4.2. *The function $\psi(x)$ defined by equation (6.4.25) satisfies the inequality*

$$\psi(x) \le e^{-x}(1+x) \tag{6.4.26}$$

for all $x \ge 0$.

Proof. It is sufficient to show that inequality (6.4.26) holds for any $\psi_n(x)$, $n \ge 0$, from sequence (6.4.24). We begin with $n = 0$. Then the elementary inequality

$$\psi_0(x) = e^{-x^2/2} \le e^{-x}(1+x)$$

or, equivalently,

$$w_1(x) = x^2/2 - x + \log(1+x) \geq 0, \quad x \geq 0,$$

can be easily verified by differentiation of $w_1(x)$. On the other hand, it follows from Lemma 6.4.1 that

$$e^{-x}(1+x) \geq R[e^{-x}(1+x)].$$

Hence, we obtain for $n = 1$

$$\psi_1(x) = R(\psi_0) \leq R(e^{-x}(1+x)).$$

Moreover, for any $n \geq 1$ we have by induction

$$\psi_{n+1}(x) = R(\psi_n) \leq R(e^{-x}(1+x)) \leq e^{-x}(1+x).$$

This proves inequality (6.4.26) for any $\psi_n(x)$, $n \geq 0$. Hence, the same inequality holds for the limit (6.4.25), and this completes the proof. □

We can formulate the final result of this section in the following way.

Theorem 6.4.1. *The integral equation* (6.4.16), *where* μ *and* $K(\psi, \psi)$ *are defined in equations* (6.4.13) *and* (6.4.7), *respectively, has a solution* $\psi(x)$ *such that*

$$e^{-x^2/2} \leq \psi(x) \leq e^{-x}(1+x), \quad x \geq 0, \tag{6.4.27}$$

provided

$$\rho(s) \geq 0, \quad 0 \leq s \leq 1, \quad \int_0^1 ds\rho(s) = 1;$$
$$0 \leq a(s) \leq 1, \quad 0 \leq b(s) \leq 1, \quad a(s) + b(s) \geq 1, \quad a^2(s) + b^2(s) \leq 1. \tag{6.4.28}$$

The function $\psi(|k|)$, $k \in \mathbb{R}^n$, *is a characteristic function in* $\mathbb{R}^n$, $n \geq 1$. *It also satisfies the integro-differential equation* (6.4.11). *The corresponding (radial) probability density* $F_n(|v|)$ *is given by the usual inverse Fourier transform,*

$$F_n(|v|) = \frac{1}{(2\pi)^n} \int_{\mathbb{R}^n} dk\psi(|k|)e^{ik\cdot v}, \quad v \in \mathbb{R}^n. \tag{6.4.29}$$

Proof. The solution $\psi(x)$ is constructed above as the limit of the monotone increasing sequence (6.4.22), (6.4.24). It follows from (6.4.24) that

$$\psi(x) \geq \psi_0(x) = e^{-x^2/2}.$$

The upper limit estimate in (6.4.27) follows from Lemma 6.4.2. Note that the equality $\psi(0) = 1$ follows from (6.4.27). The integro-differential equation (6.4.11) for $\psi(x)$ follows by differentiation from equation (6.4.16).

It remains to show that $\psi(|k|)$, $k \in \mathbb{R}^n$, is a characteristic function in $\mathbb{R}^n$, $n \geq 1$. It follows from elementary properties of characteristic functions [85] that $R(\psi)$ maps the set of characteristic functions to itself. Hence, it is sufficient to prove that $\psi_0(|k|)$ is the characteristic function. Indeed, it is known that

$$\psi_0(|k|) = e^{-|k|^2/2} = (2\pi)^{-n/2} \int_{\mathbb{R}^n} dv e^{-|v|^2/2 - ik\cdot v} , \quad k \in \mathbb{R}^n ,$$

for any $n = 1, 2, \ldots$. This completes the proof. □

In the next section we establish the uniqueness of the self-similar solution and its role for the large time asymptotics of solutions to the initial value problem.

6.5 Uniqueness of the self-similar profile

The solution $\psi(x)$ of equation (6.4.16) was obtained as a limit of monotone increasing sequence (6.4.24). On the other hand, we can choose the function

$$\psi(x) \leq e^{-x}(1 + x)$$

as the initial point for iterations (6.4.17). Lemma 6.4.1 implies that the resulting sequence $\{\psi_n(x),\ n = 1, 2, \ldots\}$ is monotone decreasing and bounded below by the function $e^{-x^2/2}$. Hence, this sequence also converges to some function $\tilde{\psi}(x)$. Obviously, we need to prove that $\tilde{\psi}(x) = \psi(x)$, where $\psi(x)$ is the function from Theorem 6.4.1.

Note that both functions $\psi(x)$ and $\tilde{\psi}(x)$ satisfy inequality (6.4.27). Hence, we obtain

$$0 \leq \psi(x) - e^{-x^2/2} \leq e^{-x}(1 + x) - e^{-x^2/2} = O(x^3) . \tag{6.5.1}$$

The same estimate holds for $\tilde{\psi}(x)$. The following considerations prove, in particular, that $\tilde{\psi}(x) = \psi(x)$.

We consider a bounded domain $\mathcal{D} \subset \mathbb{R}^3$ such that

$$\mathcal{D} = \{P = (x, y, z) \in \mathbb{R}^3 : x + y \geq 1,\ 2 \leq z \leq 3\} . \tag{6.5.2}$$

Geometrically $\mathcal{D}$ is a part of the cylinder

$$x^2 + y^2 \leq 1, \quad 2 \leq z \leq 3 ,$$

located on the right from the plane $x + y = 1$. Equations (6.5.2) imply that $0 \leq x \leq 1$, $0 \leq y \leq 1$. We introduce the function

$$F(P) = F(x, y, z) = 2(1 - x^z - y^z) - z(1 - x^2 - y^2) \tag{6.5.3}$$

and consider it in the domain $\mathcal{D}$.

Lemma 6.5.1. *If $P \in \mathcal{D}$, in the notation of equations (6.5.2), then $F(P) \geq 0$. Moreover $F(P) = 0$ for $P \in \mathcal{D}$ if and only if P belongs to the plane $z = 2$ or one of the following three straight lines:*

$$\begin{aligned} &(1) \quad x + y = 1; \quad z = 3; \\ &(2) \quad x = 0, \quad y = 1, \quad 2 \leq z \leq 3; \\ &(3) \quad x = 1, \quad y = 0, \quad 2 \leq z \leq 3. \end{aligned} \tag{6.5.4}$$

Proof. Since

$$y \geq 1 - x, \quad F_y = 2zy(1 - y^{z-2}) \geq 0, \quad (x, y, z) \in \mathcal{D}, \tag{6.5.5}$$

we obtain

$$\begin{aligned} F(x, y, z) &\geq g(x, z), \\ g(x, z) &= F(x, 1 - x, z) \\ &= 2\left[1 - x^z - (1 - x)^z\right] - z\left[1 - x^2 - (1 - x)^2\right]. \end{aligned} \tag{6.5.6}$$

It is easy to verify that

$$\begin{aligned} &g(x, 2) = g(x, 3) = 0, \quad 0 \leq x \leq 1; \\ &g(0, z) = g(1, z) = 0, \quad z > 0; \\ &g_{zz}(x, z) = -2\{x^z(\log x)^2 + (1 - x)^z\left[\log(1 - x)\right]^2\}, \quad 0 < x < 1, \quad z \geq 0. \end{aligned} \tag{6.5.7}$$

Hence, for any fixed $0 < x < 1$, $g_-(x, z) = -g(x, z)$ is a convex function of $z \geq 0$. The famous inequality for convex functions reads (see, e. g., [85])

$$g_-\left[x, z_1(1 - t) + z_2 t\right] \leq (1 - t)g_-(x, z_1) + t g_-(x, z_2), \quad z_{1,2} \geq 2, \quad 0 \leq t \leq 1.$$

Taking $z_1 = 2$, $z_2 = 3$, we obtain $g(x, z) \geq 0$, $0 \leq x \leq 1$, $2 \leq z \leq 3$. Then the inequality

$$F(x, y, z) \geq 0, \quad (x, y, z) \in \mathcal{D}, \tag{6.5.8}$$

in the notation of equations (6.5.2) and (6.5.3) follows from (6.5.7).

It remains to consider zeros of $F(x, y, z)$. The equality $F(x, y, 2) = 0$ follows directly from equation (6.5.3). Therefore, we assume that $z > 2$ and obtain by integration of equation (6.5.5) for F_y

$$F(x, y, z) = g(x, z) + 2z \int_{1-x}^{y} dt\, t(1 - t^{z-2}),$$

in the notation of equations (6.5.6). Since $g(x, z) \geq 0$ and $z > 2$, we conclude that coordinates (x, z) of zeros of F are defined by equations

$$g(x, z) = 0, \quad y = 1 - x, \quad (x, y, z) \in \mathcal{D}.$$

Three families of zeros of $g(x, z)$ for $z > 2$ are indicated in equations (6.5.7). They correspond to three straight lines mentioned in the formulation of Lemma 6.5.1. Let us assume that there is a point (x_0, z_0) such that

$$g(x_0, z_0) = 0, \quad 0 < x_0 < 1, \quad 2 < z_0 < 3.$$

Then we fix x_0 and consider the function $g(x_0, z)$ on the interval $2 \le z \le 3$. We know that

$$g(x_0, z) \ge 0, \quad g(x_0, 2) = g(x_0, 3) = 0, \quad g_{zz}(x_0, z) < 0$$

on that interval. Hence, the point (x_0, z_0) must be a local minimum of $g(x_0, z)$. This, however, contradicts the inequality $g_{zz}(x_0, z_0) < 0$. Thus, our assumption is wrong and this completes the proof of Lemma 6.5.1. □

We shall need also another estimate for the operator $K(\varphi, \varphi)$ from equation (6.4.7). This estimate is a simple generalization of similar estimate (5.4.14) for $\hat{Q}_+(\varphi, \varphi)$.

Lemma 6.5.2. *We consider the operator $K(\varphi, \varphi)$ from equation* (6.4.7), *where*

$$\rho(s) \ge 0, \quad 0 \le a(s) \le 1, \quad 0 \le b(s) \le 1, \tag{6.5.9}$$

acting on functions $\varphi(x)$, $x \ge 0$, such that

$$\|\varphi\| = \sup_{x \ge 0} |\varphi(x)| \le 1.$$

Then the following inequality holds for any pair of such functions $\varphi_{1,2}(x)$:

$$|K(\varphi_1, \varphi_1) - K(\varphi_2, \varphi_2)|(x) \le [L|\varphi_1 - \varphi_2|](x), \quad x \ge 0, \tag{6.5.10}$$

where

$$Lu(x) = \int_0^1 ds\, \rho(s)\{u[a(s)\, x] + u[b(s)\, x]\}. \tag{6.5.11}$$

The proof is omitted because it is just a repetition of the short proof of estimates (5.4.14) applied to the operator K instead of $\hat{Q}_+$.

After these preparations we can prove the uniqueness result for the self-similar solution constructed in Section 6.4.

Theorem 6.5.1. *Under conditions of Theorem* 6.4.1, *the solution $\psi(x)$ of the integral equation* (6.4.16), *except for its trivial case with $\mu = 0$ and*

$$[R(\psi)](x) = [K(\psi, \psi)](x) = \psi(0)\psi(x), \quad x \ge 0, \tag{6.5.12}$$

is unique in the class of functions satisfying conditions

$$|\psi(x)| \le 1, \quad \psi(x) - e^{-x^2/2} = O(x^{2+\varepsilon}), \tag{6.5.13}$$

for some $\varepsilon > 0$ and all $x \ge 0$.

Proof. Note that the validity of conditions (6.5.13) for a fixed $\varepsilon > 0$ implies automatically their validity for any $0 < \varepsilon_1 < \varepsilon$. Therefore, we assume below that $0 < \varepsilon < 1$ in equations (6.5.13).

We assume that there exist two different solutions $\psi_{1,2}(x)$ of equation (6.4.16) from Theorem 6.4.1. Then we obtain

$$u(x) = \psi_1(x) - \psi_2(x) = \int_0^1 d\tau [K(\varphi_1, \varphi_1) - K(\varphi_2, \varphi_2)](x\, e^{-\mu\tau}),$$

in the notation of Theorem 6.4.1. We assume that estimates (6.5.3) hold for $\psi_{1,2}(x)$ with values $0 < \varepsilon_{1,2} < 1$. Hence,

$$|u(x)| \leq 2, \quad u(x) = O(x^{2+\varepsilon}), \quad \varepsilon = \min(\varepsilon_1, \varepsilon_2). \tag{6.5.14}$$

Moreover, we can use Lemma 6.5.2 and obtain

$$|u(x)| \leq \int_0^1 d\tau\, [L|u|](x\, e^{-\mu\tau}), \quad x \geq 0, \tag{6.5.15}$$

in the notation of equation (6.5.11). Note that

$$Lx^p = [1 - \lambda(p)]\, x^p\,, \quad p > 0, \tag{6.5.16}$$

where

$$\lambda(p) = \int_0^1 ds p(s)[1 - a^p(s) - b^p(s)]\,. \tag{6.5.17}$$

On the other hand,

$$|u(x)| \leq C\, x^p, \quad C = \text{const.}, \quad p = 2 + \varepsilon, \quad x \geq 0, \tag{6.5.18}$$

in the notation of equations (6.5.14). We assume without loss of generality that $2 < p < 3$. Then we substitute estimate (6.5.18) into inequality (6.5.15) and perform elementary integration. The resulting estimate reads

$$1 \leq \frac{1 - \lambda(p)}{1 - p\mu},$$

provided $p\mu < 1$. Note that $\mu = \lambda(2)/2$ in the notation of equations (6.4.13) and (6.5.17). Conditions (6.4.28) of Theorem 6.4.1 imply that $0 \leq \lambda(2) < 1$. Therefore $p\mu < 1$ for $p = 2 + \varepsilon$ with sufficiently small $\varepsilon > 0$. Hence, we obtain the inequality

$$2\lambda(p) - 2p\mu = \int_0^1 ds p(s) F[a(s), b(s), p] \leq 0, \quad p = 2 + \varepsilon\,, \tag{6.5.19}$$

in the notation of equation (6.5.3). However, it was proved in Lemma 6.5.1 that

$$F[a(s), b(s), p] > 0 \tag{6.5.20}$$

for all $2 < p < 3$ and almost all $\{a(s), b(s)\}$ satisfying conditions (6.4.28). The only exceptional values are given by identities

$$F(1, 0, p) = F(0, 1, p) = 0.$$

We denote

$$\Delta_0 = \{s \in [0,1] : [a(s) = 1, b(s) = 0] \text{ or } [a(s) = 0, b(s) = 1]\}, \quad \Delta_1 = [0,1] \setminus \Delta_0 .$$

Then the strict inequality (6.5.20) holds for all $s \in \Delta_1$. On the other hand, equation (6.5.19) reads

$$2\lambda(p0 - 2p\mu = \int_{\Delta_1} ds\rho(s) F[a(s), b(s), p] \leq 0.$$

Hence, we obtain a contradiction if

$$\int_{\Delta_1} ds\rho(s) > 0.$$

It remains to consider the exceptional case

$$\int_{\Delta_1} ds\rho(s) = 0 \Rightarrow \int_{\Delta_0} ds\rho(s) = 1.$$

Then it is easy to see that $\mu = 0$ and $R(\psi)$ has the form (6.5.12). Hence, the integral equation (6.5.16) becomes a trivial identity $\psi(x) = \psi(x)$. Obviously, it has infinitely may solutions. This completes the proof of Theorem 6.5.1. □

6.6 Asymptotic property of self-similar solutions

In this section we continue to study solutions $\varphi(x,t)$ of equation (6.4.9), where the bilinear operator $K(\varphi_1, \varphi_2)$ is defined in equation (6.4.7). We assume below that $\rho(s)$, $a(s)$ and $b(s)$ satisfy conditions (6.4.28) of Theorem 6.4.1. It was shown in Section 6.4 that these conditions hold, in particular, for coefficients (6.4.10) of the Fourier transformed inelastic Maxwell–Boltzmann equation (6.4.2). Note that $\rho(s) = 1$ in the original form (6.2.9) of equation (6.4.2).

We want to prove that the self-similar solution $\varphi(x,t) = \psi(xe^{-\mu t})$, where $\psi(x)$ is constructed in Theorem 6.4.1, describes a large time asymptotic behavior of solutions to the initial value problem for equation (6.4.9) with initial data such that

$$\varphi|_{t=0} = \varphi_0(x) = 1 - \frac{x^2}{2} + O(x^{2+\varepsilon}), \quad \varepsilon > 0; \quad |\varphi_0(x)| \leq 1, \quad x \geq 0. \tag{6.6.1}$$

Remark. Note that these conditions on $\varphi_0(x)$ are equivalent to conditions (6.5.13) on $\psi(x)$.

The general idea of the proof is very simple. We consider equations (6.4.9) and (6.4.7) with initial data (6.6.1). The solution of that problem can be easily constructed in the form of a Wild sum. Following considerations of equation (3.3.19), we obtain

$$\varphi(x,t) = e^{-t} \sum_{n=0}^{\infty} (1 - e^{-t})^n \varphi_n(x),$$
$$\varphi_{n+1}(x) = \frac{1}{n+1} \sum_{k=0}^{n} K(\varphi_k, \varphi_{n-k}), \quad x \geq 0, \tag{6.6.2}$$

in the notation of equations (6.4.7) and (6.6.1). It is easy to check by induction that $|\varphi_n(x)| \leq 1$ for all $n = 0, 1, \ldots$. Hence, the series (6.6.2) converges uniformly for any $t \geq 0$. Therefore, it follows from equations (6.5.2) that $|\varphi(x,t)| \leq 1$, $t \geq 0$. Then we can also repeat the considerations used for the proof of Lemma 3.3.1 and conclude that the solution (6.6.2) is unique in the class of functions, satisfying inequality (3.3.23).

The second step is to consider this solution $\varphi(x,t)$ and to change variables to

$$\varphi(x,t) = \tilde{\varphi}(\tilde{x}, t), \quad \tilde{x} = xe^{-\mu t}, \tag{6.6.3}$$

where μ is given in equation (6.4.13). Then the equation (6.4.9) reads

$$\varphi_t - \mu x \varphi_x + \varphi = K(\varphi, \varphi) = \int_0^1 ds p(s) \varphi[a(s)x] \, \varphi[b(s)x], \tag{6.6.4}$$

where tildes are omitted. The initial condition (6.5.1) remains unchanged. It is convenient for our goals to transform the problem (6.6.4), (6.6.1) to the integral from:

$$\varphi(x,t) = e^{-t} \varphi_0(xe^{\mu t}) + \int_0^t d\tau \, e^{-(t-\tau)} \Phi(xe^{\mu(t-\tau)}, \tau), \tag{6.6.5}$$

where

$$\Phi(x,\tau) = [K(\varphi,\varphi)](x,\tau) = \int_0^1 ds p(s) \varphi[a(s)x, \tau] \, \varphi[b(s)x, \tau]. \tag{6.6.6}$$

Following the approach of Section 5.4, we can prove the next theorem.

Theorem 6.6.1. *We consider equation (6.6.4), where $p(s)$, $a(s)$, $b(s)$ satisfy conditions (6.4.28). The parameter μ is given in equation (6.4.13). Let $\varphi^{(1,2)}(x,t)$ be any two solutions of that equation satisfying initial conditions*

$$\varphi^{(i)}(x,0) = \varphi_0^{(i)}(x) = 1 - \frac{x^2}{2} + O(x^{p_i}), \quad 2 < p_i < 3;$$
$$|\varphi^{(i)}(x)| \leq 1, \quad i = 1, 2. \tag{6.6.7}$$

Then the following estimate is valid for all $x \geq 0$ *and* $t \geq 0$

$$|\varphi^{(1)}(x,t) - \varphi^{(2)}(x,t)| = O[x^p e^{-\gamma(p)t}],$$

where

$$p = \min(p_1, p_2), \quad \gamma(p) = \lambda(p) - p\mu \geq 0,$$

in the notation of equations (6.4.13) *and* (6.5.17). *The equality* $\gamma(p) = 0$ *is valid only in the trivial case of equation* (6.6.4), *when*

$$\mu = 0, \quad [K(\varphi,\varphi)](x) = \varphi(0)\varphi(x)$$

for any function $\varphi(x)$ *such that* $\varphi(0) = 1$.

Proof. We consider the integral equations (6.5.5) for $\varphi^{(1,2)}(x,t)$ and obtain after obvious transformations

$$\begin{aligned} u(x,t) &= \varphi^{(1}) (x,t) - \varphi^{(2)}(x,t) \\ &= e^{-t} u_0(xe^{\mu t}) + \int_0^t d\tau e^{-\tau} [K(\varphi^{(1)}, \varphi^{(1)}) - K(\varphi^2, \varphi^2)](xe^{\mu\tau}, t - \tau), \\ u_0(x) &= \varphi_0^{(1)}(x) - \varphi_0^{(2)}(x). \end{aligned} \tag{6.6.8}$$

Since $|\varphi^{(1,2)}(x,t)| \leq 1$ for all $x \geq 0$ and $t \geq 0$, we can use inequality (6.5.10) and obtain

$$\begin{aligned} &|K(\varphi^{(1)}, \varphi^{(1)}) - K(\varphi^{(2)}, \varphi^{(2)})|(x) \leq L|\varphi^{(1)} - \varphi^{(2)}|(x), \\ &Lu(x) = \int_0^1 ds p(x)\{u[a(s)x] + u[b(s)x]\}, \quad x \geq 0, \end{aligned} \tag{6.6.9}$$

where the irrelevant variable t is omitted. Then we apply this inequality to equation (6.6.8) and get the following estimate:

$$|u(x,t)| \leq |u_0(xe^{\mu t})|e^{-t} + \int_0^t d\tau\, e^{-\tau} [L|u|](xe^{\mu t}, t - \tau).$$

The standard comparison argument applied to $|u(x,t)|$ and the function $y(x,t)$ such that

$$y(x,t) = y_0(xe^{\mu t})e^{-t} + \int_0^t d\tau e^{-\tau} [Ly]\,(xe^{\mu\tau}, t - \tau) \tag{6.6.10}$$

shows that

$$\big|u(x,t)\big| \le y(x,t) \tag{6.6.11}$$

for all $x \ge t$ and $t \ge 0$, provided $|u_0(x)| \le y_0(x)$. Note that the linear equation (6.6.10) is the integral form of the Cauchy problem

$$y_t - \mu x y_x + y = Ly, \quad y|_{t=0} = y_0(x),$$

in the notation of equations (6.6.9) and (6.4.13). It is easy to verify that this equation has a class of exact solutions

$$y_p(x,t) = x^p \exp[-\gamma(p)t], \quad \gamma(p) = \lambda(p) - \mu p\,,$$

in the notation of equations (6.6.13) and (6.4.17). Conditions (6.6.7) of the theorem show that

$$u_0(x) = \varphi^{(1)}(x,0) - \varphi^{(2)}(x,0) = O\big[x^p\big], \quad p = \min(p_1, p_2),$$

where $2 < p < 3$. Note that

$$\gamma(p) = \lambda(p) - \mu p = \frac{1}{2}\int_0^1 ds p(s) F\big[a(s), b(s), p\big], \tag{6.6.12}$$

in the notation of equation (6.5.3). This integral was studied in detail in the proof of Theorem 6.5.1. It was already shown there that under conditions (6.4.28), $\gamma(p) > 0$ for any $2 < p < 3$ with one trivial exception (mentioned in Theorem 6.5.1). The exceptional value $\gamma(p) = 0$ corresponds to the trivial case of equation (6.6.4) with

$$\mu = 0, \quad \big[K(\varphi,\varphi)\big](x) = \varphi(0)\varphi(x)$$

for any function $\varphi(x)$ such that $\varphi(0) = 1$. This completes the proof. □

Corollary 4. *Let $\varphi(x,t)$ be the solution of the Cauchy problem* (6.6.4), (6.6.1) *(with some $0 < \varepsilon < 1$) constructed in Theorem* 6.6.1. *Then*

$$\big|\varphi(x,t) - \psi(x)\big| \le \min(2,\, Cx^{2+\varepsilon}\, e^{-\gamma(2+\varepsilon)}t), \quad C = \text{const.}, \tag{6.6.13}$$

where $\psi(x)$ is the function from Theorem 6.5.1 *and $\gamma(p) > 0$ is given in* (6.6.12).

Proof. We choose $\varphi^{(1)} = \varphi(x,t)$, $\varphi^{(2)} = \psi(x)$ in Theorem 6.6.1. It follows from equations (6.6.1) and (6.5.13) that

$$\big|\varphi(x,0) - \psi(x)\big| \le \min(2,\, Cx^{2+\varepsilon}), \quad 0 < \varepsilon < 1,$$

with some constant $C > 0$. Moreover, $|\psi(x)| \le 1$ and $|\varphi(x,t)| \le 1$ for all $x \ge 0$ and $t \ge 0$. Therefore, the estimate (6.6.13) follows from Theorem 6.6.1. This completes the proof. □

In the next section we consider corresponding self-similar solutions of the inelastic Boltzmann equation.

6.7 Distribution functions and power-like tails

In this section we return to the inelastic Maxwell–Boltzmann equation (6.2.5) for the distribution function $f(v, t)$, $v \in \mathbb{R}^3$. Note that we have considered in Section 6.6 only characteristic functions $\varphi(|k|, t)$ for radial solutions $f(|v|, t)$. It is obvious that we can extend this study to the general case of non-isotropic solutions $\varphi(k, t)$, $k \in \mathbb{R}^3$, of equations (6.2.7) and (6.2.8). We just need to consider a class of initial data

$$\varphi|_{t=0} = \varphi_0(k) = 1 - \frac{|k|^2}{2} + O(|k|^{2+\varepsilon}), \quad \|\varphi_0(k)\| \le 1, \quad k \in \mathbb{R}^3, \tag{6.7.1}$$

and to construct the solution of the problem (6.2.7), (6.7.1) in the form of the Wild sum (3.3.21) (see also (6.6.2)). Then, similarly to Section 6.6, we denote

$$\varphi(k, t) = \tilde{\varphi}(\tilde{k}, t), \quad \tilde{k} = k\, e^{-\mu t}, \quad k \in \mathbb{R}^3, \tag{6.7.2}$$

where μ is the same as in equations (6.6.3). Since we consider the collision term (6.2.7), the functions $a(s)$ and $b(s)$ are given in equations (6.4.10). Hence,

$$\mu = \frac{1}{2} \int_0^1 ds \rho(s)[1 - a^2(s) - b^2(s)] = z(1 - z), \quad z = \frac{1 + e}{2}, \tag{6.7.3}$$

where $0 < e < 1$ is the restitution coefficient. The equation for $\tilde{\varphi}(\tilde{k}, t)$, where tildes are omitted, reads

$$\varphi_t - \mu k \cdot \varphi_k + \varphi = \Gamma(\varphi, \varphi) = \frac{1}{2\pi} \int_{S^2} dn \left| \frac{k \cdot n}{|k|} \right| \varphi(k_+)\varphi(k_-), \quad k \in \mathbb{R}^3, \tag{6.7.4}$$

in the notation of equations (6.2.7). Note that the function $\psi(|k|)$ constructed in Theorem 6.4.1 is the stationary solution of that equation. Then we can prove that

$$\varphi(k, t) \underset{t \to \infty}{\to} \psi(k), \quad k \in \mathbb{R}^3, \tag{6.7.5}$$

following the same scheme as in Section 5.5. In order to avoid some irrelevant technical details, we presented in Section 6.6 a simplified proof (with exponential estimate of the rate of convergence in Theorem 6.6.1) for radial initial data. The results for the radial case were firstly published in [42] and then extended to the general case in [45]. We just formulate the final result of that paper without proof.

Theorem 6.7.1 ([45]). *Let $f(v, t)$ be a solution of the inelastic Maxwell–Boltzmann equation (6.2.5) with initial data $f_0(v)$ such that*

$$\langle f_0, 1 \rangle = 1, \quad \langle f_0, v \rangle = 0, \quad \langle f_0, |v|^2 \rangle = 3, \quad \langle f_0, |v|^{2+\varepsilon} \rangle < \infty.$$

Then

$$e^{-3\mu t} f(ve^{-\mu t}, t) \underset{t\to\infty}{\rightarrow} F(|v|), \quad v \in \mathbb{R}^3,$$

where the convergence is understood in the sense of convergence of probability measures. The function $F(|v|) = F_3(|v|)$ *is defined in equation* (6.4.29) *from Theorem* 6.4.1.

We note that this theorem was proved in [45] for the n-dimensional case with $n \geq 2$. For brevity we consider only $n = 3$. We also note that the estimate (6.6.13) of the rate of convergence in the Fourier space remains the same for non-isotropic solutions $\varphi(k, t)$, $k \in \mathbb{R}^3$.

The rest of this section will be devoted to some properties of the distribution function $F(|v|)$ given in the three-dimensional case by integral (6.4.29), i. e.,

$$F(|v|) = \frac{1}{(2\pi)^3} \int\limits_{\mathbb{R}^3} dk\psi(|k|)\, e^{ik\cdot v}, \quad v \in \mathbb{R}^3. \tag{6.7.6}$$

Note that

$$e^{-|k|^2/2} \leq \psi(|k|) \leq e^{-|k|}(1 + |k|),$$

in accordance with Theorem 6.4.1. Take notice that

$$\langle \psi(|k|), (|k|^N + 1)\rangle < \infty$$

for any $N > 0$. Hence, $F(|v|)$ is bounded in $\mathbb{R}^3$ and has (as a function of $v \in \mathbb{R}^3$) uniformly bounded derivatives of all orders. However, the question about the asymptotic behavior of $F(|v|)$ for large $|v|$ is not very simple. It can be guessed from the exact solution (6.3.27) for the one-dimensional model that $F(|v|)$ can have a power-like tail for high velocities.

We can clarify this question by using equations for moments of $F(|v|)$. Let us assume that

$$m_{2p}(F) = \langle F(|v|), |v|^{2p}\rangle < \infty \tag{6.7.7}$$

for some $p > 1$. It is clear that

$$\cos t = \sum_{j=0}^{[p]} (-1)^j \frac{t^{2j}}{(2j)!} + O(t^{2p}),$$

where $[p]$ denotes the integer part of $p > 1$. Then we obtain from equation (6.7.5)

$$\begin{aligned} \psi(|k|) &= \langle F(|v|), e^{-ik\cdot v}\rangle = \sum_{n=0}^{[p]} \psi_n |k|^{2n} + O(|k|^{2p}), \\ \psi_n &= \frac{m_{2n}}{(2n+1)!} = \frac{1}{(2n+1)!}\langle F, |v|^{2n}\rangle. \end{aligned} \tag{6.7.8}$$

If $\psi(x)$ is the function constructed in Theorem 6.4.1, then

$$\psi_0 = 1, \quad \psi_1 = \frac{1}{2}. \tag{6.7.9}$$

We consider the integro-differential equation (6.4.11), where

$$K(\varphi, \varphi) = \int_0^1 ds\rho(s)\varphi[a(s)x]\,\varphi[b(s)x], \tag{6.7.10}$$

in the notation of Theorem 6.4.1. Then we substitute $\psi(x)$ from equations (6.7.7) into equation (6.4.11) and obtain the following equations for coefficients ψ_n:

$$[-2\mu n + \lambda(2n)]\psi_n = \sum_{l=1}^{n-1} b_{l,n-l}\,\psi_l\,\psi_{n-l}, \quad 2 \le n \le [p];$$

$$b_{l,m} = \int_0^1 ds\rho(s)a^{2l}(s)b^{2m}(s), \quad l, m \ge 1, \tag{6.7.11}$$

in the notation of equations (6.4.13) and (6.5.17). It is proved in Theorem 6.4.1 that $F(|v|) \ge 0$. Therefore, any power moment (6.7.7) of F is positive. The same is true for coefficients ψ_n in equations (6.7.7)–(6.7.10). It is clear from equations (6.7.10) that necessary conditions for existence of moments $m_{2N}(F)$ with $N \ge 2$ (note that $m_0(F) = 1$ and $m_2(F) = 3$ are fixed) are the following:

$$\gamma(4) > 0, \quad \gamma(6) > 0, \quad \ldots, \quad \gamma(2N) > 0, \tag{6.7.12}$$

where

$$\gamma(p) = \lambda(p) - \mu p, \quad \mu = \frac{\lambda(2)}{2}. \tag{6.7.13}$$

Note that $\gamma(2) = 0$. The function $\gamma(p)$ was already considered in Section 6.6. In particular, it was proved in Theorem 6.6.1 that $\gamma(p) > 0$ if $2 < z < 3$ and $\mu \ne 0$. We can prove the following criterion for the maximal order of the moment $m_{2N}(F)$.

Theorem 6.7.2. *We consider $F(|v|)$ from Theorems 6.4.1 and 6.7.1, assuming that $\mu \ne 0$. Then the maximal integer $N \ge 1$ such that*

$$m_{2N}(F) = \langle F(|v|), |v|^{2N} \rangle < \infty$$

is equal to the maximal integer $N \ge 1$ satisfying the inequality $2N < p_0$, where $p_0 > 2$ is the biggest real root of equation

$$\gamma(p) = 0, \quad p \ne 2, \tag{6.7.14}$$

in the notation of equations (6.7.12).

Proof. We recall that $\gamma(p)$ is a concave function, since

$$\gamma''(p) = \lambda''(p) = -\int_0^1 ds\rho(s)\{a^p(s)[\log a(s)]^2 + b^p(s)[\log b(s)]^2\} < 0,$$

$$z \geq 2, \quad 0 \leq a(s), \quad b(s) \leq 1.$$

On the other hand,

$$\gamma(2) = 0; \quad \gamma(p) > 0 \quad \text{if} \quad 1 < p < 3;$$
$$\mu(2-p) \leq \gamma(p) \leq 1 - \mu p \quad \text{if} \quad p \geq 2.$$

Note also that $0 < \mu \leq 1/4$ under conditions of Theorems 6.4.1 and 6.7.2. Hence, any real root $p_0 > 2$ of equation (6.7.13) is located on the interval

$$3 \leq p_0 \leq \mu^{-1}, \quad \mu^{-1} \geq 4\,.$$

The uniqueness of this root follows from concavity of $\gamma(p)$: This function has only one positive maximum on the interval $(3, p_0)$ and no other extrema. Because of the uniqueness of p_0, the positivity conditions (6.7.11) are automatically fulfilled for all even integer numbers $2n$ such that

$$4 \leq 2n < p_0, \quad n = 2, 3, \ldots, N\,,$$

provided $p_0 > 4$. If $3 \leq p_0 \leq 4$, then the only bounded even moment of $F(|v|)$ corresponds to $N = 1$. If $N \geq 2$, the moments $m_{2n}(F)$, $n = 2, \ldots, N$, can be found in explicit form from equations (6.7.7)–(6.7.10). The asymptotic expansion of $\psi(|k|)$ from equations (6.7.7) is valid with $[p] = N$ and cannot be improved. This completes the proof. □

We consider in more detail the specific example (6.4.2) with $\rho(s) = 1$ related to the inelastic Maxwell–Boltzmann equation (6.2.8)–(6.2.10). Then

$$\lambda(p) = \int_0^1 ds[1 - z^p s^{p/2} - (1-\beta s)^{p/2}], \quad p \geq 2,$$

in the notation of equations (6.4.3). Hence

$$\lambda(p) = 1 - \Delta^{-1}(p)\{z^p + \beta^{-1}[1 - (1-\beta)\Delta(p)]\},$$
$$\Delta(p) = 1 + \frac{p}{2}; \quad \mu = \frac{\lambda(2)}{2} = \frac{z(1-z)}{2}.$$

We note that $1 - \beta = (1-z)^2$ and change variables to

$$z = 1 - t = \frac{1+e}{2}, \quad p = 2y, \quad F(y) = -(y-1)\gamma(2y)\,.$$

Then equation (6.7.13) is equivalent to

$$F(y) = y\big[(y+1)t(1-t) - 1\big] + \delta_y(t) = 0, \quad y \geq 1,$$
$$\delta_y(t) = (1-t)^{2y} + t^2(1-t^{2y})(1-t^2) \leq (1-t^2)^{-1} \leq 4/3\,. \tag{6.7.15}$$

Near the elastic limit $e \to 1$, we obtain the asymptotic formula

$$y(t) \approx t^{-1} = \frac{2}{1-e}, \quad e \to 1 \quad (t \to 0)\,,$$

or, equivalently,

$$p_0 \approx \frac{4}{1-e}, \quad e \to 1,$$

in the notation of Theorem 6.7.2. At the same time we have

$$F(1) = 0, \quad F(2) = -(1-2t^2)(1-t)^2 < 0,$$

since $t \in [0, 1/2]$. Therefore, the moment of the fourth order $m_4(F)$ is bounded for any restitution coefficient $0 \leq e < 1$.

This is just an example of application of Theorem 6.7.2. The exact root $y_0 > 2$ of equation (6.7.4) for any given value of $t \in [0, 1/2]$ can be easily found numerically.

In the next section we briefly discuss multi-linear Maxwell models and their applications.

6.8 Multi-linear Maxwell models

In Sections 6.1–6.7 we have studied a class of generalized Maxwell models of the Boltzmann equation for inelastic collisions. It was shown that the Fourier transform is the most appropriate method to deal with this equation. In fact, we mostly studied the equation for characteristic functions $\varphi(k,t)$, $k \in \mathbb{R}^3$, not the original Boltzmann equation for the distribution function $f(v,t)$, $v \in \mathbb{R}^3$. Then we need to express the results in conventional language of distribution functions. In some sense, it is similar to using two different languages. Sometimes this is not easy from the beginning, but it becomes natural after getting some experience.

These considerations help to understand our way of constructing the generalized Maxwell models. It was originally done in the papers [43, 44] jointly with C. Cercignani and I. M. Gamba. We partly follow these papers in Sections 6.8 and 6.9. Roughly speaking, we consider a large class of stochastic models of N-particle systems ("particles" can be quite different objects: from molecules to cars, people, and goods). They can be described at the formal limit $N \to \infty$ (with assumption of molecular chaos [101]) by the one-particle distribution function $f(v,t)$, where $v \in \mathbb{R}^3$, $n \geq 1$, is still called the velocity variable. The variable $t \geq 0$ denotes time. We assume the conservation of number of particles and normalize $f(v,t)$ in such a way that

$$\int_{\mathbb{R}^3} dvf(v,t) = 1\,. \tag{6.8.1}$$

We also assume that the frequency of interactions (say, multi-particle collisions) is constant. Then we can postulate the kinetic equation in the form

$$f_t = Q_+(f) - f\,. \tag{6.8.2}$$

Here

$$Q_+(f) = \alpha_1 Q_+^{(1)}(f) + \alpha_2 Q_+^{(2)}(f) + \cdots + \alpha_M Q_+^{(M)}(f), \tag{6.8.3}$$

where $Q_+^{(j)}$, $j = 1,\dots,M$, are j-linear positive operators describing interactions of $j \geq 1$ particles and $\alpha_j \geq 0$ are relative probabilities of such interactions. It is assumed that

$$\text{each} \quad \int_{\mathbb{R}^d} [Q_+^{(j)}(f)](v)dv = 1; \quad \text{and that} \quad \sum_{j=1}^{M} \alpha_j = 1, \tag{6.8.4}$$

so condition (6.8.1) always holds.

Next, we focus on what properties of operators $Q_+^{(j)}$ are needed to make them consistent with Maxwell-type interactions.

We postulate the main property of multi-particle systems with such interactions in the following way: *Temporal evolution of the system is invariant under scaling transformations of the phase space.*

That is, if S_t is the evolution operator of the above discussed N-particle system such that

$$S_t\{v_1(0),\dots,v_M(0)\} = \{v_1(t),\dots,v_M(t)\}, \quad t \geq 0,$$

then

$$S_t\{\lambda v_1(0),\dots,\lambda v_M(0)\} = \{\lambda v_1(t),\dots,\lambda v_M(t)\} \tag{6.8.5}$$

for any constant $\lambda > 0$. It is easy to see that this assumption leads to the following property of $Q_+^{(j)}$ $(j = 1,\dots,M)$:

$$Q_+^{(j)}(A_\lambda f) = A_\lambda Q_+^{(j)}(f), \quad A_\lambda f(v) = \lambda^d f(\lambda v), \quad \lambda > 0. \tag{6.8.6}$$

Note that the transformation A_λ is consistent with the normalization (6.8.1).

This property shows that it is convenient to use the Fourier transform

$$\hat{f}(k,t) = \mathcal{F}(f) = \int_{\mathbb{R}^d} f(v,t)e^{-ik\cdot v}dv, \quad k \in \mathbb{R}^d, \tag{6.8.7}$$

since the resulting equation

$$\hat{f}_t = \hat{Q}_+(\hat{f}) - \hat{f}, \quad \hat{Q}_+(\hat{f}) = \sum_{j=1}^{M} \alpha_j \hat{Q}_+^{(j)}(\hat{f}) \tag{6.8.8}$$

is invariant under scaling transformations $k \to \lambda k$, $k \in \mathbb{R}^d$. Finally, it is natural to assume (at least in the case when $v \in \mathbb{R}^d$ is a velocity of the particle) that all interactions are invariant under rotations in $\mathbb{R}^d$. Then the general problem can be simplified if we confine ourselves to a class of isotropic functions $f(|v|, t)$.

In particular, denoting

$$u(x,t) = \int_{\mathbb{R}^d} dv f(|v|, t) e^{-ik\cdot v}, \quad x = |k|^2, \tag{6.8.9}$$

we consider a particular form of equation (6.8.8) for $u(x,t)$. The resulting equation is the main mathematical object studied in this paper.

All the above considerations remain valid for $d = 1$; the only differences are that, first, equation (6.8.2) should be considered as the one-dimensional Kac equation [101], and second, rotations in $\mathbb{R}^1 = \mathbb{R}$ should be replaced by reflections. An interesting one-dimensional system, presented in [44], is based on the above discussed multi-particle stochastic model with non-negative phase variables $v = \mathbb{R}_+$, for which the Laplace transform

$$u(x,t) = \int_0^\infty f(v,t) e^{-xv} dv, \quad x \geq 0, \tag{6.8.10}$$

leads to exactly the same class of equations for $u(x,t)$, described below.

We consider equation (6.8.8) for the case of isotropic solutions $\hat{f}(k,t) = u(|k|^2, t)$. The operator $\hat{Q}_+$ is a linear combination of n-linear operators $\hat{Q}_+^{(j)}$, $j = 1, \ldots, M$, acting on the $x = |k|^2$ variable and invariant under dilations $\{x_\lambda = \lambda x,\ \lambda > 0\}$ in $\mathbb{R}_+$. A general class of such operators acting on $u(x)$ can be written in the form

$$\hat{Q}_+^{(j)}(u) = \int_0^\infty da_1 \cdots \int_0^\infty da_j Q_j(a_1, \ldots, a_j) \prod_{i=1}^{j} u(a_i x),$$

where $Q_j(a_1, \ldots, a_j)$ can be a generalized function of j non-negative variables. In our case both $u(x)$ and $\hat{Q}_+^{(j)}(u)$ are related by Fourier (or Laplace) transforms to some probability densities in $\mathbb{R}^d$ or $(\mathbb{R}_+)$ (see equations (6.8.1), (6.8.4), (6.8.9), and (6.8.10)). Hence,

$$u(0) = 1, \quad \int_0^\infty \int_0^\infty Q_j(a_1, \ldots, a_j) da_1 \ldots da_j = 1,$$

and moreover $u \in \mathcal{C}(\mathbb{R}_+)$. Finally, we note that the original (before Fourier/Laplace transforms) operators $Q_+^{(j)}$ were positive, i. e., $Q_+^{(j)}(f) \geq 0$ if $f \geq 0$. To satisfy this condition it is sufficient to assume that $Q_j(a_1, \dots, a_j) \geq 0$ in the above formulas. This follows directly from the fact that the product of two transforms is the transform of a convolution of originals. These arguments explain our choice of equations below.

We slightly change the notation and consider the following equation for $u(x,t)$:

$$u_t + u = \Gamma(u), \quad x \geq 0, \quad t \geq 0, \tag{6.8.11}$$

where

$$\Gamma(u) = \sum_{j=1}^{M} \alpha_j \Gamma^{(j)}(u), \quad \sum_{j=1}^{M} \alpha_j = 1, \quad \alpha_j \geq 0,$$
$$\Gamma^{(j)}(u) = \int_0^\infty \dots \int_0^\infty A_j(a_1, \dots, a_j) \prod_{i=1}^{j} u(a_k x) da_1 \dots da_j, \quad j = 1, \dots, M. \tag{6.8.12}$$

We assume that

$$A_j(a) = A_j(a_1, \dots, a_j) \geq 0, \quad \int_0^\infty da_1 \dots \int_0^\infty da_j A(a_1, \dots, a_j) = 1, \tag{6.8.13}$$

where $A_j(a) = A_j(a_1, \dots, a_j)$ is a generalized density of a probability measure in $\mathbb{R}_+^j$ for any $j = 1, \dots, M$. We also assume that all $A_j(a)$ have compact support, i. e.,

$$A_j(a_1, \dots, a_j) \equiv 0 \quad \text{if} \quad \sum_{k=1}^{j} a_k^2 > R^2, \quad j = 1, \dots, M, \tag{6.8.14}$$

for sufficiently large $0 < R < \infty$. In fact a much weaker assumption that

$$\int_0^\infty \dots \int_0^\infty A_j(a_1, \dots, a_j) \sum_{k=1}^{j} a_k^p \, da_1 \dots da_j < \infty, \quad j = 1, \dots, M, \tag{6.8.15}$$

for all $p > 0$ is needed for most of our results.

Classical models of elastic or inelastic particle interactions of Maxwell type are particular cases of equation (6.8.1) with, for example,

$$M = 2, \quad \alpha_1 = \int_0^1 ds H(s), \quad \alpha_2 = \int_0^1 ds G(s),$$
$$A_1(a_1) = \frac{1}{\alpha_1} \int_0^1 ds\, H(s) \delta[a_1 - c(s)],$$
$$A_2(a_1, a_2) = \frac{1}{\alpha_2} \int_0^1 ds\, G(s) \delta[a_1 - a(s)]\, \delta[a_2 - b(s)], \tag{6.8.16}$$

where the interaction law is determined by the functions $a(s)$, $b(s)$, and $c(s)$.

Then, it is clear that equation (6.8.11) can be considered as a generalized Fourier transformed isotropic Maxwell model with multiple interactions provided $u(0,t) = 1$. The case $M = \infty$ in equations (6.8.12) can be treated in the same way.

Therefore, the general problem we consider below can be formulated in the following way: We consider the initial value problem

$$u_t + u = \Gamma(u), \quad u|_{t=0} = u_0(x), \quad x \geq 0, \quad t \geq 0, \tag{6.8.17}$$

in the Banach space $B = C(\mathbb{R}_+)$ of continuous functions $u(x)$ with the norm

$$\|u\| = \sup_{x \geq 0} |u(x)|. \tag{6.8.18}$$

It is usually assumed that $\|u_0\| \leq 1$ and that the operator Γ is given by equations (6.8.12). On the other hand, there are just a few properties of $\Gamma(u)$ that are essential for existence, uniqueness, and large time asymptotics of the solution $u(x,t)$ of the problem equation (6.8.17). Therefore, in many cases the results can be applied to many general classes of operators Γ in equation (6.8.17) and more general functional space, for example $B = C(\mathbb{R}^d)$ (anisotropic models). That is why we study below the class (6.8.12) of operators Γ as the most important example, but simultaneously indicate which properties of Γ are relevant in each case. In particular, most results do not use a specific form (6.8.12) of Γ and, in fact, are valid for a more general class of operators.

Following this way of study, we first consider the problem (6.8.17) with Γ given by equations (6.8.12) and point out the most important properties of Γ.

We simplify notations and omit in most of the cases below the argument x of the function $u(x,t)$. The notation $u(t)$ (instead of $u(x,t)$) means then the function of the real variable $t \geq 0$ with values in the space $B = C(\mathbb{R}_+)$.

Remark 1. We shall omit below the argument $x \in \mathbb{R}_+$ of functions $u(x)$, $v(x)$, etc., in some cases when this does not cause a misunderstanding. In particular, inequalities of the kind $|u| \leq |v|$, for functions $u(x)$ and $v(x)$, should be understood as a pointwise control in absolute value, i. e., "$|u(x)| \leq |v(x)|$ for any $x \geq 0$."

We start by giving the following general definition for operators acting in a unit ball of a Banach space B denoted by

$$U = \{u \in B : \|u\| \leq 1\}. \tag{6.8.19}$$

Definition 6.8.1. The operator $\Gamma = \Gamma(u)$ is called an L-Lipschitz operator if there exists a linear bounded operator $L\colon B \to B$ such that the inequality

$$|\Gamma(u_1) - \Gamma(u_2)|(x) \leq (L|u_1 - u_2|)(x), \quad x \geq 0, \tag{6.8.20}$$

holds for any pair of functions $u_{1,2}$ in U.

Remark 2. Note that the L-Lipschitz condition (6.8.20) holds, by definition, at any point $x \in \mathbb{R}_+$.

Thus, condition (6.8.20) is much stronger than the classical Lipschitz condition

$$\|\Gamma(u_1) - \Gamma(u_2)\| < C\|u_1 - u_2\| \quad \text{if} \quad u_{1,2} \in U, \tag{6.8.21}$$

which obviously follows from (6.8.20) with the constant $C = \|L\|_B$, the norm of the operator L in the space of bounded operators acting in B. In other words, the terminology "L-Lipschitz condition" means the pointwise Lipschitz condition with respect to a specific linear operator L. A generalization to the case $B = C(\mathbb{R}^d)$ is obvious: We just need to change $x \geq 0$ to $x \in \mathbb{R}^d$ in equations (6.8.17), (6.8.18), and (6.8.20).

The next lemma shows that the operator $\Gamma(u)$ defined in (6.8.12), which satisfies $\Gamma(1) = 1$ (mass conservation) and maps U into itself, satisfies an L-Lipschitz condition, where the linear operator L is the one given by the linearization of Γ near unity.

We assume without loss of generality that the kernels $A_j(a_1, \ldots, a_j)$ in equations (6.8.12) are symmetric with respect to any permutation of the argument $(a_1, \ldots, a_j)$, $j = 2, 3, \ldots, M$.

Theorem 6.8.1. *The operator* $\Gamma(u)$ *defined in equations* (6.8.12) *maps* U *into itself and satisfies the* L*-Lipschitz condition* (6.8.20), *where the linear operator* L *is given by*

$$Lu(x) = \int_0^\infty da K(a) u(ax), \tag{6.8.22}$$

with

$$K(a) = \sum_{j=1}^{M} j\alpha_j K_j(a),$$

where

$$K_j(a) = \int_0^\infty \cdots \int_0^\infty A_j(a, a_2, \ldots, a_j) da_2 \ldots da_j, \quad \sum_{j=1}^{M} \alpha_j = 1 \tag{6.8.23}$$

for symmetric kernels $A_j(a_1, a_2, \ldots, a_j)$, $j = 2, \ldots$.

Proof. First, the operator $\Gamma(u)$ in (6.8.12)–(6.8.14) maps B into itself and also satisfies

$$\|\Gamma(u)\| \leq \sum_{j=1}^{M} \alpha_j \|u\|^j, \quad \sum_{j=1}^{M} \alpha_j = 1. \tag{6.8.24}$$

Hence,

$$\|\Gamma(u)\| \leq 1 \quad \text{if} \quad \|u\| \leq 1, \tag{6.8.25}$$

and then $\Gamma(U) \subset U$, so it maps U into itself.

Since $\Gamma(1) = 1$, we introduce the linearized operator $B \to B$ such that formally

$$\Gamma(1 + \varepsilon u) = 1 + \varepsilon Lu + O(\varepsilon^2). \tag{6.8.26}$$

By using the symmetry of kernels $A_j(a)$, $j = 2, 3, \ldots, M$, one can easily check that L is given by equations (6.8.22) and (6.8.26).

In order to prove the L-Lipschitz property (6.8.20) for the operator Γ given in equations (6.8.12), we make use of the multi-linear structure of the integrand associated with the definition of $\Gamma(u)$. Indeed, from the elementary identity

$$ab - cd = \frac{a + c}{2}(b - d) + \frac{b + d}{2}(a - c).$$

We obtain provided $\|u_{1,2}\| \le 1$

$$\left|\Gamma^{(2)}(u_1) - \Gamma^{(2)}(u_2)\right| \le 2 \int_0^\infty daK_2(a) \left|u_1(ax) - u_2(ax)\right| da \tag{6.8.27}$$

for $3 \le j \le M$ (the case $j = 1$ is trivial). This problem can be obviously reduced to an elementary inequality

$$\left|\prod_{k=1}^{j} x_k - \prod_{k=1}^{j} y_k\right| \le \sum_{k=1}^{j} |x_k - y_k|, \quad j = 3, \ldots, \tag{6.8.28}$$

provided $|x_k| \le 1$, $|y_k| \le 1$, $k = 1, \ldots, j$. Since this is true for $j = 2$, we can use the induction. Let

$$a = x_{j+1}, \quad c = y_{j+1}, \quad b = \prod_{k=1}^{j} x_k, \quad d = \prod_{k=1}^{j} y_k.$$

Then

$$\left|\prod_{k=1}^{j+1} x_k - \prod_{k=1}^{j+1} y_k\right| = |ab - cd| \le |a - c| + |b - d|$$

$$\le |x_{j+1} - y_{j+1}| + \sum_{k=1}^{j} |x_k - y_k|,$$

and inequality (6.8.28) is proved for any $j \ge 3$. Then we proceed exactly as in case $j = 2$ and prove the estimate (6.8.28) for arbitrary $j \ge 3$. Inequality (6.8.20) follows directly from the definition of operators Γ and L. □

Corollary 5. *The L-Lipschitz condition* (6.8.21) *is fulfilled for* $\Gamma(u)$ *given in equations* (6.8.2) *with the constant*

$$C = \|L\| = \sum_{j=1}^{M} j\alpha_j, \quad \sum_{j=1}^{M} \alpha_j = 1, \tag{6.8.29}$$

where $\|L\|$ *is the norm of* L *in* B.

Proof. The proof follows directly from inequality (6.8.20) and equations (6.8.22) and (6.8.23). □

It is also easy to prove that the L-Lipschitz condition holds in $B = C(\mathbb{R}^d)$ for "gain-operators" in the Fourier transformed Boltzmann equations for both elastic and inelastic Maxwell models.

6.9 Self-similar asymptotics

We present below some results of the paper [44], omitting proofs for the sake of brevity. The proofs can be found in that paper. Our goal in this section is to solve the problem (6.8.17) for the general multi-linear operator $\Gamma(u)$, which is very similar to the case of the relatively simple bilinear operator (6.4.7) studied in Sections 6.4–6.7.

Theorem 6.9.1 ([44]). *Consider the Cauchy problem* (6.8.17) *with* $\|u_0\| \leq 1$ *and assume that the operator* $\Gamma\colon B \to B$:

(a) *maps the closed unit ball* $U \subset B$ *into itself; and*
(b) *satisfies an L-Lipschitz condition* (6.8.20) *for some positive bounded linear operator* $L\colon B \to B$.

Then, for any $t \geq 0$:

(i) *there exists a unique solution* $u(t)$ *of the problem* (6.8.17) *such that* $\|u(t)\| \leq 1$; *and*
(ii) *any two solutions* $u(t)$ *and* $w(t)$ *of problem* (6.8.17) *with initial data in the unit ball* U *satisfy the pointwise in* x *inequality*

$$|u(t) - w(t)| \leq \exp\{t(L - I)\}(|u_0 - w_0|). \tag{6.9.1}$$

The proof is based on the Picard iteration scheme and Theorem 6.8.1.

We note that the operator Γ given in (6.8.2) has the following properties:

(a) Γ maps the unit ball U of the Banach space $B = C(\mathbb{R}_+)$ into itself, that is,

$$\|\Gamma(u)\| \leq 1 \quad \text{for any} \quad u \in C(\mathbb{R}_+) \quad \text{such that} \quad \|u\| \leq 1;$$

(b) Γ is an L-Lipschitz operator with L given by

$$Lu(x) = \int_0^\infty K(a)\, u(a\, x) da, \quad x \geq 0, \tag{6.9.2}$$

where $K(a)$ is a generalized density of a positive measure in $\mathbb{R}_+$ satisfying

$$0 < \int_0^\infty K(a)\, a^p da < \infty, \quad \text{for any} \quad p \geq 0; \tag{6.9.3}$$

that means

$$|\Gamma(u_1) - \Gamma(u_2)|(x) \le (L|u_1 - u_2|)(x) = \int_0^\infty K(a)\,|u_1(ax) - u_2(ax)|\,da, \tag{6.9.4}$$

for all $x \ge 0$ for any two functions $u_{1,2} \in C(\mathbb{R}_+)$ such that $\|u_{1,2}\| \le 1$;

(c) Γ is invariant under dilations:

$$e^{\tau\mathcal{D}}\Gamma(u) = \Gamma(e^{\tau\mathcal{D}}u), \quad \mathcal{D} = x\frac{\partial}{\partial x}, \quad e^{\tau\mathcal{D}}u(x) = u(xe^{\tau}), \quad \tau \in \mathbb{R}. \tag{6.9.5}$$

No specific information about Γ beyond these three conditions will be used in this section.

It was already shown in Theorem 6.9.1 that the conditions (a) and (b) guarantee existence and uniqueness of the solution $u(x,t)$ to the initial value problem (6.8.27)–(6.8.28). The property (b) yields the estimate (6.9.1), which is very important for large time asymptotics, as we shall see below. Property (c) suggests a special class of self-similar solutions to equation (6.8.17). Note that the operator L in property (b) has a general form of linear positive operator invariant under dilations, i. e., its specific form is connected with property (c).

We introduce some new notations.

Definition 6.9.1. Let L be the integral operator given in equation (6.9.4). Then

$$Lx^p = k(p)x^p, \quad 0 < k(p) = \int_0^\infty K(a)a^p\,da < \infty, \quad p \ge 0, \tag{6.9.6}$$

and the spectral function $\mu(p)$ is defined by

$$\mu(p) = \frac{k(p) - 1}{p}. \tag{6.9.7}$$

An immediate consequence of properties (a) and (b), as stated in (6.9.4), is that one can obtain a criterion for a pointwise in x estimate of the difference of two solutions to the initial value problem (6.8.17).

Lemma 6.9.1. *Let $u_{1,2}(x,t)$ be any two classical solutions of the problem* (6.8.17) *with* Γ (a) *and* (b), *and let initial data satisfy the conditions*

$$|u_{1,2}(x,0)| \le 1, \quad |u_1(x,0) - u_2(x,0)| \le Cx^p, \quad x \ge 0, \tag{6.9.8}$$

for some positive constant C and p. Then

$$|u_1(x,t) - u_2(x,t)| \le Cx^p\, e^{-t[1-k(p)]}, \quad \textit{for all} \quad t \ge 0. \tag{6.9.9}$$

The proof of Lemma 6.9.1 is based on inequality (6.9.4) (see [43] for details). Then we can investigate a convergence to stationary solution $\pi(x)$ of the problem (6.8.17), satisfying the equation

$$\pi = \Gamma(\pi), \quad \|u\| \le 1. \tag{6.9.10}$$

If the stationary solution $\bar{u}(x)$ does exist (note, for example, that $\Gamma(0) = 0$ and $\Gamma(1) = 1$ for Γ given in equations (6.8.12)), then the large time asymptotics of some data $u_0(x)$ in (6.8.17) can be studied directly on the basis of Lemma 6.9.1. It is enough to assume that $u_0(x,t) \to \bar{u}(x)$ as $t \to \infty$ for any $x \ge 0$.

This simple consideration, however, does not answer the following question:

What happens with $u(x,t)$ if inequality (6.9.8) for $|u_0(x) - \bar{u}(x)|$ is satisfied with such p that $k(p) > 1$?

In order to address this and similar questions we consider a class of self-similar solutions of equation (6.8.17). Indeed, property (c) of Γ shows that equation (6.8.17) admits a class of formal solutions $u_s(x,t) = w(x\,e^{\mu_* t})$ with some real μ_*. It is convenient for our goals to use a terminology that slightly differs from the usual one.

Definition 6.9.2. The function $w(x)$ is called a self-similar profile associated with the initial value (6.8.17) if it satisfies the problem

$$\mu_* \mathcal{D}w + w = \Gamma(w), \quad \mathcal{D} = x\frac{\partial}{\partial x}, \quad \|w\| \le 1. \tag{6.9.11}$$

Note that the convergence of solutions $u(x,t)$ of the initial value problem (6.8.17) to a stationary solution $\bar{u}(x)$ can be considered as a special case of the self-similar asymptotics with $\mu_* = 0$.

Under the assumption that self-similar solutions exist (the existence is proved in the next section), we prove a fundamental result on the convergence of solutions $u(x,t)$ of the initial value (6.8.17) to self-similar ones (sometimes called in the literature *self-similar stability*).

Lemma 6.9.2. *We assume that:*

(i) *for some $\mu_* \in \mathbb{R}$, there exists a classical (continuously differentiable if $\mu_* \neq 0$) solution $w(x)$ of the problem* (6.9.11)*;*

(ii) *the initial data $u(x,0) = u_0$ in the problem* (6.8.17) *satisfy*

$$u_0 = w + O(x^p), \quad \|u_0\| \le 1, \quad \textit{for } p > 0 \textit{ such that } \mu(p) < \mu_*, \tag{6.9.12}$$

where $\mu(p)$ defined in (6.9.7) *is the spectral function associated to the operator L. Then*

$$\left|u(x\,e^{-\mu_* t}, t) - w(x)\right| \le O\left[x^p\,e^{-pt(\mu_* - \mu(p))}\right], \tag{6.9.13}$$

and therefore

$$\lim_{t\to\infty} u(x\,e^{-\mu_* t}, t) = w(x), \quad x \ge 0. \tag{6.9.14}$$

The proof is based on Lemma 6.9.1. It remains to construct the self-similar profile $w(x)$. We consider equation (6.9.11) written in the form

$$\mu_* x w'(x) + w(x) = g(x), \quad g = \Gamma(w), \quad \mu_* \in \mathbb{R}, \tag{6.9.15}$$

and, assuming that $\|w\| < \infty$, transform this equation to the following integral form. Multiplying the equation by the integrating factor $\mu_*^{-1} x^{\frac{1-\mu_*}{\mu_*}}$, integrating and, changing coordinates in the resulting right hand side integral, it is easy to verify that the resulting integral equation reads

$$w(x) = \int_0^1 g(x\tau^{\mu_*})d\tau. \tag{6.9.16}$$

We prove the following result, formulated in terms of the spectral function $\mu(p)$ from (6.9.7).

Theorem 6.9.2. *Consider equation* (6.9.15) *with arbitrary* $\mu_* \in \mathbb{R}$ *and the operator* Γ *satisfying the conditions* (a) *and* (b) *from equations* (6.9.2)–(6.9.4). *Assume that there exists a continuous function* $w_0(x)$, $x \geq 0$, *such that*

(i) $\|w_0(x)\| \leq 1$; *and*

(ii)

$$\int_0^1 g_0(x\tau^{\mu_*})d\tau = w_0(x) + O(x^p), \quad g_0 = \Gamma(w_0), \tag{6.9.17}$$

with some $p > 0$ *satisfying the inequality* $\mu(p) < \mu_*$.

Then there exists a classical solution $w(x)$ *of equation* (6.9.15) *such that*

$$\|w\| \leq 1, \quad w(x) = w_0(x) + O(x^p), \quad \textit{with the same } p > 0.$$

The solution is unique in the class of continuous functions satisfying conditions

$$\|w\| < \infty, \quad w(x) = w_0(x) + O(x^{p_1}),$$

with any positive p_1 *such that* $\mu(p_1) < \mu_*$.

The existence is proved by the following iteration procedure. We choose an initial approximation $w_0 \in U$ such that $\|w_0\| \leq 1$ and consider the iteration scheme

$$w_{n+1}(x) = \int_0^1 g_n(x\tau^{\mu_*})d\tau, \quad g_n = \Gamma(w_n), \quad n = 0, 1, \ldots.$$

The convergence of iterations is proved in [44] on the basis of properties (a) and (b) of the operator Γ. In that paper one can find a lot of extra information on properties of self-similar solutions and other related applications of Maxwell models. An interesting application to multi-linear models to economic games with large numbers of participants can be found in [54]. Our approach to self-similar asymptotics was used in the recent paper [52] for investigation of the classical (elastic) Boltzmann equation related to shear flow and similar problems.

In the next section we briefly consider other applications of Fourier transform.

6.10 Other applications of Fourier transform to the Boltzmann equation

In this section we return to the classical Boltzmann equation considered in Sections 3.2 and 3.3. The simplest generalization of equation (3.3.1) is obviously a set of equations for spatially uniform mixtures of Maxwell gases. The relaxation process in such a mixture, containing N kinds of molecules, is given by the equation (see also Section 2.5)

$$\frac{\partial f_i}{\partial t} = \sum_{j=1}^{N} \int_{\mathbb{R}^3\times S^2} dw dn\, g_{ij}\Big(\frac{u\cdot n}{n}\Big)[f_i(v')f_j(w') - f_i(v)f_j(w)],$$

$$v' = \frac{m_i v + m_j w + m_j u\cdot n}{m_i + m_j}, \qquad w' = \frac{m_i v + m_j w + m_i u\cdot n}{m_i + m_j},$$

$$u = v - w; \quad i,j = 1,\dots,N, \tag{6.10.1}$$

where $f_i(v,t)$ is the distribution function, m_i is the mass of molecules of the kind $i = 1,\dots,N$, and $\{g_{ij}(\cos\theta);\ i,j = 1,\dots,N\}$ is the matrix of the products of the differential scattering cross-sections of molecules of type j at corresponding relative velocities, while the remaining notation is the same as that used previously. The Fourier transformation is applied to system (6.10.1),

$$\varphi_i(k,t) = \int_{\mathbb{R}^3} dv\, f_i(v,t)e^{-ik\cdot v}, \quad i = 1,\dots,N,$$

in the same way as in Section 3.2, and gives the following result:

$$\frac{\partial \varphi_i}{\partial t} = \sum_{j=1}^{N} \int_{S^2} dn\, g_{ij}\Big(\frac{k\cdot n}{k}\Big)\Big\{\varphi_i\Big[\frac{m_i k + m_j k\cdot n}{m_i + m_j}\Big]\varphi_j\Big[\frac{m_j(k - k\cdot n)}{m_i + m_j}\Big]$$
$$- \varphi_j(0)\varphi_i(k)\Big\}, \quad i = 1,\dots,N.$$

Hence, we obtain here the same simplifications as for a single equation, and a theory of relaxation in the mixture of Maxwell gases can be constructed using the same method as for a simple gas (see, in particular, [47]).

Another simple generalization of equation (3.3.1), first pointed out by Ernst (see [80] and references therein), is the analog of this equation in Euclidean space

$\mathbb{R}^d$ of arbitrary dimension $d = 2, 3, \ldots$. The change to the Fourier representation in Section 3.2 in no way uses the dimensionality of velocity space, and hence in $\mathbb{R}^d$ the Fourier transform of the collision integral will have the same form as in (3.2.8), where the integration is carried out over the surface of the unit sphere in $\mathbb{R}^d$ $(d = 2, \ldots)$. This kind of equation is encountered in physical applications, unconnected with the kinetic theory of gases.

The problem of using the Fourier representation to investigate the Boltzmann equation in the case of non-Maxwell molecules is of more interest. This problem has already been touched on in Section 3.2. We will now dwell on it in somewhat more detail. We will rewrite the inner integral in (3.2.7) in the form

$$\int\limits_{\mathbb{R}^d\times\mathbb{R}^d} dv dw\, f(v) f(w) g\Big(|u|, \frac{k\cdot n}{k}\Big)\big[e^{-iv\cdot k_+ - iw\cdot k_-} - e^{-ik\cdot v}\big], \tag{6.10.2}$$

where

$$k_\pm = \frac{k \pm |k| n}{2}, \quad g(|u|, \cos\theta) = |u|\,\sigma(|u|, \cos\theta), \quad u = u - w, \quad k \in \mathbb{R}^d.$$

Considering $g(|u|, \mu)$ (μ occurs in (6.10.2) parametrically) as an arbitrary function of $|u| = |v - w|$, we formulate the following problem: In what cases can (6.10.2) be expressed without integration in terms of the Fourier transform

$$\varphi(k) = \int\limits_{\mathbb{R}^d} dv\, f(v) e^{-ik\cdot v} \tag{6.10.3}$$

of the distribution function $f(v)$? It is clear that this is possible in at least two cases:
(1) when $g(|u|, \mu) \equiv g(\mu)$, i. e., in the case of Maxwell molecules; and
(2) when $g(|u|, \mu)$ is a polynomial in $|u|^2$ with coefficients which depend on μ,

$$g(|u|, \mu) = g_0(|u|) + |u|^2 g_1(\mu) + \cdots + |u|^{2n} g_n(\mu).$$

Unfortunately, the second case is unreal since the scattering cross-sections $\sigma(u, \mu)$ in actual cases are limited as $u \to \infty$ (we assume the parameter $-1 < \mu < 1$ to be fixed). For hard spheres, for example, $g(|u|, \mu) = C\,|u|$, $C = \text{const}$.

Nevertheless, it makes sense to consider the model Boltzmann equation

$$f_t = \int\limits_{\mathbb{R}^d\times S^{d-1}} dw dn\, |u|^2 g\Big(\frac{u\cdot n}{u}\Big)\big[f(v')f(w') - f(v)f(w)\big], \quad u = v - w, \tag{6.10.4}$$

since a comparison of the result for (6.10.4) with the results for Maxwell molecules can reveal qualitative changes in the relaxation process, due to the increase in the collision frequency as the relative velocity increases. This in turn enables us to obtain a better understanding of the relaxation pattern in a gas of hard spheres or molecules which interact with a potential $U(r)\, r^{-n}$, $n > 4$.

It is convenient now to consider (6.10.4) in the space $\mathbb{R}^d$ of arbitrary dimensions $d = 2, \ldots$. Integral (6.10.2) takes the form

$$-g\left(\frac{k\cdot n}{k}\right)\left\{\left(\frac{\partial}{\partial k_+}-\frac{\partial}{\partial k_-}\right)^2\varphi(k_+)\varphi(k_-)\right.$$
$$\left.-\left(\frac{\partial}{\partial k_+^0}-\frac{\partial}{\partial k_-^0}\right)^2\varphi(k_+^0)\varphi(k_-^0)\right\},$$

where we have assumed that the differentiation is first carried out, and we then substitute the values of the arguments

$$k_\pm = \frac{k \pm |k|n}{2}, \quad k_+^0 = k, \quad k_-^0 = 0.$$

Hence, we obtain the following form for the Fourier representation of equation (6.10.4):

$$\varphi_t = -\int_{S^{d-1}} dn\, g\left(\frac{k\cdot n}{k}\right)\left\{\left(\frac{\partial}{\partial k_+}-\frac{\partial}{\partial k_-}\right)^2\varphi(k_+)\varphi(k_-)\right.$$
$$\left.-\left(\frac{\partial}{\partial k_+^0}-\frac{\partial}{\partial k_-^0}\right)^2\varphi(k_+^0)\varphi(k_-^0)\right\}. \qquad (6.10.5)$$

We will now consider the following specific case of equation (6.10.4):

(1)

$$d = 2 \Rightarrow \int_{S^1} dn \cdots = \int_{-\pi}^{\pi} d\theta;$$

(2)

$$g(\cos\theta) = \frac{1}{8}|\sin\theta|, \quad -\pi < \theta < \pi;$$

(3) the function $f(v,t)$ is isotropic with respect to $v \in \mathbb{R}^2$, and we assume the normalization

$$\int_{\mathbb{R}^2} dv f(|v|,t) = \frac{1}{2}\int_{\mathbb{R}^2} dv f(|v|,t)\, v^2 = 1 \quad \Rightarrow \quad f_M(v) = \frac{1}{2\pi}e^{-v^2/2}.$$

Then, putting, as usual, $x = |k|^2/2$, after simple calculations we reduce equation (6.10.5) for the function $\varphi(|k|^2/2, t)$ to the form

$$\varphi_t = -\int_{-\pi}^{\pi} dy\, \varphi'(y)\, \varphi'(x-y) + [-\varphi(x) - \varphi'(x) + x\varphi''(x)],$$

where the prime denotes differentiation with respect to x or y, while the argument t is omitted. It is clear that this equation can be simplified using a Laplace transformation. Putting

$$\Phi(p,t) = \int_0^{\infty} dx e^{-px} \varphi(x,t),$$

we obtain the partial differential equation

$$\Phi_t + p\Phi\,(p\Phi - 1) = -\frac{\partial}{\partial p} p^2\Phi - \Phi + 1\,. \tag{6.10.6}$$

Note now the function $\Phi(p,t)$ can be expressed very simply in terms of the Laplace transformation of the distribution function $f(|v|,t)$, since

$$F(z,t) = \int dv f(|v|,t)\, e^{-z\frac{v^2}{2}} = \frac{1}{z}\Phi\left(\frac{1}{z}\right), \tag{6.10.7}$$

and consequently, the Fourier transformation can be regarded simply as an intermediate step to obtain the final equation for $F(z,t)$. This equation, according to (6.10.6) and (6.10.7), has the form

$$F_t - F_z + F + \frac{F^2 - 1}{z} = 0.$$

Separating here the equilibrium solution $F(z) = 1/(z+1) + G(z,t)$, we obtain the simple equation

$$G_t - G_z + G\left(1 + \frac{2}{z} - \frac{2}{z+1}\right) + \frac{G^2}{z} = 0,$$

which converges, by an obvious replacement, to a linear form. The solution of the Cauchy problem for this equation has the form

$$G(z,t) = \frac{z^2(s+1)\,G_0(s)}{(z+1)\{s^2(z+1)e^t + G_0(s)(s+1)[(z+1)e^t - (s+1)]\}},$$

where

$$s = z + t, \quad G_0(s) = G(z+t, 0).$$

Hence, we have constructed an accurately solvable model of the two-dimensional non-linear Boltzmann equation with a collision frequency that depends on the velocity. The construction of a relaxation theory for this model reduces simply to investigating the explicit formulas written above.

The model described was obtained independently by Ernst and Hendrics [83] and by the author [24].

Remarks on Chapter 6

1. Chapter 6 concerns non-classical versions of the Boltzmann equation for Maxwell-type interactions. The presentation is partly based on papers [36, 38, 42, 44, 45]. There is a large number of related results on inelastic Maxwell models; only a small part of them is cited in the book (see, e. g., [15, 50] and references therein).
2. We did not discuss in Chapter 6 a relatively new and interesting field of applications of Maxwell models, namely, applications to socio-economic problems like a distribution of wealth (see, e. g., [54, 75]). It is perhaps too early to make some definite conclusions about perspectives of that direction of research, but kinetic Maxwell models seem to be very well suited for it. Even applications of Maxwell models to pure mathematics sometimes can be discussed (see, e. g., [48]).

7 Boltzmann equation and hydrodynamics beyond Navier–Stokes

7.1 Boltzmann equation for small Knudsen numbers

We consider in this chapter the full Boltzmann equation for the distribution function $f(x, v, t)$

$$f_t + v \cdot f_x = Q(f, f), \quad x \in \mathbb{R}^3, \quad v \in \mathbb{R}^3, \quad t \in \mathbb{R}_+, \tag{7.1.1}$$

where the collision term $Q(f, f)$ is given in equations (2.1.20). We shall use below some properties of this equation discussed in Sections 2.2–2.4. In particular, we are interested in first moments of $f(x, v, t)$, namely,

$$\rho(x, t) = \langle f, 1 \rangle, \quad u(x, t) = \frac{1}{\rho} \langle f, v \rangle, \quad T(x, t) = \frac{m}{3\rho} \langle f, |v - u|^2 \rangle, \tag{7.1.2}$$

in the notation of equation (2.1.9). We remind the reader the physical meaning of these quantities: $\rho \in \mathbb{R}_+$, $u \in \mathbb{R}^3$, and $T \in \mathbb{R}_+$ denote the number density, the bulk velocity and the absolute temperature, respectively. The constant parameter $m > 0$ stands for molecular mass (usually we assume below that $m = 1$ in appropriate units). The absolute temperature is expressed in equations (7.1.2) in energy units. Then the corresponding local Maxwell distribution reads

$$f_M(x, v, t) = \rho (2\pi T)^{-3/2} \exp\left[-\frac{|c|^2}{2T} \right], \quad c = v - u, \tag{7.1.3}$$

where f_M depends on x and t through variables $\rho(x, t)$, $u(x, t)$, and $T(x, t)$.

Formal equations for (ρ, u, T) can be obtained from equation (7.1.1) in the following way (see Section 2.2). We multiply this equation by 1, v, $|v|^2$ and obtain equations (2.2.10) by integration in velocities. Then we note that

$$\begin{aligned} \langle f, |v|^2 \rangle &= \langle f, |c|^2 + |u|^2 \rangle = \rho(3T + |u|^2), \\ \langle f, v_\alpha v_\beta \rangle &= \langle f, c_\alpha c_\beta + u_\alpha u_\beta \rangle = \rho u_\alpha u_\beta + \pi_{\alpha\beta}, \\ \langle f, v_\alpha |v|^2 \rangle &= \langle f, (c_\alpha + u_\alpha)|c + u|^2 \rangle \\ &= \rho u_\alpha (|u|^2) + 3T) + 2\pi_{\alpha\beta} u_\beta + 2q_\alpha; \quad \alpha, \beta = 1, 2, 3, \end{aligned}$$

where

$$\begin{aligned} \pi_{\alpha\beta} &= \left\langle f,\ c_\alpha c_\beta - \frac{|c|^2}{3} \delta_{\alpha\beta} \right\rangle, \quad c = v - u, \\ q_\alpha &= \frac{1}{2} \langle f, c_\alpha, |c|^2 \rangle, \quad \alpha, \beta = 1, 2, 3, \end{aligned} \tag{7.1.4}$$

https://doi.org/10.1515/9783110550986-008

and substitute these equalities into equations (2.2.10). Finally, we obtain the following set of equations

$$\rho_t + \operatorname{div} \rho u = 0;$$
$$\rho(\partial_t + u \cdot \partial_x)\, u_\alpha + \frac{\partial p}{\partial x_\alpha} + \frac{\partial \pi_{\alpha\beta}}{\partial x_\beta} = 0; \quad p = \rho T,$$
$$\frac{3}{2}\rho(\partial_t + u \cdot \partial_x)\, T + p \operatorname{div} u + \left(\pi_{\alpha\beta}\frac{\partial u_\alpha}{\partial x_\beta} + \operatorname{div} q\right) = 0, \quad \alpha, \beta = 1, 2, 3, \tag{7.1.5}$$

where p denotes the pressure. These equations are obviously unclosed, since they include unknown terms (fluxes): tensor π and vector q.

In the general case there is no way to find $\rho(x,t)$, $u(x,t)$, and $T(x,t)$ without solving the full kinetic equation (7.1.1) for $f(x,v,t)$. There are, however, certain conditions, when we can do it with sufficient accuracy. In order to clarify these conditions it is convenient to pass to dimensionless variables in equation (7.1.1) like we did in Section 3.3. (see equation (3.3.3)). For simplicity we assume that the initial condition is given in the form

$$f|_{t=0} = \rho_0 T_0^{-3/2} \tilde{f}_0(\tilde{x}, \tilde{v}), \quad \tilde{v} = \frac{v}{\sqrt{T_0}}, \quad \tilde{x} = \frac{x}{L}. \tag{7.1.6}$$

We can also assume that the kernel $g(|v-w|, \mu)$ of the collisional integral (2.1.20) reads

$$g(|v-w|, \mu) = g_0(T_0)\tilde{g}(|\tilde{v} - \tilde{w}|, \mu). \tag{7.1.7}$$

Then we assume that

$$f(x, v, t) = \rho_0 T_0^{-3/2} \tilde{f}(\tilde{x}, \tilde{v}, \tilde{t}), \quad \tilde{t} = \rho_0 g_0 t, \tag{7.1.8}$$

and substitute this equality into equation (7.1.1). Finally, we obtain the dimensionless version of the Boltzmann equation (7.1.1)

$$f_t + v \cdot f_x = \frac{1}{\varepsilon} Q(f, f), \tag{7.1.9}$$

where

$$\varepsilon = \mathrm{Kn} = \frac{l}{L}, \quad l = \frac{\sqrt{T_0}}{\rho_0 g_0}. \tag{7.1.10}$$

Here l denotes the mean free pass of gas molecules. For example, $l = (\rho_0 d^2)^{-1}$ for hard spheres of diameter d. The parameter ε is called the Knudsen number.

It is important for applications to know that the typical values of l for air are of order 10^{-4} cm under normal conditions on the Earth. The typical macroscopic length L can obviously be much bigger. This explains the importance of gas flows in the region of small Knudsen numbers. In particular, in the formal limit $\varepsilon \to 0$ we obtain from equation (7.1.6) that $Q(f,f) = 0$ in that limit. It follows from the well-known results on uniqueness of solution of this equation (see, e. g., [72]) that the limiting function f is the local Maxwellian (7.1.3). Since integrals (7.1.4) vanish for $f = f_M$, the limiting equations (7.1.5) for $\varepsilon = 0$ read

$$\begin{aligned} \rho_t + \operatorname{div} \rho u &= 0, \\ \rho \mathcal{D}_0 u + \operatorname{grad} p &= 0, \\ \frac{3}{2}\mathcal{D}_0 T + T \operatorname{div} u &= 0, \quad \mathcal{D}_0 = \partial_t + u \cdot \partial_x, \quad p = \rho T. \end{aligned} \tag{7.1.11}$$

These are classical Euler equations for monoatomic ideal gas. The Euler equations are also very important in mathematics, since they were a starting point for the development of contemporary theory of quasi-linear hyperbolic equations (see, e. g., [113]).

In the next section we discuss two classical methods of solving equation (7.1.1) for small ε.

7.2 Hilbert and Chapman–Enskog methods

A natural way to solve any equation with small parameter ε is to use formal series in powers of ε. Therefore, we consider equation (7.1.9) and assume that

$$f(x, v, t) = \sum_{n=0}^{\infty} \varepsilon^n f_n(x, v, t). \tag{7.2.1}$$

Note that coefficients f_n, $n \geq 0$, of the series do not depend on ε. Then we obtain the following set of equations for f_n:

$$Q(f_0, f_0) = 0; \quad (\partial_t + v \cdot \partial_x) f_{n-1} = \sum_{k=0}^{n} Q(f_k, f_{n-k}), \quad n = 1, 2, \dots, \tag{7.2.2}$$

where

$$\begin{aligned} Q(f_k, f_l) &= \int_{\mathbb{R}^3 \times S^2} dw\, d\omega\, G(|V|, \hat{V} \cdot \omega)[f_k(v')f_l(w') - f_k(v)f_l(w)], \\ \omega &\in S^2, \quad V = v - w, \quad \hat{V} = \frac{V}{|V|}, \\ v' &= \frac{1}{2}(v + w + |V|\omega), \quad w' = \frac{1}{2}(v + w - |V|\omega), \quad k, l = 0, 1, \dots . \end{aligned} \tag{7.2.3}$$

Remark. The notations in equations (7.2.3) are slightly different as compared, e. g., with equation (2.1.20). The reason is that the notations $u(x, t)$ and $g(v)$ are used in this chapter for some other functions.

The equation for f_0 was already discussed above. We obtain

$$f_0 = f_M = \rho_0 T_0^{-3/2} \exp\left[-\frac{|v - u_0|^2}{2T_0}\right], \tag{7.2.4}$$

where functions $\rho_0(x,t)$, $u_0(x,t)$, and $T_0(x,t)$ are not defined yet. Then we consider the next equation for f_1 and obtain

$$(\partial_t + v \cdot \partial_x) f_M = Q(f_M, f_1) + Q(f_1, f_M). \tag{7.2.5}$$

It is easy to verify that

$$\langle \Psi(v),\ Q(f_1, f_2) + Q(f_2, f_1) \rangle = 0, \quad \Psi(v) \in \mathrm{Span}(1, v, |v|^2), \tag{7.2.6}$$

for any pair of functions $f_{1,2}(v)$. Hence, the equation

$$\partial_t \langle f_M, \Psi(v) \rangle + \partial_x \cdot \langle f_M, v\Psi(v) \rangle = 0 \tag{7.2.7}$$

is a necessary condition for existence of solution $f_1(x,v,t)$ of equation (7.2.5). On the other hand, the considerations of Section 7.1 show that equation (7.2.7) is equivalent to Euler equations (7.1.11) for unknown "parameters" $\rho_0(x,t)$, $u_0(x,t)$, and $T_0(x,t)$ of the Maxwellian (7.2.4). The Euler equations are well-posed (at least for a short time interval). Therefore, the initial value problem with given "nice" data

$$\rho_0|_{t=0} = \rho_0^{(0)}(x), \quad u_0|_{t=0} = u_0^{(0)}(x), \quad T_0|_{t=0} = T_0^{(0)}(x) \tag{7.2.8}$$

has a unique solution on that interval. Then the limiting function f_0 (7.2.4) is completely known. At the same time the necessary condition for existence of solution f_1 of equation (7.2.5) is the linear integral equation. It can be transformed to

$$L_{f_M} \varphi = (\partial_t + v \cdot \partial_x) \log f_M(x,v,t), \tag{7.2.9}$$

where

$$L_{f_M} \varphi = f_M^{-1} [Q(f_M\, \varphi, f_M) + Q(f_M, f_M\, \varphi)] \tag{7.2.10}$$

is the linearized Boltzmann collision integral. The operator L_{f_M} can be reduced to its standard form that was studied in detail in Sections 3.5–3.7 for the particular case of Maxwell molecules. This operator in the general case of intermolecular forces will be discussed below. At the moment we just want to avoid technical details.

It is important to mention that this approach to construction of asymptotic solutions to the Boltzmann equation (7.1.9) was proposed by D. Hilbert in 1912 [96]. In particular, he derived equation (7.2.9) and showed that the general solution of this equation reads

$$\varphi = \varphi^{\perp} + \alpha + \beta \cdot v + \gamma |v|^2, \tag{7.2.11}$$

where $\varphi^{\perp}(x,v,t)$ is a unique solution of equation (7.2.9) such that

$$\langle f_M\,\varphi^{\perp},\Psi\rangle = 0,\quad \Psi\in \mathrm{Span}(1,v,|v|^2). \tag{7.2.12}$$

The functions $\alpha(x,t)$, $\beta(x,t)$, and $\gamma(x,t)$ in equation (7.2.11) cannot be obtained from equation (7.2.7). A natural way to find them is to consider equations (7.2.5) for $n=2$ and to use the conditions of their solvability. Then we obtain similarly to equation (7.2.7)

$$\partial_t\langle f_M(\alpha+\beta\cdot v+\gamma|v|^2),\ \Psi(v)\rangle + \partial_x\cdot\langle f_M(\varphi^{\perp}+\alpha+\beta\cdot v+\gamma|v|^2),\ v\Psi(v)\rangle = 0,$$

in the notation of equation (7.2.12) for $\Psi(v)$. This linear equation is obviously equivalent to five equations for five unknowns (two scalars α, γ and one vector $\beta\in\mathbb{R}^3$). Assuming that these equations are well-posed and the corresponding initial data for α, β, and γ at $t=0$ are known, we can find a unique solution for $f_1(x,v,t)$ in Hilbert series (7.2.1). This procedure can be obviously extended by induction to all $n\geq 2$. The reader can find more detailed discussion on Hilbert solutions in [92] and [144].

The Hilbert method allows in principle to construct any “coefficient” $f_n(x,v,t)$ of the formal series (7.2.1). However, the series contains terms proportional to $\varepsilon^k t^l$ with $k,l=1,2,\ldots$ for standard initial data. This is an usual difficulty for equations with a small parameter near time derivative. There are also some other reasons to modify the Hilbert expansion. The earliest and probably most important step in this direction was made independently by S. Chapman [73] and D. Enskog [79] in 1916–1917. For the first time a systematic presentation of the Chapman–Enskog method was given in the book [74], firstly published in 1939. Since then this famous method is discussed in almost all books on kinetic theory (see, e. g., [70, 104, 112, 144]).

Below we present a slightly different version of this method following the author’s paper [33]. We simplify notations by denoting

$$\mathcal{D}=\partial_t+v\cdot\partial_x,\quad M(|c|)=(2\pi T)^{-3/2}\exp\left(-\frac{|c|^2}{2T}\right),\quad c=v-u, \tag{7.2.13}$$

and represent the solution of equation (7.1.9) as a sum

$$f=\rho M+\varepsilon F, \tag{7.2.14}$$

where $\rho(x,t;\varepsilon)$, $u(x,t;\varepsilon)$, and $T(x,t;\varepsilon)$ are “true” hydrodynamic moments of f. Hence,

$$\langle\Psi,F\rangle=0,\quad \langle\Psi,\ \mathcal{D}\rho M\rangle+\varepsilon\langle\Psi,\ \mathcal{D}F\rangle=0, \tag{7.2.15}$$

for any $\Psi(v)\in\mathrm{Span}(1,v,|v|^2)$. This leads to the usual set of unclosed equations of hydrodynamics (7.1.11).

We shall use these equations in slightly different notations, which clearly show the dependence on ε and F. The equations read

$$\begin{aligned}
&\mathcal{D}_0\rho + \rho \text{ div } u = 0,\\
&\rho\,\mathcal{D}_0 u_\alpha + \frac{\partial p}{\partial x_\alpha} + \varepsilon \frac{\partial \pi_{\alpha\beta}}{\partial x_\beta} = 0,\\
&\frac{3}{2}\rho\,\mathcal{D}_0 T + p \text{ div } u + \varepsilon\left(\pi_{\alpha\beta}\frac{\partial u_\alpha}{\partial x_\beta} + \text{div } q\right) = 0, \quad \mathcal{D}_0 = \partial_t + u\cdot\partial_x,
\end{aligned} \tag{7.2.16}$$

where

$$p = \rho T, \quad \pi_{\alpha\beta} = \langle F, c_\alpha c_\beta\rangle, \quad q = \frac{1}{2}\langle F, |c|^2 c\rangle, \quad \alpha,\beta = 1,2,3. \tag{7.2.17}$$

The equation for $F(x, v, t; \varepsilon)$ reads

$$\mathcal{D}\rho M + \varepsilon\,\mathcal{D}F = \rho\, ML(F/M) + \varepsilon Q(F,F), \tag{7.2.18}$$

where the linearized collision operator L is defined by equality

$$MLg = Q(Mg, M) + Q(M, Mg). \tag{7.2.19}$$

Note that $L = L_M$, in the notation of equation (7.2.10).

In accordance with equations (7.2.19), (7.2.13), and (7.2.3), we obtain for any "nice" function $g(v)$ the following explicit formula:

$$\begin{aligned}
&L\varphi(v)\\
&\quad = \int_{\mathbb{R}^3\times S^2} dw d\omega\, G(|V|, \hat{V}\cdot\omega)\, M(|v-u|)[g(v') + g(w') - g(v) - g(w)],\\
&V = v - w, \quad \hat{V} = \frac{V}{|V|}, \quad \omega \in S^2,\\
&v' = \frac{v + w + |V|\omega}{2}, \quad w' = \frac{v + w - |V|\omega}{2}.
\end{aligned} \tag{7.2.20}$$

The operator L acts only on the velocity $v \in \mathbb{R}^3$. It is self-evident that the bulk velocity $u \in \mathbb{R}^3$ and the temperature $T \in \mathbb{R}_+$ play a role of constant parameters. Moreover, the obvious change of variable v to $c = v - u$ reduces equation (7.2.15) to its particular case with $u = 0$. This case was studied in detail in Sections 3.5–3.7 for Maxwell molecules, i. e., for such kernel $G(|V|, \mu)$ in the integral (7.2.15) that depends only on $\mu \in [-1, 1]$. General properties of L do not depend on that assumption. They are also independent of parameters u and T of the Maxwellian M in (7.2.15).

We denote by H the Hilbert space with the scalar product

$$\langle g_1, g_2\rangle_M = \langle Mg_1,\ g_2\rangle \tag{7.2.21}$$

and consider L as an operator acting from H to H. The following basic properties of L will be used below:

A $Lg = 0$ if and only if $g \in N(L) = \mathrm{Ker} L = \mathrm{Span}(1, v, |v|^2)$;

B L is a self-adjoint and semi-negative operator, i. e.,

$$\langle g_1, Lg_2 \rangle_M = \langle Lg_1, g_2 \rangle_M, \quad \langle g, Lg \rangle_M \le 0; \tag{7.2.22}$$

the equality $\langle g, Lg \rangle_M = 0$ is possible only for $g \in N(L)$;

C $H = N(L) \oplus R(L)$, where $R(L) = LH$ is a range of the operator L in H and $\oplus$ denotes the orthogonal sum (with respect to the scalar product (7.2.21)).

Then the problem

$$Lg = \varphi, \quad g \in R(L), \quad \varphi \in R(L) \tag{7.2.23}$$

has a unique solution $g = L^{-1}\varphi$. We extend L^{-1} to the linear operator $\tilde{L}^{-1}$ such that

$$\tilde{L}^{-1}\varphi = \begin{cases} L^{-1}\varphi, & \text{if } \varphi \in R(L), \\ 0, & \text{if } \varphi \in N(L). \end{cases} \tag{7.2.24}$$

Then equation (7.2.18) can be transformed to

$$F = M\left[\tilde{L}^{-1}(\mathcal{D} \log \rho M) + \frac{\varepsilon}{\rho}\tilde{L}^{-1}\frac{\mathcal{D}F - Q(F,F)}{M}\right]. \tag{7.2.25}$$

We denote

$$\tilde{Q}(F) = Q(F,F),$$

$$F_0 = M\tilde{L}^{-1}(\mathcal{D} \log \rho M) = -M\tilde{L}^{-1}\left(c_\alpha \cdot \frac{\partial}{\partial x_\alpha}\frac{|c|^2}{2T}\right), \quad c = v - u. \tag{7.2.26}$$

The same notation $c = v - u$ for the thermal velocity is often used below. Then, omitting tildes, we obtain

$$F = F_0 + \frac{\varepsilon}{\rho} ML^{-1}\frac{\mathcal{D}F - Q(F)}{M}, \tag{7.2.27}$$

where

$$F_0 = M\left[\frac{1}{T}\overline{\frac{\partial u_\alpha}{\partial x_\beta}}\varphi_{\alpha\beta}(c) + \frac{1}{T^2}\frac{\partial T}{\partial x_\alpha}\varphi_\alpha(c)\right],$$

$$\varphi_{\alpha\beta} = L^{-1}\left(c_\alpha c_\beta - \frac{|c|^2}{3}\delta_{\alpha\beta}\right), \quad \varphi_\alpha = L^{-1}\left[\frac{c_\alpha}{2}(|c|^2 - 5T)\right],$$

$$\overline{a_{\alpha\beta}} = \frac{1}{2}\left(a_{\alpha\beta} + a_{\beta\alpha} - \frac{2}{3}\delta_{\alpha\beta} \operatorname{Tr} a\right), \quad \operatorname{Tr} a = a_{11} + a_{22} + a_{33}. \tag{7.2.28}$$

Note that

$$\overline{\frac{\partial u_\alpha}{\partial x_\beta}}\varphi_{\alpha\beta}(c) = \frac{\partial u_\alpha}{\partial x_\beta}\varphi_{\alpha\beta}(c),$$

since $\varphi_{\alpha\alpha} = 0$, $\varphi_{\alpha\beta} = \varphi_{\beta\alpha}$. The equations (7.2.27) coupled with equations (7.2.16) are obviously exact. They present the shortest way from the Boltzmann equation (7.1.9) to Navier–Stokes equations considered in the next section.

7.3 Navier–Stokes equations

We begin with the formal limit of equation (7.2.27) at $\varepsilon = 0$. Then $F = F_0(x, v, t)$, in the notation of equations (7.2.28). Hence, we obtain asymptotic formulas for tensor $\pi(x, t; \varepsilon)$ and vector $q(x, t; \varepsilon)$ in equations (7.2.16):

$$\pi_{\alpha\beta}|_{\varepsilon=0} = \pi_{\alpha\beta}^{\mathrm{NS}} = \langle F_0, c_\alpha c_\beta \rangle,$$
$$q_\alpha|_{\varepsilon=0} = q_\alpha^{\mathrm{NS}} = \frac{1}{2}\langle F_0, c_\alpha |c|^2 \rangle, \quad \alpha, \beta = 1, 2, 3. \tag{7.3.1}$$

Then straightforward calculations lead to formulas

$$\pi_{\alpha\beta}^{\mathrm{NS}} = -2\mu(T)\overline{\frac{\partial u_\alpha}{\partial x_\beta}}, \quad q_\alpha^{\mathrm{NS}} = -\lambda(T)\frac{\partial T}{\partial x_\alpha}, \tag{7.3.2}$$

where

$$\mu(T) = -\frac{1}{10T}\langle M\varphi_{\alpha\beta}, c_\alpha c_\beta \rangle, \quad \lambda(T) = -\frac{1}{6T^2}\langle M\varphi_\alpha, c_\alpha |c|^2 \rangle \tag{7.3.3}$$

are respectively the coefficients of viscosity and heat conductivity. The resulting equations (7.2.16) are of the Navier–Stokes type. This is the reason for upper indices NS in equations (7.3.1) and (7.3.2). It is easy to show that both coefficients $\mu(T)$ and $\lambda(T)$ are positive. Note that

$$L\varphi_{\alpha\beta}(c) = c_\alpha c_\beta - \frac{|c|^2}{3}\delta_{\alpha\beta}, \quad \varphi_{\alpha\alpha} = 0,$$

in the notation of equations (7.2.28). Hence,

$$\langle M\varphi_{\alpha\beta}, c_\alpha c_\beta \rangle = \langle \varphi_{\alpha\beta}, L\varphi_{\alpha\beta} \rangle_M < 0, \tag{7.3.4}$$

as it follows from property B of L. Then $\mu(T) > 0$. Similar considerations for $\varphi_\alpha(c)$ show that $\lambda(T) > 0$.

It is also easy to give a formal proof of the H-theorem for Navier–Stokes equations. Indeed, the H-function for local Maxwellian reads

$$H(\rho, T) = \rho\langle M, \log \rho M \rangle = \rho\left[\log \rho(2\pi T)^{-3/2} - \frac{3}{2}\right]. \tag{7.3.5}$$

Since the total mass is preserved, we can consider the simplified function

$$\Phi(\rho, T) = \rho \log \rho T^{-3/2} \tag{7.3.6}$$

and obtain from equations (7.2.16) with $\pi = \pi^{\mathrm{NS}}$ and $q = q^{\mathrm{NS}}$ the following equation for Φ:

$$\Phi_t + \operatorname{div}\left[u\Phi + \frac{\varepsilon\lambda(T)}{T}\operatorname{grad} T\right]$$
$$= -\frac{\varepsilon}{T^2}\left[2T\mu(T)\left(\overline{\frac{\partial u_\alpha}{\partial x_\beta}}\right)^2 + \lambda(T)\,|\operatorname{grad} T|^2\right], \tag{7.3.7}$$

where the summation over all $\alpha, \beta = 1, 2, 3$ is assumed. Then we formally obtain that the functional

$$\Gamma(t) = \int_{\mathbb{R}^3} dx\{\Phi[\rho(x,t), T(x,t)] - \Phi(\rho_\infty, T_\infty)\} \tag{7.3.8}$$

is a monotone decreasing function of time t for “nice” solutions of the Navier–Stokes equation with constant values ρ_∞ and T_∞ at infinity. It is obvious that $\Gamma(t) = \text{const.}$ for solutions of the Euler equations (7.1.11).

General properties of Navier–Stokes equations are independent of the specific form of positive dissipative coefficients $\mu(T)$ and $\lambda(T)$. Nevertheless, there are many applied problems where we need to know exact numerical values of these coefficients. They are expressed in equations (7.3.3) in terms of some integrals that contain two functions $\varphi_{\alpha\beta}(c)$ and $\varphi_\alpha(c)$ given in equations (7.2.28) for $\alpha, \beta = 1, 2, 3$. These equations mean that

$$\begin{gathered} L\varphi_{\alpha\beta}(c) = c_\alpha c_\beta - \frac{|c|^2}{3}\delta_{\alpha\beta}, \quad L\varphi_\alpha(c) = \frac{1}{2}c_\alpha(|c|^2 - 5T), \\ \langle M\varphi_{\alpha\beta}(c), \Psi(c)\rangle = 0, \quad \langle M\varphi_\alpha(c), \Psi(c)\rangle = 0; \quad \alpha, \beta = 1, 2, 3, \end{gathered} \tag{7.3.9}$$

for any $\Psi(c) \in \operatorname{Span}(1, c, |c|^2)$.

How can we solve equations (7.3.9)? From the beginning we consider the case of Maxwell molecules, i. e., the case of the kernel $G = G(\mu)$, $\mu \in [-1, 1]$ in the integral operator L (7.2.3). It is easy in this case to use Theorem 3.6.1(ii) or simply make straightforward calculations and obtain

$$\begin{aligned} &L\left(c_\alpha c_\beta - \frac{|c|^2}{3}\delta_{\alpha\beta}\right) = \lambda_{02}\left(c_\alpha c_\beta - \frac{|c|^2}{3}\delta_{\alpha\beta}\right), \\ &\quad Lc_\alpha(|c|^2 - 5T) = \lambda_{11}c_\alpha(|c|^2 - 5T), \\ &\qquad \lambda_{ln} = 2\pi\int_0^\pi d\theta\,\sin\theta\,G(\cos\theta)\left[P_l\left(\cos\frac{\theta}{2}\right)\cos^{2n+l}\frac{\theta}{2}\right. \\ &\qquad\quad \left. + P_l\left(\sin\frac{\theta}{2}\right)\sin^{2n+l}\frac{\theta}{2} - 1 - \delta_{n0}\delta_{l0}\right], \quad n, l = 0, 1, \ldots. \end{aligned} \tag{7.3.10}$$

The expression for eigenvalues can be simplified and we obtain

$$\begin{gathered} \varphi_{\alpha\beta} = -\eta\left(c_\alpha c_\beta - \frac{|c|^2}{3}\delta_{\alpha\beta}\right), \quad \varphi_\alpha = -\frac{3}{4}\eta(c_\alpha - 5T), \\ \eta^{-1} = \frac{3\pi}{2}\int_{-1}^{1} d\mu G(\mu)(1 - \mu^2). \end{gathered} \tag{7.3.11}$$

Substituting these formulas into equations (7.3.3), we obtain for Maxwell molecules

$$\mu(T) = \eta T, \quad \lambda(T) = \frac{15}{4}\eta T. \tag{7.3.12}$$

Other important molecular models are hard spheres and power-like potentials $\Phi(r) = \alpha r^n$, $\alpha > 0$, considered in Section 2.4. For any fixed $n > 1$ the kernel $G(|V|, \cos\theta)$ in equation (7.2.20) reads

$$G(|V|, \cos\theta) = |V|^{\gamma_n} G_n(\cos\theta), \quad \gamma_n = 1 - \frac{4}{n}, \tag{7.3.13}$$

where $\theta \in [0, \pi]$ is the scattering angle. The case of hard spheres with diameter d can be formally considered as the limit, as $n \to \infty$. Then we obtain

$$G_{\mathrm{HS}}(|V|, \cos\theta) = |V|\frac{d^2}{4}, \tag{7.3.14}$$

i. e., the same formula with $\gamma_\infty = 1$ and $G_\infty = d^2/4$. Obviously, $n = 4$ for Maxwell molecules and equations (7.3.9) can be solved explicitly in that case. If $n \neq 4$ we cannot obtain exact solutions, but some important information can be found. Let us fix some $n \neq 4$ and consider equations (7.3.9) and (7.2.20) with kernel (7.3.13). Then it is natural to look for solutions of equations (7.3.9) in the dimensionless form

$$\varphi_{\alpha\beta}(c) = T^m \tilde{\varphi}_{\alpha\beta}\left(\frac{c}{\sqrt{T}}\right), \quad \varphi_\alpha(c) = T^l \tilde{\varphi}_\alpha\left(\frac{c}{\sqrt{T}}\right). \tag{7.3.15}$$

The numbers m and l must be chosen in such a way that functions $\tilde{\varphi}_{\alpha\beta}(\tilde{c})$ and $\tilde{\varphi}_\alpha(\tilde{c})$ do not depend on T. Then, after simple transformations, we obtain

$$m = 1 - \frac{\gamma_n}{2}, \quad l = \frac{3 - \gamma_n}{2}.$$

Finally, equations (7.3.3) yield the following general formulas:

$$\mu(T) = \mu(1)T^{1-\gamma_n/2}, \quad \lambda(T) = \lambda(1)T^{1-\gamma_n/2}, \quad \gamma_n = 1 - \frac{4}{n}, \tag{7.3.16}$$

for the intermolecular potential $\Phi(r) \sim r^{-n}$, $n > 1$. The case of hard spheres formally corresponds to $n = \infty$. Hence, it is sufficient for such potentials to find constant values $\mu(1)$ and $\lambda(1)$. The best way to do it is to use expansions in orthogonal series in eigenfunctions for Maxwell molecules (see Theorem 3.6.1(ii)). This is a traditional way to solve approximately equations (7.3.9); see, e. g., the classical book [74].

What can be done beyond the Navier–Stokes level? This is the main question we intend to discuss. First we briefly formulate the whole formal scheme of the Chapman–Enskog method applied to equations (7.2.27) and (7.2.16). We assume that

$$F = \sum_{n=0}^{\infty} \varepsilon^n F_n(x, v, t), \tag{7.3.17}$$

where F_0 is given in equations (7.2.28). Then we need to expand in similar series all terms of equations (7.2.27). It is obvious that

$$Q(F) = \sum_{n=0}^{\infty} \varepsilon^n Q_n(F), \quad Q_n(F) = \sum_{k=0}^{n} Q(F_k, F_{n-k}),$$

in the notation of equations (7.2.3). The expansion of the derivative $\mathcal{D}F$ in (7.2.27) is a bit tricky. We note that

$$\mathcal{D}F = \mathcal{D}_0 F + c \cdot \partial_x F, \quad \mathcal{D}_0 = \partial_t + u \cdot \partial_x .$$

Hence,

$$\mathcal{D}F = \mathcal{D}_0 F + \sum_{n=0}^{\infty} \varepsilon^n \, (c \cdot \partial_x F_n).$$

The last step is to expand $\mathcal{D}_0 F$ in powers of ε in order to get rid of time derivatives. It is possible because any $F_n(x, v, t)$ in the series (7.3.17) depends on x and t only through hydrodynamic variables (ρ, u, T) and their spatial derivatives of finite orders. The formal expansion of $\mathcal{D}_0 \rho$, $\mathcal{D}_0 u$, and $\mathcal{D}_0 T$ follows from equations (7.3.17). We obtain

$$\begin{aligned} \mathcal{D}_0 \rho &= -\rho \ \mathrm{div}\ u, \quad \rho\, \mathcal{D}_0 u_\alpha = -\partial_{x_\alpha} p - \sum_{n=1}^{\infty} \varepsilon^n \partial_{x_\beta} \pi_{\alpha\beta}^{(n)}, \\ \frac{3}{2} \rho\, \mathcal{D}_0 T &= -p \ \mathrm{div}\ u - \sum_{n=1}^{\infty} \varepsilon^n (\pi_{\alpha\beta}^{(n)} \partial_{x_\beta} u_\alpha + \mathrm{div}\ q^{(n)}), \end{aligned} \tag{7.3.18}$$

where

$$\begin{aligned} p &= \rho T, \quad \pi_{\alpha\beta}^{(n)} = \langle F_{n-1}, c_\alpha c_\beta \rangle, \\ q_\alpha^{(n)} &= \frac{1}{2} \langle F_{n-1}, c_\alpha |c|^2 \rangle, \quad n \geq 1; \quad \alpha, \beta = 1, 2, 3. \end{aligned} \tag{7.3.19}$$

It is clear that these formulas allow to construct the operator $\mathcal{D}_0$ acting on F in the form of operator series

$$\mathcal{D}_0 = \sum_{n=0}^{\infty} \varepsilon^n \mathcal{D}_{0n},$$

and therefore we obtain

$$\mathcal{D}F = \sum_{n=0}^{\infty} \varepsilon^n (\mathcal{D}F)_n, \quad (\mathcal{D}F)_n = \sum_{k=0}^{n} \mathcal{D}_{0k} F_{n-k} + c \cdot \partial_x F.$$

Hence, we get from equations (7.2.27) and (7.2.28) the following recursive formulas:

$$F_{n+1} = \frac{1}{\rho} M\, L^{-1} \frac{1}{M} \left\{ c \cdot \partial_x F + \sum_{k=0}^{n} [\mathcal{D}_{0k} F_{n-k} - Q(F_k, F_{n-k})] \right\}, \tag{7.3.20}$$

where $n = 0, 1, \ldots$. In principle, these formulas allow to construct "exact" equations of hydrodynamics in the form of power series (7.3.18). In the next section we try to make the first step in this direction in order to improve the Navier–Stokes approximation.

7.4 Burnett equations

Since we are interested mainly in the correction terms $\pi^{(2)}$ and $q^{(2)}$ from equations (7.3.19), we do not need to use formulas (7.3.20). Instead we consider equation (7.2.7) and derive by integration the following identities:

$$\begin{aligned}\pi_{\alpha\beta} &= \left\langle F,\, c_\alpha c_\beta - \frac{|c|^2}{3}\,\delta_{\alpha\beta}\right\rangle = \pi_{\alpha\beta}^{\mathrm{NS}} + \varepsilon I(\varphi_{\alpha\beta}),\\ q_\alpha &= \frac{1}{2}\langle F,\, c_\alpha\,(|c|^2 - 5T)\rangle = q_\alpha^{\mathrm{NS}} + \varepsilon I(\varphi_\alpha), \quad \alpha,\beta = 1,2,3,\end{aligned} \tag{7.4.1}$$

where

$$I(h) = \frac{1}{\rho}\,\langle h,\, \mathcal{D}F - Q(F)\rangle, \tag{7.4.2}$$

in the notation of equations (7.2.28) and (7.3.2). These identities follow from the fact that L^{-1} is a self-adjoint operator in H; see equation (7.2.2). Since

$$\mathcal{D} = \partial_t + v\cdot\partial_x = \mathcal{D}_0 + c\cdot\partial_x, \quad \mathcal{D}_0\rho = -\rho\ \mathrm{div}\ u,$$

we can transform equation (7.4.2) to

$$\begin{aligned}I(h) &= \mathcal{D}_0\frac{1}{\rho}\,\langle F,\, h\rangle + \frac{1}{\rho}\left[\frac{\partial}{\partial x}\cdot\langle F,\, ch\rangle - \Delta(h)\right],\\ \Delta(h) &= \frac{1}{\rho}[\langle F,\, \mathcal{D}h\rangle + \langle Q(F),\, h\rangle].\end{aligned} \tag{7.4.3}$$

Equations (7.4.1) and (7.4.3) with $h = \varphi_{\alpha\beta}$ and $h = \varphi_\alpha$ suggest the most convenient way to derive the Burnett equations. These are equations of hydrodynamics (7.2.16), where

$$\pi_{\alpha\beta} = \pi_{\alpha\beta}^{\mathrm{NS}} + \varepsilon\pi_{\alpha\beta}^{B}, \quad q_\alpha = q_\alpha^{\mathrm{NS}} + \varepsilon q_\alpha^{B}, \quad \alpha,\beta = 1,2,3. \tag{7.4.4}$$

The upper index B is related to D. Burnett, who firstly studied these correction terms in 1935 [58, 59]. The terms π^B and q^B can be calculated by formulas (7.4.1) and (7.4.3) with $F = F_0$ from equations (7.2.28) for operator $\mathcal{D}_0$, i. e.,

$$\begin{aligned}&\mathcal{D}_0\rho = -\rho\,\mathrm{div}\ u, \quad \mathcal{D}_0 u = -\frac{1}{\rho}\mathrm{grad}\, p, \quad \mathcal{D}_0 T = -\frac{2}{3}T\,\mathrm{div}\ u,\\ &\mathcal{D}_0 = \partial_t + u\cdot\partial_x, \quad p = \rho T.\end{aligned} \tag{7.4.5}$$

We evaluate below only some terms with higher derivatives which are important for our goals. The complete Burnett equations are well known (see, e. g., books [74, 86, 144]) and therefore we present them below without details of calculation. The details can be found, in particular, in [55].

Coming back to equations (7.4.1), we denote

$$\pi_{\alpha\beta}^{\mathrm{NS}} = I(\varphi_{\alpha\beta}) = \mathcal{D}_0 P_{\alpha\beta} + \frac{1}{\rho}\left[\frac{\partial}{\partial x_\gamma}\langle c_\gamma \varphi_{\alpha\beta}, F_0\rangle - \Delta_{\alpha\beta}\right],$$

$$q_\alpha = I(\varphi_\alpha) = \mathcal{D}_0 Q_\alpha + \frac{1}{\rho}\left[\frac{\partial}{\partial x_\gamma}\langle c_\gamma \varphi_\alpha, F_0\rangle - \Delta_\alpha\right], \tag{7.4.6}$$

where

$$Q_\alpha = \frac{1}{\rho}\langle F_0, \varphi_\alpha, \rangle, \quad \Delta_{\alpha\beta} = \langle F_0, \mathcal{D}\varphi_{\alpha\beta}\rangle + \langle \varphi_{\alpha\beta}, Q(F_0)\rangle,$$

$$P_{\alpha\beta} = \frac{1}{\rho}\langle F_0, \varphi_{\alpha\beta}, \rangle, \quad \Delta_\alpha = \langle F_0, \mathcal{D}\varphi_\alpha\rangle + \langle \varphi_\alpha, Q(F_0)\rangle. \tag{7.4.7}$$

Note that the remainder terms $\Delta_{\alpha\beta}$ and Δ_α depend only on first derivatives of ρ, T, and u. They do not influence higher derivatives in the non-linear equations. Therefore, we first concentrate on the "main" terms in equations (7.4.6) and then consider the remaining terms with first derivatives.

The straightforward calculation of the integrals with F_0 given in equation (7.2.28) yields

$$P_{\alpha\beta} = \frac{A}{\rho}\overline{\frac{\partial u_\alpha}{\partial x_\beta}}, \quad Q_\alpha = \frac{B}{\rho}\frac{\partial T}{\partial x_\alpha}, \quad \frac{\partial}{\partial x_\gamma}\langle c_\gamma \varphi_{\alpha\beta}, F_0\rangle = \overline{\frac{\partial}{\partial x_\alpha} C \frac{\partial T}{\partial x_\beta}},$$

$$\frac{\partial}{\partial x_\beta}\langle c_\beta \varphi_\alpha, F_0\rangle = \frac{\partial}{\partial x_\beta} C\, T \overline{\frac{\partial u_\alpha}{\partial x_\beta}}, \tag{7.4.8}$$

where

$$A = A(T) = \frac{1}{5T}\langle \varphi_{\alpha\beta}, \varphi_{\alpha\beta}\rangle_M = \frac{1}{5T}\langle \varphi_{\alpha\beta}^2\rangle_M,$$

$$B = B(T) = \frac{1}{3T^2}\langle \varphi_\alpha, \varphi_\alpha\rangle_M = \frac{1}{3T^2}\langle \varphi_\alpha^2\rangle_M,$$

$$C = C(T) = \frac{1}{5T^2}\langle c_\alpha \varphi_\alpha, \varphi_{\alpha\beta}\rangle_M\,. \tag{7.4.9}$$

Then we use the identity

$$\mathcal{D}_0 \frac{\partial}{\partial x_\beta} = \frac{\partial}{\partial x_\beta}\mathcal{D}_0 - \frac{\partial u_\gamma}{\partial x_\beta}\frac{\partial}{\partial x_\gamma}$$

and the above described rule for calculating $\mathcal{D}_0\rho$, $\mathcal{D}_0 u$, and $\mathcal{D}_0 T$ by equations (7.4.5). Thus we obtain

$$\mathcal{D}_0 P_{\alpha\beta} = -\frac{A}{\rho}\left[\overline{\frac{\partial}{\partial x_\alpha}\left(\frac{1}{\rho}\frac{\partial p}{\partial x_\beta}\right)} + \overline{\frac{\partial u_\alpha}{\partial x_\gamma}\frac{\partial u_\gamma}{\partial x_\beta}} - \alpha(T)\,\mathrm{div}\,u\,\overline{\frac{\partial u_\alpha}{\partial x_\beta}}\right],$$

$$\mathcal{D}_0 Q_\alpha = -\frac{B}{\rho}\left[\frac{2}{3}\frac{\partial}{\partial x_\alpha} T\,\mathrm{div}\,u + \frac{\partial u_\beta}{\partial x_\alpha}\frac{\partial T}{\partial x_\beta} - \beta(T)\,\mathrm{div}\,u\,\frac{\partial T}{\partial x_\alpha}\right],$$

$$\alpha(T) = 1 - \frac{2}{3}\frac{TA'(T)}{A(T)}, \quad \beta(T) = 1 - \frac{2}{3}\frac{TB'(T)}{B(T)}. \tag{7.4.10}$$

Hence,

$$\pi^B_{\alpha\beta} = -\frac{1}{\rho}[\overline{\Pi^{(1)}_{\alpha\beta}} + \overline{\Pi^{(2)}_{\alpha\beta}} + \Delta_{\alpha\beta}], \quad q^B_\alpha = -\frac{1}{\rho}[\Omega^{(1)}_\alpha + \Omega^{(2)}_\alpha + \Delta_\alpha], \tag{7.4.11}$$

where

$$\begin{aligned}
\Pi^{(1)}_{\alpha\beta} &= A\frac{\partial}{\partial x_\alpha}\left(\frac{1}{\rho}\frac{\partial p}{\partial x_\beta} - X_\beta\right) - \frac{\partial}{\partial x_\alpha}C\frac{\partial T}{\partial x_\beta}, \quad p = \rho T,\\
\Pi^{(2)}_{\alpha\beta} &= A\left[\frac{\partial u_\alpha}{\partial x_\gamma}\frac{\partial u_\gamma}{\partial x_\beta} - \alpha(T)\frac{\partial u_\alpha}{\partial x_\beta}\,\mathrm{div}\,u\right],\\
\Omega^{(1)}_\alpha &= \frac{2B}{3}\frac{\partial}{\partial x_\alpha}T\,\mathrm{div}\,u - \frac{\partial}{\partial x_\beta}C\,T\overline{\frac{\partial u_\alpha}{\partial x_\beta}},\\
\Omega^{(2)}_\alpha &= B\frac{\partial T}{\partial x_\beta}\left[\frac{\partial u_\beta}{\partial x_\alpha} - \beta(T)\delta_{\alpha\beta}\,\mathrm{div}\,u\right];
\end{aligned} \tag{7.4.12}$$

other notations are given in equations (7.4.7)–(7.4.10).

The other terms can be evaluated in a similar way. We omit these calculations (see, e. g., [55] for details) and formulate the final result.

Proposition 7.4.1. *The Burnett terms in equations* (7.4.4) *can be written in the standard form* [74, 144]

$$\begin{aligned}
q^B_\alpha = \frac{\mu^2}{p}\bigg[&\gamma_1\frac{\partial T}{\partial x_\alpha}\,\mathrm{div}\,u + 2\gamma_2\left(\frac{1}{3}\frac{\partial}{\partial x_\alpha}T\,\mathrm{div}\,u + \frac{\partial T}{\partial x_\beta}\frac{\partial u_\beta}{\partial x_\alpha}\right)\\
&+\left(\frac{\gamma_3}{\rho}\frac{\partial p}{\partial x_\beta} + \gamma_4 T\frac{\partial}{\partial x_\beta} + \gamma_5\frac{\partial T}{\partial x_\beta}\right)\overline{\frac{\partial u_\alpha}{\partial x_\beta}}\bigg], \quad p = \rho T,
\end{aligned} \tag{7.4.13}$$

$$\begin{aligned}
\pi^B_{\alpha\beta} = \frac{\mu^2}{p}\bigg\{&\omega_1\overline{\frac{\partial u_\alpha}{\partial x_\beta}}\,\mathrm{div}\,u\\
&-\omega_2\left[\overline{\frac{\partial}{\partial x_\alpha}\frac{1}{\rho}\frac{\partial p}{\partial x_\beta}} + \overline{\frac{\partial u_\alpha}{\partial x_\gamma}\frac{\partial u_\gamma}{\partial x_\beta}} + 2\overline{\frac{\partial u_\alpha}{\partial x_\gamma}\frac{\partial u_\gamma}{\partial x_\beta}}\right]\\
&+\omega_3\overline{\frac{\partial^2 T}{\partial x_\alpha \partial x_\beta}} + \frac{\omega_4}{\rho T}\overline{\frac{\partial p}{\partial x_\alpha}\frac{\partial T}{\partial x_\beta}} + \frac{\omega_5}{T}\overline{\frac{\partial T}{\partial x_\alpha}\frac{\partial T}{\partial x_\beta}} + \omega_6\overline{\overline{\frac{\partial u_\alpha}{\partial x_\gamma}}\,\overline{\frac{\partial u_\gamma}{\partial x_\beta}}}\bigg\},
\end{aligned} \tag{7.4.14}$$

where $p = \rho T$ *and* $\mu(T)$ *is a viscosity coefficient given in equations* (7.3.3). *The dimensionless coefficients* $\{\gamma_i,\ i = 1, \dots, 5;\ \omega_k,\ k = 1, \dots, 6\}$ *are given by the formulas*

$$\begin{aligned}
\gamma_1 &= \frac{T}{\mu^2}\left[D + \frac{5}{3}(B + H) - \frac{2}{3}T\frac{dB}{dT}\right],\\
\gamma_2 &= -\frac{T}{\mu^2}B, \quad \gamma_3 = -\frac{T}{\mu^2}G, \quad \gamma_4 = \frac{T}{\mu^2}C,\\
\gamma_5 &= \frac{T}{\mu^2}\left[2(B + H) - I - E + \frac{d}{dT}CT\right],
\end{aligned}$$

$$
\begin{aligned}
\omega_1 &= \frac{T}{\mu^2}\left[\frac{1}{3}\left(7A - 2T\frac{dA}{dT}\right) + P + K\right],\\
\omega_2 &= \frac{T}{\mu^2}A, \quad \omega_3 = \frac{T}{\mu^2}C, \quad \omega_4 = -\frac{T}{\mu^2}Q,\\
\omega_5 &= \frac{T}{\mu^2}\left(T\frac{dC}{dT} - R - J\right),\\
\omega_6 &= \frac{T}{\mu^2}\left(4A + \frac{12}{7}K - Y\right),
\end{aligned} \tag{7.4.15}
$$

where

$$
\begin{aligned}
A &= \frac{1}{5T}\langle\varphi_{\alpha\beta}^2, 1\rangle_M, \quad B = \frac{1}{3T^2}\langle\varphi_\alpha^2, 1\rangle_M, \quad C = \frac{1}{5T^2}\langle\varphi_{\alpha\beta}\, c_\alpha\, \varphi_\beta\rangle_M\,,\\
D &= \frac{2}{9T}\left\langle\varphi_\alpha, \frac{\partial\varphi_\alpha}{\partial T}\right\rangle_M, \quad E = \frac{1}{5T}\left\langle\varphi_{\alpha\beta}, c_\alpha\frac{\partial\varphi_\beta}{\partial T}\right\rangle_M,\\
G &= \frac{1}{5T}\left\langle\varphi_{\alpha\beta}, \frac{\partial\varphi_\alpha}{\partial c_\beta}\right\rangle_M, \quad P = \frac{2}{15}\left\langle\varphi_{\alpha\beta}, \frac{\partial\varphi_{\alpha\beta}}{\partial T}\right\rangle_M,\\
Q &= \frac{1}{5T}\left\langle\varphi_\alpha, \frac{\partial\varphi_{\alpha\beta}}{\partial c_\beta}\right\rangle_M, \quad R = \frac{1}{5T}\left\langle\varphi_\alpha, c_\beta\frac{\partial\varphi_{\alpha\beta}}{\partial T}\right\rangle_M,\\
H &= \frac{2}{15T^2}\langle\, |c|^4 g(|c|^2), g'(|c|^2)\,\rangle_M, \quad \varphi_\alpha(c) = c_\alpha g(|c|^2);\\
K &= \frac{4}{45T}\langle|c|^6 h((|c|^2), h'((|c|^2))\rangle_M, \quad \varphi_{\alpha\beta}(c) = \overline{c_\alpha c_\beta}\, h(|c|^2)\,.
\end{aligned} \tag{7.4.16}
$$

The integrals I, J, and Y depend on the collision term and read

$$
\begin{aligned}
I(T) &= \frac{1}{5T^3}\langle[Q(M\varphi_{\alpha\beta}, M\varphi_\beta) + Q(M\varphi_\beta, M\varphi_{\alpha\beta})], \varphi_\alpha\rangle\,,\\
J(T) &= \frac{1}{5T^3}\langle Q(M\varphi_\alpha, M\varphi_\beta), \varphi_{\alpha\beta}\rangle\,,\\
Y(T) &= \frac{9}{2T^2}\langle Q(M\varphi^{(2)}, M\varphi^{(2)}), \varphi^{(2)}\rangle\,.
\end{aligned} \tag{7.4.17}
$$

Finally, we note that equations (7.4.13) and (7.4.14) are given usually in the literature without general formulas for the coefficients (γ, ω). The coefficients are also absent in original papers [58, 59] by Burnett. He considered only terms related to the pressure tensor and did not consider the heat flux. To our knowledge, the only book that contains such formulas is the book [144], by Truesdell and Muncaster. Their notation and the way of calculation are quite different from ours, but the formulas are, of course, equivalent to our equations (7.4.13)–(7.4.16). For brevity we omit details of comparison.

The Burnett coefficients $\{\gamma_i,\ i = 1, \dots, 5;\ \omega_k,\ k = 1, \dots, 6\}$ are in the general case functions of temperature T. However, they are simply numbers in important cases of hard spheres and power-like potentials. These numbers are given by some integrals that can be evaluated approximately. The exact values are well known only for

Maxwell molecules [74]:

$$\gamma_1 = \frac{75}{8}, \quad \gamma_2 = -\frac{45}{8}, \quad \gamma_3 = -3, \quad \gamma_4 = 3, \quad \gamma_5 = \frac{117}{4};$$
$$\omega_1 = \frac{10}{3}, \quad \omega_2 = 2, \quad \omega_3 = 3, \quad \omega_4 = 0, \quad \omega_5 = 3, \quad \omega_6 = 8\,.$$

More information about properties of Burnett coefficients can be found in [55] and [136].

The complexity of the classical Burnett equations is not a big obstacle if one uses modern computers. Unfortunately there is another serious problem with these equations. It is discussed in the next section.

7.5 Ill-posedness of Burnett equations

In fact, formulas (7.4.11) and (7.4.12) are sufficient to explain why Burnett equations are ill-posed. Considering just the terms with higher derivatives, we transform equations (7.2.16), (7.4.4), (7.3.2), and (7.4.11) to

$$\rho_t = \cdots, \qquad \frac{\partial u_\alpha}{\partial t} = \frac{\varepsilon^2}{\rho^2}\overline{\frac{\partial \Pi^{(1)}_{\alpha\beta}}{\partial x_\beta}} + \cdots, \qquad T_t = \frac{2\,\varepsilon^2}{3\,\rho^2}\,\mathrm{div}\, S^{(1)} + \cdots, \tag{7.5.1}$$

where dots denote terms that do not contain third derivatives. Then, after simple calculations, we obtain

$$\rho_t = \cdots, \qquad u_t = \frac{2\,\varepsilon^2}{3\,\rho^2}\left[\frac{AT}{\rho}\Delta(\nabla\rho) + (A-C)\Delta(\nabla T)\right] + \cdots,$$
$$T_t = \left(\frac{2}{3}\right)^2 \frac{\varepsilon^2}{\rho^2}\, T(B-C)\,\Delta\,\mathrm{div}\, u + \cdots. \tag{7.5.2}$$

It is sufficient to consider one-dimensional solutions

$$\rho = \rho(x_1, t), \quad u = \{u_1(x_1,t), 0, 0\}, \quad T = T(x_1, t).$$

Then the matrix M of the coefficients for third derivatives reads

$$\begin{pmatrix} 0 & 0 & 0 \\ \frac{AT}{\rho} & 0 & A-C \\ 0 & \frac{2(B-C)T}{3} & 0 \end{pmatrix}. \tag{7.5.3}$$

Its non-zero eigenvalues are

$$\lambda_\pm = \pm\left[\frac{2}{3}T(B-C)(A-C)\right]^{1/2}.$$

Hence, under the obvious assumption $T > 0$, the hyperbolicity condition reads

$$\Im\lambda_{\pm} = 0, \quad \Leftrightarrow \quad (B - C)(A - C) \geq 0\,. \tag{7.5.4}$$

It is easy to verify that this condition is not fulfilled in most typical molecular models for the Boltzmann equation. In order to do this we use temporary notations

$$\Psi_{\alpha\beta}(c) = c_\alpha c_\beta - \frac{|c|^2}{3}\delta_{\alpha\beta}, \quad \Psi_\alpha(c) = \frac{c_\alpha}{2}(|c|^2 - 5T)$$

and assume that $\mu(T)$ and $\lambda(T)$ denote respectively the coefficients of viscosity and heat conductivity. It follows from equation (7.3.3) that

$$\mu(T) = -\frac{1}{10T}\langle\varphi_{\alpha\beta}, \Psi_{\alpha\beta}\rangle_M, \quad \lambda(T) = -\frac{1}{3T^2}\langle\varphi_\alpha, \Psi_\alpha\rangle_M.$$

The usual approximation [74] for functions $\varphi_{\alpha\beta}$ and φ_α from equations (7.2.8) is given by

$$\varphi_{\alpha\beta} \approx \gamma(T)\Psi_{\alpha\beta}(c), \quad \varphi_\alpha \approx \delta(T)\Psi_\alpha(c). \tag{7.5.5}$$

Then

$$\gamma(T) = \frac{\mu(T)}{T}, \quad \delta(T) = -\frac{2\lambda(T)}{5T}, \tag{7.5.6}$$

since

$$\langle\Psi_{\alpha\beta}, \Psi_{\alpha\beta}\rangle_M = 10T^2, \quad \langle\Psi_\alpha, \Psi_\alpha\rangle_M = \frac{15}{2}T^3.$$

The approximation (7.5.5) is exact for Maxwell molecules; moreover, $\lambda(T) = 15\mu(T)/4$ in that case. By using equations (7.5.5) and (7.5.6), we evaluate the integrals (7.4.9) and obtain

$$A = 2\frac{\mu^2}{T}, \quad B = \frac{2\lambda^2}{5T}, \quad C = \frac{4\lambda\mu}{5T}.$$

Therefore,

$$A - C = \frac{4\lambda\mu}{5T}\left(\frac{5\mu}{2\lambda} - 1\right), \quad B - C = \frac{2\lambda^2}{5T}\left(1 - \frac{\mu}{\lambda}\right).$$

The ratio

$$\mathrm{Pr} = \frac{5}{2}\frac{\mu(T)}{\lambda(T)}$$

is called in fluid mechanics the Prandtl number (for monoatomic gases) [103]. It is well known that the approximate equality $\mathrm{Pr} \simeq 2/3$ (exact for Maxwell molecules) holds for all typical molecular models: hard spheres, etc. On the other hand, the hyperbolicity condition (7.5.4) can be approximately (under assumption (7.5.5)) written as

$$1 \le \mathrm{Pr} \le 5/4.$$

The realistic value $\mathrm{Pr} = 2/3$ obviously violates this condition. Therefore, the Burnett equations *are probably ill-posed for all typical molecular models*, though our proof is rigorous just for Maxwell molecules. All above formulas are exact in this case. We note that $\mathrm{Pr} = 1$ for the Bhatnagar–Gross–Krook (BGK) model; however, this model is too unrealistic (see e. g. the book [70]).

It should be stressed that we need something more than hyperbolicity for any reasonable modification of Burnett equations. It is clear that any constant solution $\{\rho = \rho_0 > 0,\ \ u = 0,\ \ T = T_0 > 0\}$ must be stable with respect to small perturbations. Such stable modifications are discussed in Sections 7.6–7.10.

7.6 General method of regularization

We begin with formulation of the problem. It is well known (see, in particular, Sections 7.2 and 7.3) that the classical Chapman–Enskog method allows to obtain formally closed "equations of hydrodynamics" in the form of symbolic power series

$$\frac{\partial}{\partial t}\begin{Bmatrix}\rho\\ u\\ T\end{Bmatrix} = (A_0^E + \varepsilon A_1^{\mathrm{NS}} + \varepsilon^2 A_2^B + \cdots)\begin{Bmatrix}\rho\\ u\\ T\end{Bmatrix}, \tag{7.6.1}$$

where all $A_k^{\cdots}, k = 0, 1, 2, \ldots$, are non-linear operators applied to the "hydrodynamic vector" $(\rho, u, T)^{\mathrm{Tr}}$. The first two operators A_0^E and A_1^{NS} correspond to Euler and Navier–Stokes equations. These are most important for gas dynamics sets of equations. However, if we add the third operator A_2^B we are in trouble, since Burnett equations are ill-posed. This was noticed in 1982 [23], and since then several interesting general approaches to deal with this problem have been proposed. We confine ourselves in this section to an alternative procedure of truncation of the Chapman–Enskog series (7.6.1). This approach was proposed in [31, 32]; its idea is very simple. We consider an abstract evolution equation for a vector y from some Banach space Y (or simply a vector ordinary differential equation [ODE] in the case of finite-dimensional space Y) having the form similar to equation (7.6.1):

$$\frac{dy}{dt} = T(y; \varepsilon) = A(y) + \varepsilon B(y) + \varepsilon^2 C(y) + \cdots, \tag{7.6.2}$$

where $y \in Y$; $A(y), B(y), \ldots$, are differentiable non-linear operators acting in Y. In other words, we assume the existence of Frechet derivatives (linear operators) $A'(y)$ such that $A(y + h) - A(y) = A'(y)h + O(\|h\|^2)$ with standard notation for a norm in Y [106].

Let us assume that the usual way of truncation of equations (7.6.2) based on the rule "Neglect all terms of order $O(\varepsilon^{n+1})$" does not work for $n = 2$, since the corresponding approximate equation

$$y_t = A(y) + \varepsilon B(y) + \varepsilon^2 C(y) \tag{7.6.3}$$

does not have any reasonable solutions. What then? An alternative way of truncation is to use a change of variables (all operators are assumed to be time-independent)

$$z = y + \varepsilon^2 R(y). \tag{7.6.4}$$

Then $y \simeq z - \varepsilon^2 R(z)$ and simple calculations yield

$$z_t = A(z) + \varepsilon B(z) + \varepsilon^2 \{C(z) + [R, A](z)\} + O(\varepsilon^3),$$

where $[R, A] = R'A - A'R$ (do not confuse with the standard notation for commutators of two operators when the operators A and R are non-linear). In fact, this notation becomes the usual one $[A, R] = AR - RA$ in the case of linear operators A and R, since $A' = A$ for linear operator A. The result can be expressed in the following form.

Proposition 7.6.1. *Any truncated equation*

$$y_t = A(y) + \varepsilon B(y) + \varepsilon^2 C(y) + \cdots \tag{7.6.5}$$

is formally equivalent to a family of equations

$$z_t = A(z) + \varepsilon B(z) + \varepsilon^2 \tilde{C}(z) + \cdots, \tag{7.6.6}$$

where

$$z = y + \varepsilon^2 R(y) \iff y = z - \varepsilon^2 R(z) + \cdots, \quad \tilde{C}(z) = C(z) + R'(z)A(z) - A'(z)R(z)$$

with any differentiable operator R.

Remark. Note that even classical Navier–Stokes equations are not defined uniquely. Only Euler equations are unique.

Thus we can apply this scheme to Chapman–Enskog series (7.6.1) truncated for $n = 2$ (classical Burnett equations) and obtain a family of Burnett-type equations which depend on an arbitrary operator R. It was shown in the first publication [31] of this approach that it is possible to choose such R that the corresponding Burnett-type equations become well-posed. However, the choice of R is obviously not unique. Then the next question is to choose the regularizing operator R in some "optimal" way. An attempt to do it in papers [32, 33] resulted in the derivation of *new Burnett-type equations that were called the Generalized Burnett equations (GBEs)*. GBEs were later applied to some physical problems: shock waves with moderate Mach numbers [35], half-space problems [16], and others (see Section 7.10). Yet it is difficult to prove that GBEs represent the best possible variant of hydrodynamic equations at the Burnett level. It is even more difficult to give a mathematically rigorous proof that these equations do improve the Navier–Stokes results for small Knudsen numbers. On the other hand, all these problems can be solved if we assume that the physical system under consideration (rarefied gas) is close to equilibrium. Then we can replace formally the Boltzmann equation (7.6.1) by its linearized version and apply the above approach. This will be done in the next section.

7.7 Linearized problem

The notations in Sections 7.7–7.9 based on the paper [34] are slightly different from the rest of the book.

We consider equation (7.1.9) in dimensionless form, make a substitution

$$f = m + g\sqrt{m}, \quad m = (2\pi)^{-3/2}\exp\left(-\frac{|v|^2}{2}\right), \tag{7.7.1}$$

and formally neglect the non-linear term. Then we obtain the linearized Boltzmann equation and consider the corresponding initial value problem for $g(x, v, t)$ for $t > 0$:

$$\begin{aligned} g_t + v\cdot g_x &= \frac{1}{\varepsilon}Lg, \quad Lg = m^{-1/2}[Q(m, gm^{1/2}) + Q(gm^{1/2}, m)], \\ g|_{t=0} &= g_0;\ x \in \mathbb{R}^3,\ v \in \mathbb{R}^3. \end{aligned} \tag{7.7.2}$$

We introduce the standard notation for scalar product and norm in the (complex) Hilbert space $H = L_2(\mathbb{R}^3)$, adding a natural restriction on the initial data:

$$\langle g_1, g_2\rangle = \int_{\mathbb{R}^3} dv g_1(v) g_2^*(v), \quad \|g\|^2 = \langle g, g\rangle, \quad \int_{\mathbb{R}^3} dx \|g_0\|^2 < \infty, \tag{7.7.3}$$

where the star denotes the complex conjugate value. This problem is very well studied in the literature (see in particular [7, 78]). We shall see below a connection of its solution with our approach based on Proposition 7.6.1 applied to the corresponding Chapman–Enskog series. For the sake of simplicity we consider below only one-dimensional flows, assuming that $x \in \mathbb{R}^1$ instead of $x \in \mathbb{R}^3$ in (7.7.2) and (7.7.3). We also consider only axially symmetric flows such that g depends on $v \in \mathbb{R}^3$ only through v_x and $|v|$. Note that the null-space of L is given by

$$\begin{aligned} N(L) &= \{g \in L_2(\mathbb{R}^3),\ Lg = 0\} \\ &= \operatorname{Span}\{\sqrt{m},\ v_x\sqrt{m},\ |v|^2\sqrt{m},\}, \quad \dim N(L) = 3. \end{aligned} \tag{7.7.4}$$

Here and below, we ignore irrelevant "perpendicular" modes and consider only functions of v_x and $|v|$. Then we introduce hydrodynamic variables

$$\tilde{\rho} = \langle g, \sqrt{m}\rangle, \quad \tilde{u} = \langle g, v_x\sqrt{m}\rangle, \quad \tilde{T} = \frac{1}{3}\langle g, (|v|^2 - 3)\sqrt{m}\rangle, \tag{7.7.5}$$

which are in agreement with initial hydrodynamic variables (7.1.2) and equation (7.7.1) such that $\rho \approx 1 + \tilde{\rho}$, $u \approx \tilde{u}$, $T \approx 1 + \tilde{T}$. Tildes are omitted below. It is easy to check that in the limit $\varepsilon = 0$, we obtain the linearized Euler equations

$$\rho_t + u_x = 0, \quad u_t + \rho_x + T_x = 0, \quad T_t + \frac{2}{3} u_x = 0. \tag{7.7.6}$$

The standard Chapman–Enskog expansion for equation (7.7.2) yields the formal linear partial differential equation for the hydrodynamic vector $z = (\rho, u, T)^{\mathrm{Tr}}$. We obtain

$$\underbrace{z_t + M_0 z_x}_{\text{Euler}} = \underbrace{\varepsilon M_1 z_{xx}}_{\text{Navier–Stokes}} + \underbrace{\varepsilon^2 M_2 z_{xxx}}_{\text{Burnett}} + \cdots, \tag{7.7.7}$$

where M_i, $i = 0, 1, 2, \ldots$, are constant real (3×3) matrices, defined uniquely by the Chapman–Enskog method. The matrix M_0 corresponds to Euler equations (7.7.6); M_0 is independent of intermolecular forces and has the following entries $m_{ij}^{(0)}$ $(i, j = 1, 2, 3)$:

$$m_{11}^{(0)} = m_{22}^{(0)} = m_{33}^{(0)} = m_{13}^{(0)} = m_{31}^{(0)} = 0,$$
$$m_{12}^{(0)} = m_{21}^{(0)} = m_{23}^{(0)} = 1, \quad m_{32}^{(0)} = 2/3\,.$$

Note that matrices M_i with $i = 1, 2, \ldots$ depend on intermolecular forces. If we replace dots in equation (7.7.7) by zero, then the result coincides with the classical (ill-posed) Burnett equations. Nevertheless, the following statement explains how we can use these "bad" equations.

Proposition 7.7.1. *There exists a transformation*

$$y = R_0 z + \varepsilon R_1 z_x + \varepsilon^2 R_2 z_{xx}, \tag{7.7.8}$$

where $R_{0,1,2}$ are constant 3×3 matrices, that leads (formally) to diagonal Burnett equations

$$\begin{aligned} &y_t + c_0 \Lambda_0 (y_x + \alpha \varepsilon^2 y_{xxx}) = \beta \varepsilon \Lambda_1 y_{xx} + O(\varepsilon^3), \\ &c_0 = \sqrt{5/3}, \quad \Lambda_0 = \operatorname{diag}(1, -1, 0), \quad \Lambda_1 = (1, 1, \gamma), \end{aligned} \tag{7.7.9}$$

with some positive numbers α, β, γ.

Proof. We perform the Fourier transform of truncated equations (7.7.7) and obtain the following vector ODE:

$$\hat{z}_t + ikM_0 \hat{z} = -\varepsilon k^2 M_1 \hat{z} + \varepsilon^2 (ik)^3 M_2 \hat{z}, \tag{7.7.10}$$

where

$$\hat{z} = (\hat{\rho}, \hat{u}, \hat{T})^{\mathrm{Tr}}, \quad \hat{\rho}(k,t) = \mathfrak{F}(\rho) = \int_{-\infty}^{\infty} dx e^{-ikx} \rho(x,t), \quad \hat{u} = \mathfrak{F}(u), \quad \hat{T} = \mathfrak{F}(T). \tag{7.7.11}$$

Note that the matrix M_0 has three distinct eigenvalues:

$$\lambda_0 = 0, \quad \lambda_\pm = \pm\sqrt{5/3} = \pm c_0. \tag{7.7.12}$$

We introduce new variables

$$s = \hat{\rho} - \frac{3}{2}\hat{T}, \quad w_\pm = \hat{\rho} + \hat{T} \pm c_0\hat{u} \tag{7.7.13}$$

and obtain the following equation for vector $w = (w_+, w_-, s)^{\mathrm{Tr}}$:

$$w_t + ikc_0\Lambda_0 w = -\varepsilon k^2 \bar{M}_1 w + \varepsilon^3 (ik)^3 \bar{M}_2 w, \quad \Lambda_0 = \mathrm{diag}\,(1, -1, 0), \tag{7.7.14}$$

with some new constant real matrices $\bar{M}_{1,2}$. To complete the proof of Proposition 7.7.1 we use the following statement.

Lemma 7.7.1. *If $n \ge 2$, A is any $(n \times n)$ matrix and $w(t) \in \mathbb{R}^n$ satisfies the equation*

$$w_t + \Lambda w = \varepsilon A w, \quad \Lambda = \mathrm{diag}(\lambda_1, \dots, \lambda_n), \quad \lambda_i \ne \lambda_j \quad \textit{for} \quad i \ne j, \quad n \ge 2,$$

then there is a unique $(n \times n)$ matrix B such that (a) $\mathrm{diag}\, B = 0$ *and (b) the substitution $w = y + \varepsilon B y$ leads to the equation*

$$y_t + \Lambda y = \varepsilon(\mathrm{diag}\, A) y + O(\varepsilon^2).$$

Proof. Note that $y = (1 + \varepsilon B)^{-1}$ and therefore $y_t + \Lambda y = \varepsilon(A + [B, \Lambda]) y + O(\varepsilon^2)$.

Let $A = \{a_{ij}\}$, $B = \{b_{ij}\}$. Then $[B, \Lambda]_{ij} = (B\Lambda - \Lambda B)_{ij} = b_{ij}(\lambda_j - \lambda_i)$. Hence, we obtain

$$B = \left\{ b_{ij} = \frac{a_{ij}}{\lambda_i - \lambda_j},\ i \ne j;\ b_{ii} = 0;\ i, j = 1, \dots, n \right\},$$

and this completes the proof. □

Now we can easily complete the proof of Proposition 7.7.1 by applying the same lemma to the equation with a new diagonal operator $\Lambda_1 = \Lambda - \varepsilon\ \mathrm{diag}\ A$. Then we can "kill" all non-diagonal terms of order $O(\varepsilon^2)$. For brevity we omit straightforward calculations which show that the resulting diagonal equations coincide precisely with (7.7.9) (see also Section 7.9 below). The non-diagonal part of modified equations (7.7.14) will have the third order in ε and therefore can be neglected. This completes the proof of the Fourier transformed version of Proposition 7.7.1 for sufficiently small values of $\varepsilon|k|$. On the other hand, it is sufficient for the proof of Proposition 7.7.1, since it considers the diagonalizing transformation only at the formal level. Therefore, the proof of Proposition 7.7.1 is complete. □

Remark. It is worth to note that Proposition 7.7.1 automatically solves the above discussed problem of non-uniqueness of regularizing transformation for Burnett equations. The "optimal" transformation is supposed to not only make the equations well-posed, but also simplify them as much as possible. In our case it makes them diagonal,

i. e., transforms them to three independent equations: two Burgers–KdV equations for two sound modes and one heat equation for the thermal mode. Nothing simpler can be expected at the Burnett level.

In the next section we shall clarify a connection of the above transformation with asymptotics for small ε of solutions of the Cauchy problem (7.7.1).

7.8 Asymptotic expansion for small Knudsen numbers

We consider the problem (7.7.2) and pass to the Fourier representation

$$\hat{g}(k,v,t) = \int_{-\infty}^{\infty} dx e^{-ikx} g(x,v,t), \quad \hat{g}_t + ikv_x\hat{g} = \frac{1}{\varepsilon}L\hat{g},$$
$$\hat{g}|_{t=0} = \hat{g}_0, \quad k \in \mathbb{R}. \tag{7.8.1}$$

Note that we consider solutions $\hat{g}(k,v,t)$ that depend on v only through v_x and $|v|$. The null-space $N(L)$ (7.7.4) of the self-adjoint operator L is three-dimensional. It is convenient to introduce a special orthonormal basis in $N(L)$:

$$e_1 = \sqrt{\frac{m}{30}}(|v|^2 + v_x\sqrt{15}), \quad e_2 = \sqrt{\frac{m}{30}}(|v|^2 - v_x\sqrt{15}),$$
$$e_3 = \sqrt{\frac{m}{6}}(|v|^2 - 3). \tag{7.8.2}$$

It is easy to verify that

$$\langle e_i, e_j\rangle = \delta_{ij}, \quad \langle v_x\, e_i,\, e_j\rangle = \gamma_i\,\delta_{ij}, \quad i,j = 1,2,3;$$
$$\gamma_1 = -\gamma_2 = \sqrt{\frac{5}{3}}, \quad \gamma_3 = 0. \tag{7.8.3}$$

The solution of the Cauchy problem (7.8.1) can be written as

$$\hat{g}(k,v,t) = \exp\left[\frac{t}{\varepsilon}A(i\varepsilon k)\right]\hat{g}_0(k,v), \quad A(\mu) = L - \mu v_x, \quad \mu \in \mathcal{C}. \tag{7.8.4}$$

The existence and uniqueness of the solution for any fixed $\varepsilon > 0$ and real k follows from general theory of semi-groups under very general restrictions on the scattering cross-section in the Boltzmann collision integral (see [7, 8] for hard spheres and hard potentials with compact support and [78] for the more general class of cross-sections). Moreover, the corresponding semi-group is contracting (it follows from dissipativity of L: $\langle g, Lg\rangle \le 0$ for any "nice" real functions $g(v)$) and therefore

$$\|\hat{g}(k,v,t)\| \le \|\hat{g}_0(k,v)\|, \tag{7.8.5}$$

in the notation of (7.7.3). We shall consider below such initial conditions that $\|\hat{g}_0(k,v)\|$ is bounded on any compact set in k-space and

$$\int_{-\infty}^{\infty} dk(1+|k|^N)\,\|\hat{g}_0(k,v)\| < \infty \tag{7.8.6}$$

with some $N > 0$ (to be fixed below). Then we can use the inverse Fourier transform and obtain

$$g(x,v,t) = \frac{1}{2\pi}\int_{-\infty}^{\infty} dke^{ikx}\hat{g}(k,v,t), \quad x \in \mathbb{R}. \tag{7.8.7}$$

For any fixed real positive a we can split this integral into two parts,

$$g(x,v,t) = I^{<}(a) + I^{>}(a), \quad I^{<}(a) = \int_{\varepsilon|k|<a} \cdots, \quad I^{>}(a) = \int_{\varepsilon|k|\geq a} \cdots. \tag{7.8.8}$$

The second integral can be estimated as follows:

$$\begin{aligned}\|I^{>}(a)\| &\leq \frac{1}{2\pi}\int_{\varepsilon|k|\geq a} dk\|\hat{g}_0(k,v)\| \\ &\leq \frac{\varepsilon^N}{2\pi a^N}\int_{-\infty}^{\infty} dk|k|^N\|\hat{g}_0(k,v)\| = O(\varepsilon^N),\end{aligned} \tag{7.8.9}$$

where N is the same as in inequality (7.8.6). For our goals it is sufficient to have $N = 4$. It remains to evaluate the first integral. For brevity we assume that we consider the case of intermolecular potential with compact support (e. g., hard spheres). Then we can use the following result by A. A. Arsen'ev [7] (simplified for our axially symmetric case Lemmas 13 and 15; see also [102]).

Lemma 7.8.1. *There exists a number $r > 0$ such that*
(i) *the eigenvalue problem*

$$\lambda\phi(v;\mu) = A(\mu)\varphi(v;\mu), \quad |\mu| \leq r, \tag{7.8.10}$$

has exactly three linearly independent solutions λ_j, ϕ_j, $j = 1,2,3$, having the following form of convergent power series:

$$\begin{aligned}&\lambda_j = \lambda_j(\mu) = \sum_{n=1}^{\infty}\mu^n\lambda_n^{(j)}, \quad \varphi_j(v;\mu) = e_j(v) + \sum_{n=1}^{\infty}\mu^n\varphi_n^{(j)}(v), \\ &\langle e_j, \varphi_n^{(j)}\rangle = 0, \quad j = 1,2,3,\end{aligned} \tag{7.8.11}$$

in the notation of (7.8.2);

(ii) *the solution* $\hat{g}(k,v,t)$ *of the problem* (7.8.1), (7.8.6) *for real* k, *such that* $\varepsilon|k| < r$, *reads*

$$\hat{g}(k,v,t) = \sum_{j=1}^{3} \exp\left\{\lambda_j(i\varepsilon k)\frac{t}{\varepsilon}\right\} P_j(i\varepsilon k)\, g_0(k,v) + \exp\left\{-b(r)\frac{t}{\varepsilon}\right\} B(\varepsilon k,t)\, g_0(k,v), \tag{7.8.12}$$

where $b(r) > 0$, B *is a bounded operator such that* $\|B(\varepsilon,k)\| < C(r) < \infty$. *Operators* P_j, $j = 1,2,3$, *are orthogonal projectors on corresponding subspaces of A (convergent power series in* $\varepsilon|k|$*).*

Hence, we obtain the following general asymptotic formula valid for almost all $x \in \mathbb{R}^1$:

$$g(x,v,t) = g_{as}(x,v,t) + O\left(\exp\left\{-b(r)\frac{t}{\varepsilon}\right\}\right) + O(\varepsilon^N),$$
$$g_{as}(x,v,t) = \sum_{j=1}^{3} \int_{\varepsilon|k|<r} \frac{dk}{2\pi} \exp\left\{ikx + \frac{t}{\varepsilon}\lambda_j(ik\varepsilon)\right\} P_j(ik\varepsilon)\, g_0(k,v), \tag{7.8.13}$$

where N is the same as in (7.8.6) and the constants in O-symbols are bounded in norm (7.7.3) uniformly with respect to $x \in \mathbb{R}^1$. It is also possible to prove (see Lemma 12 in [7]) that the second term from (7.8.8) is in fact of order $O[\exp(-ct/\varepsilon)]$, $c > 0$, but the simplified estimate (7.8.9) is sufficient for our goals. The constant r is not very important; the same formula (7.7.13) is valid for any independent of ε fixed constant $r' < r$. The final step is to consider various approximations of the asymptotic formula (7.7.13). We consider equations (7.7.11) and truncate the power series there.

Let

$$\lambda_j^{[n]}(\mu) = \sum_{k=1}^{n} \mu^k \lambda_k^{(j)}, \quad P_j^{[l]}(\mu) = \sum_{k=1}^{l} \mu^k P_k^{(j)},$$
$$g_{n,l}(x,v,t) = \sum_{j=1}^{3} \int_{\varepsilon|k|<r} \frac{dk}{2\pi} \exp\left\{ikx + \frac{t}{\varepsilon}\lambda_j^{[n]}(ik\varepsilon)\right\} P_j^{[l]}(ik\varepsilon)\, \hat{g}_0(k,v), \tag{7.8.14}$$

where the connection of series (7.7.11) for eigenfunctions φ_j and corresponding projectors P_j will be explained below. Then by repeating almost without changes, but with straightforward calculations from [7] (pp. 878–882), we obtain the following result.

Proposition 7.8.1. *If* $g_0(x,v)$ *satisfies conditions*

$$\int_{-\infty}^{\infty} dx \|g_0(k,v)\|^2 < \infty,$$

$$\int_{-\infty}^{\infty} dk(1+|k|^N)\,\|\hat{g}_0(k,v)\|, \quad \hat{g}_0 = \mathfrak{F}(g_0), \quad N \geq 1, \tag{7.8.15}$$

then for all integer $n < N$ and $l < N$ the following approximate formula is valid for the solution $g(x,v,t)$ of the Cauchy problem (7.7.2)*:*

$$\|g(x,v,t) - g_{n,l}(x,v,t)\| \leq C\left(\varepsilon^{n-1} + \varepsilon^{l+1} + e^{-b\frac{t}{\varepsilon}}\right), \tag{7.8.16}$$

where $C > 0$ and $b > 0$ are independent of x and t.

In order to clarify the series for projectors $P(\mu)$ we consider the eigenvalue problem (7.8.10) for real values of μ (note that the dependence of μ is analytic in some neighborhood of zero for potentials with compact support [7]). Then the operator $A = L - \mu v_x$ is real and self-adjoined in H. The eigenfunctions $\phi_j(v)$, $j = 1,2,3$, are pairwise orthogonal. Therefore, we have a simple formula

$$P_j(\mu)g(v) = \langle g, \varphi_j\rangle \frac{\varphi_j(v;\mu)}{\|\varphi_j\|^2}, \quad \Im\,\mu = 0, \quad j = 1,2,3. \tag{7.8.17}$$

The series for $P_j(ik\varepsilon)$ can be obtained by complex continuation in the μ variable. We shall see below that the solutions of diagonal Burnett equations (7.7.10) approximate the hydrodynamic quantities with uniform in time error $O(\varepsilon^2)$, whereas the corresponding error for diagonal Navier–Stokes equations has the lower order $O(\varepsilon)$.

7.9 Accuracy of equations of hydrodynamics and connection with the Chapman–Enskog expansion

We introduce a hydrodynamic vector

$$w = (w_1, w_2, w_3))^{\mathrm{Tr}}, \quad w_i(x,t) = \langle g(x,v,t), e_i\rangle, \quad i = 1,2,3, \tag{7.9.1}$$

where e_1, e_2, e_3 are vectors (7.8.2) of the orthonormal basis in $N(L)$ satisfying conditions (7.8.3). Note that all components of w are linear combinations of initial hydrodynamic variables (7.7.6). Since $|\langle h_1, h_2\rangle| \leq \|h_1\|\|h_2\|$ for any functions $h_{1,2}(v)$, we can make the following conclusions from Proposition 7.8.1: Under conditions of Proposition 7.8.1 we can construct corresponding approximations of the hydrodynamic vector w in the form

$$w^{n,l} = \left(\langle g_{n,l}, e_1\rangle, \langle g_{n,l}, e_2\rangle, \langle g_{n,l}, e_3\rangle\right)^{\mathrm{Tr}}, \tag{7.9.2}$$

where n and l are the same as in Proposition 7.8.1 and the functions $g_{n,l}(x,v,t)$ are given in (7.8.14). Then of course we have also the same order of error (see (7.8.16)), i. e.,

$$|w(x,t) - w^{n,l}(x,t)| \leq C(\varepsilon^{n-1} + \varepsilon^{l+1} + e^{-b\frac{t}{\varepsilon}}). \tag{7.9.3}$$

Here is a delicate point which shows a principal difference between Burnett's level and higher levels for equations of hydrodynamics. It is clear from above considerations that first we need to transform hydrodynamic equations of any level to diagonal form (like it was done with Burnett equations in Proposition 7.7.1). For brevity we assume that we use for these equations correct initial conditions of right order $l = \mathrm{Max}(n-2, 0)$. Then we can expect that the solutions of the diagonal equations will be very close to $w^{(n,l)}(x,t)$ and have the same order of approximation as in (7.8.16). Generally speaking, this is not true. The matter is that the integrals in (7.8.16) depend on a constant $r > 0$ and do not coincide with usual formulas for inverse Fourier transform (see, e. g., (7.8.7)). To estimate this additional error in the modified formula (7.8.16) we need to estimate integrals like

$$J_m^{(n)}(r) = \int_{r/\varepsilon}^{\infty} dk \; \exp\left[\frac{t}{\varepsilon}\mathbb{R}\,\lambda_m^{[n]}(i\varepsilon k)\right] \|\hat{g}_0(k,v)\|,$$

$$\mathbb{R}\,\lambda_m^{[n]}(i\varepsilon k) = \sum_{p=1}^{[n/2]} (i\varepsilon k)^{2p}\lambda_m^{(2p)}, \tag{7.9.4}$$

where $n \geq 2$, $m = 1,2,3$. It is well known that $\lambda_{1,2,3}^{(2)} > 0$ for any molecular model in the Boltzmann collision operator L. Therefore, we obtain $J_m^{(2)} = O[\exp(-ct/\varepsilon)]$, $c > 0$. Hence, this additional error is negligible at the Navier–Stokes and Burnett levels. However, it may happen for some molecular model that $\lambda_m^{(4)} > 0$ for some $m = 1, 2$, or 3. Then the corresponding (diagonal!) super-Burnett equations are ill-posed and cannot be used directly. On the other hand, their formal Fourier transform remains useful since the vector $w^{4,2}$ given in equations (7.8.14) and (7.9.2) (with integrals over a compact domain in k-space) approximates the true solution with error of order $O(\varepsilon^3)$.

After some standard calculations, we obtain the following explicit formulas for components w_j, $j = 1,2,3$, of the hydrodynamic vector $w(x,t)$ at Navier–Stokes ($n = 2, l = 0$) and Burnett ($n = 3, l = 1$) levels:

$$w_j^{\mathrm{NS}} = \frac{1}{2\pi}\int_{-\infty}^{\infty} dk e^{ikx}\langle \hat{g}_{2,0}(k,t), e_j\rangle,$$

$$w_j^{B} = \int_{-\infty}^{\infty} dk e^{ikx}\langle \hat{g}_{3,1}(k,t), e_j\rangle; \quad j = 1,2,3, \tag{7.9.5}$$

where

$$\langle \hat{g}_{2,0}(k,t), e_j\rangle = \exp\left\{\lambda_j^{[2]}\frac{t}{\varepsilon}\right\}\langle \hat{g}_0(k), e_j\rangle,$$

$$\lambda_j^{[2]} = -i\varepsilon k\gamma_j + \varepsilon^2|k|^2\langle (v_x - \gamma_j)e_j, \psi_j\rangle, \quad \psi_j = L^{-1}(v_x - \gamma_j)e_j, \tag{7.9.6}$$

with usual definition of the inverse operator L^{-1} acting in the orthogonal complement H_1 of $N(L)$. At the Burnett level, we obtain

$$\langle \hat{g}_{3,1}(k,t), e_j\rangle = (\langle \hat{g}_0, e_j\rangle + ik\varepsilon\langle \hat{g}_0, \varphi_j^{(1)}\rangle)\exp\left\{\lambda_j^{[3]}\frac{t}{\varepsilon}\right\}$$
$$+ ik\varepsilon \sum_{l=1}^{3\,(l\neq j)} \langle \hat{g}_0, e_l\rangle \rho_{l,j} \exp\left\{\lambda_l^{[3]}\frac{t}{\varepsilon}\right\}, \qquad (7.9.7)$$

where

$$\rho_{l,j} = \frac{1}{\gamma_l - \gamma_j}\langle v_x \psi_l, e_j\rangle, \quad \varphi_j^{(1)} = \psi_j + \sum_{l=1}^{3\,(l\neq j)} \rho_{j,l} e_l,$$
$$\lambda_j^{[3]} = \lambda_j^{[2]} + i\varepsilon^3 k^3 \langle \psi_j, (v_x - \gamma_j)\psi_j^{(1)}\rangle; \quad l, j = 1,2,3, \quad l \neq j. \qquad (7.9.8)$$

It is clear that functions (7.9.5) can be expressed through solutions of diagonal equations of hydrodynamics with appropriate initial data. The next statement directly follows from equations (7.9.3) and (7.9.5).

Proposition 7.9.1. *The above solutions of diagonal equations of hydrodynamics satisfy the estimates*

$$|w^{N\text{-}S}(x,t) - w(x,t)| < C_1\varepsilon,$$
$$|w^{B}(x,t) - w(x,t)| < C_2\varepsilon^2, \quad C_{1,2} = \text{const.}, \qquad (7.9.9)$$

where $w(x,t)$ is given in equations (7.9.2) and $t \geq \varepsilon^{1+\delta}$ with any $\delta > 0$.

Proof. Indeed it follows from (7.8.14), (7.9.2), and (7.9.5) that

$$|w^{N\text{-}S}(x,t) - w^{2,0}(x,t)| \leq O\Big[\max_{1\leq m\leq 3} J_m^{(2)}(r)\Big] = O\Big[\exp\Big(-\frac{ct}{\varepsilon}\Big)\Big], \quad c > 0,$$

in the notation of (7.9.4). Exactly the same estimate is valid for $|w^B(x,t) - w^{3,1}(x,t)|$. Then we combine these estimates with inequalities (7.9.3), and this completes the proof. □

Inequalities (7.9.9) clearly show that the Burnett equations do improve the results obtained at the Navier–Stokes level provided that we use them in the correct way. Finally, we briefly discuss the general structure of the Chapman–Enskog expansion (7.7.7) in the linearized case considered above. If we consider the problem (7.7.2), where $x \in \mathbb{R}^3$ and g depends on $v \in \mathbb{R}^3$ only through v_x and $|v|$, and choose the hydrodynamic variables in the form (7.9.1), then we obtain the Fourier transformed Chapman–Enskog series in the form

$$\hat{w}_t = U\hat{w} = \left(\sum_{n=0}^{\infty} \varepsilon^n U_n\right)\hat{w}, \quad U_0 = i\sqrt{\frac{5}{3}}\,|k|\ \operatorname{diag}(1,-1,0),$$
$$U_1 = -|k|^2 U^{\mathrm{NS}}, \quad U_2 = (i|k|)^3 U^B, \dots. \qquad (7.9.10)$$

What can be said about the general structure of the matrix U?

Proposition 7.9.2. *The following identities are valid:*

$$U = B\Lambda B^{-1}, \quad \Lambda = \frac{1}{\varepsilon}\operatorname{diag}\{\lambda_1(i\varepsilon k), \dots, \lambda_3(i\varepsilon k)\},$$
$$B = \{b_{ij};\, i,j = 1,\dots,3\}, \quad b_{ij} = \langle \varphi_j(i\varepsilon k), e_i \rangle. \tag{7.9.11}$$

The series

$$\lambda_j = \sum_{j=1}^{\infty} \lambda_n^{(j)}(i\varepsilon k)^n; \quad b_{ij} = \delta_{ij} + \sum_{n=1}^{\infty} b_n^{ij}(i\varepsilon k)^n; \quad i,j = 1,\dots,3; \tag{7.9.12}$$

converge for such $k \in \mathbb{R}$ that $|k| \le r/\varepsilon$.

Proof. It follows directly from differentiation of the sum in equations (7.8.12) multiplied scalarly by unit vectors e_j, $j = 1,2,3$. The convergence of series in equations (7.9.12) follows from convergence of similar series in equations (7.8.11). This completes the proof. □

Hence, the Chapman–Enskog equations of hydrodynamics read as

$$w_t = B\Lambda B^{-1}w, \quad w|_{t=0} = w_0^{(as)}, \tag{7.9.13}$$

where $B(\varepsilon)$, $\Lambda(\varepsilon)$, and $w_0^{(as)}$ are analytical at $\varepsilon = 0$. Diagonal equations can be obtained by the obvious transformation

$$w = By \Rightarrow y_t = \Lambda y, \quad y|_{t=0} = y_0 = B^{-1}w_0^{(as)}.$$

The connection with “changes of variables” from Proposition 7.6.1 which lead to diagonal hydrodynamic equations is obvious. A similar diagonalizing transformation can also be made for the general class solutions of the linearized Boltzmann equation. In the general case the null-space $N(L)$ is five-dimensional and there are some minor technical difficulties which demand a longer presentation. For the sake of brevity and clarity we decided to confine ourselves in this chapter to a simpler class of axially symmetric solutions.

We briefly summarize below the results of the section.

We have considered the problem of regularization of the Chapman–Enskog expansion (in particular, at the Burnett level). It was shown that the regularization can be done by transformation to new hydrodynamic variables. It was also shown that the way of truncation of the Chapman–Enskog series is not unique (even at the Navier–Stokes level). What is the meaning of these transformations? How can we find the optimal transformation? Does it exist? All these questions were considered in detail for the linearized Boltzmann equation. For brevity only one-dimensional (in space) axially symmetric flows were considered.

It was shown that the Chapman–Enskog expansion has a special structure which allows to pass to diagonal equations of hydrodynamics (three independent equations in the case under consideration or five independent equations in the general case). The diagonal Navier–Stokes equations are: two linearized Burgers (viscous) equations for two sound modes and the heat equation for one thermal mode. It is proved (Proposition 7.8.1) that the uniform in time and space accuracy of the Navier–Stokes approximation is $O(\varepsilon)$, where ε is the Knudsen number. The diagonal Burnett equations are: two independent linearized Burgers–KdV (viscous) equations for sound modes and the same heat equation for the thermal mode. It is proved (Proposition 7.8.1) that the uniform in time and space accuracy of the Burnett approximation is $O(\varepsilon^2)$. These estimates are valid for $t > \varepsilon^{1+\delta}$, $\delta > 0$, i. e., outside of the initial layer.

Thus, for the linearized Boltzmann equation the modified Burnett equations really improve the Navier–Stokes approximation. On the other hand, the above stated questions remain open for the non-linear case. This case is considered in the next section.

7.10 The non-linear case: generalized Burnett equations

Our goal in this section is to apply the general method of regularization from Section 7.6 to non-linear Burnett equations, i. e., to general equations of hydrodynamics (7.2.16), where fluxes π and q are given by equalities

$$\pi_{\alpha\beta} = \pi_{\alpha\beta}^{\mathrm{NS}} + \varepsilon\pi_{\alpha\beta}^{B}, \quad q_\alpha = q_\alpha^{\mathrm{NS}} + \varepsilon q_\alpha^{B}; \quad \alpha, \beta = 1, 2, 3, \tag{7.10.1}$$

in the notation of equations (7.3.2), (7.4.13), and (7.4.14). The abstract form of equations (7.2.16) reads

$$y_t = A(y) + \varepsilon B(y) + \varepsilon^2 C(y), \tag{7.10.2}$$

where $y = (\rho, u, T)^{\mathrm{Tr}}$ is the "hydrodynamic vector." The non-linear operators A, B, and C stand for Euler, Navier–Stokes, and Burnett terms, respectively.

Our task is to find an appropriate transformation

$$z = y + \varepsilon^2 R(y) \tag{7.10.3}$$

such that the resulting equations for $z(x, t)$ are well-posed and stable.

It is clear from considerations of the linearized Burnett equations that there exist infinitely many regularizing transformations. Moreover, it was shown in Section 7.7 that these transformations can significantly simplify equation (7.10.2) in the case of linear operators A, B, and C. Unfortunately the non-linear case is much more complicated. It is not easy to define what is the optimal transformation in that case. We consider below just one possible approach to this problem.

Obviously, we should restrict a class of operators R in equation (7.10.3) by using some reasonable considerations. In particular, the following two conditions seem quite natural:

A Modified equations should have a general form of equations of hydrodynamics (7.2.16) with some new fluxes $\Pi_{\alpha\beta}$ and Q_α instead of $\pi_{\alpha\beta}$ and q_α, respectively.

B Modified equations should preserve all "principal properties" of original equations.

Condition (B) will be discussed below; here we concentrate on condition (A).

There is also another argument, which is not easy to formalize. It is obviously preferable to choose "the simplest" class of transformations R (change of variables) provided this class is sufficient to achieve our goals. Therefore, we shall not consider *all* possible equivalent transformations of (7.2.16). Instead we specify a particular class which seems to be the simplest one.

Proposition 7.10.1. *The transformation of equations* (7.2.16) *with* $\varepsilon = 1$

$$\rho' = \rho, \quad u' = u, \quad T' = T + \frac{1}{\rho}\operatorname{div} R \tag{7.10.4}$$

with arbitrary (smooth) vector function $R(x,t)$ *leads to the following equations for* ρ', u', *and* T' *(primes are omitted below):*

$$\rho_t + \operatorname{div}\rho u = 0, \quad \rho\mathcal{D}_0 u_\alpha + \frac{\partial p}{\partial x_\alpha} + \frac{\partial \Pi_{\alpha\beta}}{\partial x_\beta} = 0, \quad p = \rho T,$$
$$\frac{3}{2}\rho\mathcal{D}_0 T + p\operatorname{div} u + \Pi_{\alpha\beta}\frac{\partial u_\alpha}{\partial x_\beta} + \operatorname{div} Q = 0, \quad \mathcal{D}_0 = \partial_t + u\cdot\partial_x, \tag{7.10.5}$$

where

$$\Pi = \tilde{\pi} - I\operatorname{div} R, \quad Q = \tilde{q} - \frac{3}{2}[\mathcal{D}_0 R + R\operatorname{div} u - (R\cdot\nabla)u], \tag{7.10.6}$$

where I denotes the identity tensor and tildes mean that π and q are expressed in new variables.

Proof. The proof is given by a straightforward calculation. □

The above transformation is a particular (simplified) version of more general transformations from [31]. Note that $\Pi_{\alpha\beta} = \Pi_{\beta\alpha}$, but

$$\Pi_{\alpha\alpha} = \operatorname{Tr}\Pi = -3\operatorname{div} R \neq 0 \tag{7.10.7}$$

in (7.10.5). Thus, the transformation (7.10.4) preserves the form of the original equations (7.2.16); however, it does not preserve the identity $\operatorname{Tr}\Pi = 0$.

Our goal now is to specify the vector function $R(x,t)$ in order to obtain a stable generalization of Burnett equations and satisfy the above condition (B).

We use the notation of Section 7.4 and present Burnett terms in the following form:

$$\pi_{\alpha\beta} = -\frac{1}{\rho}\left\{\frac{TA(T)}{\rho}\overline{\frac{\partial^2 \rho}{\partial x_\alpha \partial x_\beta}} + [A((T) - C(T)]\overline{\frac{\partial^2 T}{\partial x_\alpha \partial x_\beta}} + G_{\alpha\beta}\right\}, \tag{7.10.8}$$

$$q_\alpha^B = -\frac{T}{2\rho}\left\{\frac{1}{3}[4B(T) - C(T)]\frac{\partial}{\partial x_\alpha}\operatorname{div} u - C(T)\Delta u_\alpha + G_\alpha\right\}, \tag{7.10.9}$$

where $A(T)$, $B(T)$, and $C(T)$ are given in equations (7.4.9), and $G_{\alpha\beta}$ and G_α are bilinear forms on the vector space of first derivatives

$$\nabla \log \rho, \quad \nabla \log T; \quad \frac{\partial u_i}{\partial x_j}, \quad i, j = 1, 2, 3. \tag{7.10.10}$$

The following formulas (exact for Maxwell molecules) are convenient for various estimates:

$$A \approx 2\frac{\mu^2}{T}, \quad B \approx \frac{45\mu^2}{8T}, \quad C \approx 3\frac{\mu^2}{T}, \quad \lambda \approx \frac{15}{4}\mu, \tag{7.10.11}$$

in the notation of equations (7.3.2) and (7.3.3).

Coming back to the general scheme of the regularization of Burnett equations by transformation (7.10.3), we can now specify the vector $R(x,t)$ in (7.10.4). In accordance with condition (B), the symmetric tensor Π and the vector Q in (7.10.6) must have the same structure as corresponding Burnett terms. This leads to a unique choice of $R(x,t)$ in the following form:

$$R(x,t) = \varepsilon^2\frac{S(x,t)}{\rho}, \quad S = a(T)\,\nabla \log \rho + b(T)\nabla \log\, T, \tag{7.10.12}$$

with indefinite coefficients $a(T)$ and $b(T)$.

Indeed the vector function $R(x,t)$ must be: (1) a vector; (2) a linear form on derivatives; and (3) a homogeneous function of ρ of order $m = -1$. The only vector linear on derivatives of u is rot$u$, which is in fact a pseudo-vector. Therefore, we reject it and obtain (7.10.12). Note that a and b cannot depend on u because of Galilei invariance.

Hence, the uncertainty in regularizing transformation (7.10.4) is reduced to two unknown functions $a(T)$ and $b(T)$. As we shall see below, it is convenient to introduce two dimensionless parameters $\theta_{1,2}$ in the following way:

$$\begin{aligned} a(T) &= \frac{2}{3}[-TA(T) + \theta_1 g(T)], \\ b(T) &= \frac{2}{3}T[C(T) - A(T)] + \theta_2 g(T), \\ g &= \frac{T}{3}(5T + 2B - 4C). \end{aligned} \tag{7.10.13}$$

Generally speaking, this is just a simple replacement of $a(T)$ and $b(T)$ by new unknown functions $\theta_{1,2}$. Let us consider, however, a particular case of power-like intermolecular potentials (or hard spheres). Then $A(T)$, $B(T)$, and $C(T)$ are proportional to T^γ (for example, $\gamma = 1$ for hard spheres) and $g(T)$ is proportional to $T^{\gamma+1}$. Hence, the only way to preserve properties of original Burnett equations is to choose $\theta_{1,2}(T)$ in (7.10.3) independently of T. Thus, the parameters $\theta_{1,2}$ are simply numbers for all power-like potentials. Therefore, the most natural choice of unknown $a(T)$ and $b(T)$ is given by (7.10.3), where $\theta_{1,2}$ are numbers. All these considerations can be completely formalized by demanding that the new equations (7.10.5) preserve all group-invariant transformations of the original Burnett equations.

Hence, the uncertainty in the transformation (7.10.4), (7.10.12) is reduced to two numerical parameters $\theta_{1,2}$ in (7.10.13). The parameters should be chosen from conditions of hyperbolicity and stability (see Section 7.5). As we shall see below, it is convenient to introduce a third parameter θ_3 in such a way that

$$\theta_1 + \theta_2 + \theta_3 = 1. \tag{7.10.14}$$

Then we can formulate a definition of the whole class of GBEs. In order to avoid complicated notations with primes (see (7.10.4)), we do this in the following way.

Definition 7.10.1. GBEs, denoted by symbols $(\theta_1, \theta_2, \theta_3)$, are equations for *auxiliary* variables (ρ, u, T) such that the *true* hydrodynamical variables are expressed through (ρ, u, T) by equalities

$$\rho^{\mathrm{Tr}} = \rho, \quad u^{\mathrm{Tr}} = u, \quad T^{\mathrm{Tr}} = T - \frac{\varepsilon^2}{\rho} \operatorname{div} \frac{S}{\rho} \tag{7.10.15}$$

(see (7.10.4) and (7.10.12)).

Any two parameters of $\theta_{1,2,3}$ can be chosen arbitrarily. Then the third parameter can be found from (7.10.14); the functions $a(T)$ and $b(T)$ in equations (7.10.12) are given (for fixed $\theta_{1,2}$) in (7.10.13). The GBEs $(\theta_1, \theta_2, \theta_3)$ are, by definition, equations (7.10.5), where

$$\Pi_{\alpha\beta} = \varepsilon \pi_{\alpha\beta}^{\mathrm{NS}} + \varepsilon^2 \left(\pi_{\alpha\beta}^{B} - \delta_{\alpha\beta} \operatorname{div} \frac{S}{\rho} \right), \tag{7.10.16}$$

$$\begin{aligned} Q_\alpha = {} & \varepsilon q_\alpha^{\mathrm{NS}} + \varepsilon^2 q_\alpha^{B} + \frac{\varepsilon^2}{\rho} \left\{ 3 S_\beta \overline{\frac{\partial u_\alpha}{\partial x_\beta}} + \left(\frac{3}{2} a + b \right) \frac{\partial}{\partial x_\alpha} \operatorname{div} u \right. \\ & \left. + (\operatorname{div} u) \left[(a' T - 2a) \frac{\partial}{\partial x_\alpha} \log \rho + (b' T - 2b) \frac{\partial}{\partial x_\alpha} \log T \right] \right\}, \quad \alpha, \beta = 1, 2, 3, \end{aligned} \tag{7.10.17}$$

where primes denote differentiation with respect to T.

Remark. The above formula for Q_α follows by straightforward calculations from the corresponding formula in equation (7.10.6) provided the term $\mathcal{D}_0(S/\rho)$ is computed in the Euler approximation, i. e., for $\varepsilon = 0$.

7.11 Hyperbolicity and stability of GBEs

Our goal in this section is to determine a domain of parameters (θ_1, θ_2) in $\mathbb{R}^2$ such that the corresponding GBEs $(\theta_1, \theta_2, \theta_3)$ are hyperbolic and stable. By stability we mean the following property: Any constant stationary solution $\{\rho = \rho_0 > 0,\ T = T_0 > 0,\ u = 0\}$ is stable with respect to small perturbations (it is sufficient to consider $u = 0$, since the equations are Galilei-invariant). In fact, the stability implies hyperbolicity, as we shall see below.

In order to study the hyperbolicity it is sufficient to consider just terms with higher derivatives in GBEs $(\theta_1, \theta_2, \theta_3)$. By using equations (7.10.4) and (7.10.8)–(7.10.17), we obtain

$$\rho_t = \cdots, \quad \rho u_t = \varepsilon^2 \frac{g(T)}{\rho}\left[\frac{2\theta_1}{3\rho}\Delta(\nabla\rho) + \frac{\theta_2}{T}\Delta(\nabla T)\right] + \cdots,$$
$$\frac{3}{2}\rho T_t = \varepsilon^2 \frac{g(T)}{\rho}\theta_3 \,\Delta \operatorname{div} u + \cdots; \quad \theta_1 + \theta_2 + \theta_3 = 1, \tag{7.11.1}$$

where dots denote terms with lower derivatives. The reader can now see that our way of parameterization leads to a simple form of terms with third derivatives. The hyperbolicity condition reads

$$\theta_2\theta_3 = \theta_2(1 - \theta_1 - \theta_2) \geq 0. \tag{7.11.2}$$

Note that the Burnett equations are also a particular case of GBEs $(\theta_1, \theta_2, \theta_3)$ with parameters (generally speaking, parameters are functions of T)

$$\theta_1^B = \frac{TA(T)}{g(T)}, \quad \theta_2^B = \frac{2}{3}T\frac{A(T) - C(T)}{g(T)}, \tag{7.11.3}$$

which corresponds to $a = b = 0$ in equation (7.10.13).

In the approximation (7.10.11), we obtain

$$g(T) \approx \frac{37}{12}\mu^2, \quad \theta_1^B \approx \frac{24}{37}, \quad \theta_2^B \approx -\frac{24}{111}, \tag{7.11.4}$$

i. e., these equations are not hyperbolic, as we already knew.

We assume that the condition (7.11.2) is satisfied and consider the question of stability. The Knudsen number ε is just a formal parameter in our equations; we can assume that $\varepsilon = 1$ without loss of generality. Then we should consider GBEs $(\theta_1, \theta_2, \theta_3)$ in the neighborhood of the constant solution $\{\rho = \rho_0 > 0,\ T = T_0 > 0,\ u = 0\}$. It is important that all other Burnett terms (i. e., terms of order $O(\varepsilon^2)$), which are omitted in equations (7.11.1), are quadratic with respect to derivatives. Therefore, such terms do not contribute to linearized equations.

Hence, the linearized GBEs contain all usual terms of linearized Navier–Stokes equations plus Burnett terms with third derivatives (see equations (7.11.1)).

We denote

$$T = T_0(1+\tilde{T}), \quad \rho = \rho_0(1+\tilde{\rho}), \quad u = T_0^{1/2}\tilde{u},$$
$$x = l\tilde{x}, \quad t = \tau\tilde{t}; \quad \tau = \frac{\mu(T_0)}{\rho_0(T_0)}, \quad l = \tau T_0^{1/2}, \tag{7.11.5}$$

and obtain the following linearized equations (all variables below are dimensionless, tildes are omitted):

$$\rho_t + \operatorname{div} u = 0,$$
$$u_t + \nabla\rho + \nabla T = \Delta u + \frac{1}{3}\nabla \operatorname{div} u + g_0\left[\frac{2\theta_1}{3}\Delta(\nabla\rho) + \theta_2\Delta(\nabla T)\right],$$
$$T_t + \frac{2}{3}\operatorname{div} u = \frac{5}{2}\gamma_0\Delta T + \frac{2}{3}g_0\theta_3\Delta \operatorname{div} u, \quad \theta_1+\theta_2+\theta_3 = 1, \tag{7.11.6}$$

where

$$\gamma_0 = \frac{4\lambda(T_0)}{15\mu(T_0)}, \quad g_0 = \frac{T_0}{3\mu^2(T_0)}[5A(T_0) + 2B(T_0) - 4C(T_0)]. \tag{7.11.7}$$

Then we make the usual substitution (the standard notation for eigenvalues $\lambda(k)$ should not be confused with similar notation $\lambda(T)$ for the heat conduction coefficient)

$$(\rho, u, T) \sim \exp[\lambda(k)t + ik\cdot x], \quad k \in \mathbb{R}^3, \tag{7.11.8}$$

and obtain a dispersion relation, a cubic equation, for $\lambda(k)$. We omit elementary calculations and present the resulting equation:

$$\lambda^3 + a_1(z)\lambda^2 + a_2(z)\lambda + a_3(z) = 0, \quad z = |k|^2,$$
$$a_1(z) = \left(\frac{4}{3} + \frac{5}{2}\gamma_0\right)z, \quad a_2(z) = \frac{z}{3}[5 + 2(5+g_0)z + 2g_0^2\theta_2\theta_3 z^2],$$
$$a_3(z) = \frac{5}{2}\gamma_0 z^2\left(1 + \frac{2}{3}\theta_1\gamma_0 z\right). \tag{7.11.9}$$

The stability condition reads

$$\Re\lambda < 0 \quad \text{for all} \quad z > 0.$$

This is equivalent to the following conditions on coefficients [128]:

$$\{a_i > 0,\ i = 1,2,3;\ a_1a_2 > a_3\} \quad \text{for all} \quad z > 0. \tag{7.11.10}$$

The inequality $a_2 > 0$ obviously implies (for large z) the hyperbolicity condition (7.11.2). The domain of stability in the plane of parameters (θ_1, θ_2) depends on γ_0 and g_0 (note that $\gamma_0 > 0$ for any molecular model). A realistic estimate (exact for Maxwell molecules)

$$\gamma_0 \approx 1, \quad g_0 \approx \frac{37}{12} \tag{7.11.11}$$

follows from equation (7.10.11). Therefore, it seems reasonable to assume that $g_0 > 0$ for realistic models. Then we obtain necessary conditions of stability

$$\theta_1 \geq 0, \quad \theta_2(1 - \theta_1 - \theta_2) \geq 0, \tag{7.11.12}$$

which guarantee that $a_{1,2,3} > 0$. The remaining condition $a_1 a_2 > a_3$ leads to the quadratic inequality

$$\begin{aligned} &P(z) = z^2 + 2\alpha z + \beta > 0, \quad z > 0, \\ &\alpha = \frac{1}{2\Delta}\left[(5 + g_0)(8 + 15\gamma_0) - 15\gamma_0\, g_0\, \theta_1\right], \quad \beta = \frac{5(4 + 3\gamma)}{\Delta}, \\ &\Delta = (8 + 15\gamma_0)\, g_0^2\, \theta_2\,(1 - \theta_1 - \theta_2), \end{aligned} \tag{7.11.13}$$

provided $\Delta > 0$. If $\Delta = 0$, i. e., $\theta_1 = 0$ or $\theta_2 = 1 - \theta_1$, then necessary and sufficient conditions are

$$\theta_1 \geq 0, \quad (5 + g_0)(8 + 15\gamma_0) - 15\gamma_0\, g_0\, \theta_1 \geq 0. \tag{7.11.14}$$

The same inequalities for θ_1 yield the sufficient conditions if $\Delta > 0$. In this case, however, there is one more region of stability in the plane (θ_1, θ_2) for such θ_1 that

$$2\alpha\Delta = (5 + g_0)(8 + 15\gamma_0) - 15\,\gamma_0\, g_0\, \theta_1 < 0. \tag{7.11.15}$$

Then a simple analysis of the equality (7.11.13) shows that it reduces to $\alpha^2 < \beta$. Thus, we obtain

$$\theta_2^2 + \theta_2\,(\theta_1 - 1) + \frac{\left[(5 + g_0)(8 + 15\gamma_0) - 15\gamma_0\, g_0\, \theta_1\right]^2}{20\,(4 + 3\gamma_0)\,(8 + 15\gamma_0) g_0^2} < 0$$

for θ_1 satisfying the equality (7.11.14). This is a complete description of the stable domain in the plane (θ_1, θ_2) for arbitrary g_0 and γ_0. It is easy to show that such a domain always has an infinite measure (area). Having in mind (7.11.11), we can assume that $g_0 \geq -5$ for any realistic potential. Then the simple triangle region

$$\theta_{1,2} \geq 0, \quad \theta_1 + \theta_2 \leq 1 \tag{7.11.16}$$

is always stable and sufficient for practical applications.

Finally, we describe the whole domain of stability for Maxwell molecules, when equations (7.10.11) are exact. Then we obtain

$$\begin{aligned} &(a) \quad 0 \leq \theta_1 \leq 1, \quad 0 \leq \theta_2 \leq 1 - \theta_1; \\ &(b) \quad 1 < \theta_1 \leq \theta_*, \quad 1 - \theta_2 \leq \theta_2 < 0; \\ &(c) \quad \theta_1 > \theta_*, \quad \theta_2^- \leq \theta_2 \leq \theta_2^+; \\ &\theta_* = \frac{2231}{555} \approx 4.02, \quad \theta_2^\pm = \frac{1}{2}\left[1 - \theta_1 \pm \sqrt{(\theta_1 - 1)^2 - \frac{45}{161}(\theta_1 - \theta_*)^2}\right]. \end{aligned}$$

We also note that the domain

$$\theta_1 < 0, \quad 0 \le \theta_2 \le 1 - \theta_1$$

is hyperbolic, but unstable. The domain of hyperbolicity (7.11.2) for $0 \le \theta_1 \le \theta_*$ coincides with the domain of stability. A difference between these two domains remains very small for $\theta_* < \theta_1 \le 10$. Burnett equations correspond to the point (7.11.4).

It was already mentioned that the triangle (7.11.16) seems to be sufficient for practical applications with any typical molecular model, for example, hard spheres. We just need to verify the condition of stability

$$g_0 = \frac{T}{3\mu^2(T)}[5A(T) + 2B(T) - 4C(T)] \ge -5$$

for the triangle (7.11.16). Note that g_0 is a number for hard spheres and power-like potentials.

Finally, we present GBEs $(\theta_1, \theta_2, \theta_3)$ in the form convenient for applications. In order to do this we pass to the dimensionless quantities in (7.10.8)–(7.10.13) by denoting

$$\{A(T), B(T), C(T)\} = \frac{\mu^2(T)}{T}\{\tilde{A}, \tilde{B}, \tilde{C}\},$$
$$\{a(T), b(T), g(T)\} = \mu^2(T)\{\tilde{a}, \tilde{b}, \tilde{g}\},$$

where $\mu(T)$ is the viscosity coefficient (7.3.2). In the sequel of the section these notations are used *without tildes*. Then the dimensionless coefficients $\{A, B, C, a, b, g\}$ where

$$g = \frac{1}{3}(5A + 2B - 4C)$$

are numbers for any power-like and hard sphere intermolecular potential. Moreover,

$$\mu(T) = \text{const.}\, T^{\omega} \tag{7.11.17}$$

in such cases, $\omega = 1$ for Maxwell molecules, and $\omega = 1/2$ for hard spheres. Otherwise, we assume that $\{A, B, C, a, b, g\}$ are slowly varying functions of T and use (7.11.17) below, with $\omega = \omega(T)$.

Then the two-parameter family of GBEs

$$\{\text{GBEs}(\theta_1, \theta_2, \theta_3), \theta_1 + \theta_2 + \theta_3 = 1\}$$

is described in the following way. We choose the Burnett equations as a starting point. Then we fix a pair of real parameters (θ_1, θ_2) and denote

$$a = \frac{2}{3}(-A + \theta_1 g), \quad b = \frac{2}{3}(C - A) + \theta_2 g, \tag{7.11.18}$$

where the dimensionless coefficients A, C, g (see equation (7.11.17)) are known from the initial Burnett equations. For example,

$$A = 2, \quad C = 3, \quad g = \frac{37}{12} \tag{7.11.19}$$

for Maxwell molecules; see equation (7.10.11).

As explained in Section 7.10, GBEs $(\theta_1, \theta_2, \theta_3)$ are equations for *auxiliary* variables (ρ, u, T), which are connected with true hydrodynamical variables by the equalities

$$\begin{aligned} &\rho^{\mathrm{Tr}} = \rho, \quad u^{\mathrm{Tr}} = u, \quad \rho T^{\mathrm{Tr}} = P(\rho, T) = \rho T - \varepsilon^2 \operatorname{div} \frac{\mu^2(T)}{\rho} S, \\ &S = a\nabla \log \rho + b\nabla \log T, \end{aligned} \tag{7.11.20}$$

in the notation of equations (7.11.18). The equations for (ρ, u, T) read

$$\begin{aligned} \rho_t + \operatorname{div} \rho u &= 0, \\ \rho \mathcal{D} u_\alpha + \nabla_\alpha P(\rho, T) + \varepsilon \frac{\partial \pi_{\alpha\beta}}{\partial x_\beta} &= 0, \\ \frac{3}{2} \rho \mathcal{D}_0 T + P(\rho, T) \operatorname{div} u + \varepsilon \left(\pi_{\alpha\beta} \frac{\partial u_\alpha}{\partial x_\beta} + \operatorname{div} Q \right) &= 0, \end{aligned}$$

where

$$\begin{aligned} Q_\alpha = q_\alpha + \varepsilon \frac{\mu^2(T)}{\rho} \Big[& 3 S_\beta \overline{\frac{\partial u_\alpha}{\partial x_\beta}} \\ & + 2(\omega - 1) S_\alpha \operatorname{div} u + \Gamma_\alpha + \left(\frac{3}{2} a + b \right) \nabla_\alpha \operatorname{div} u \Big], \end{aligned} \tag{7.11.21}$$

q_α and $\pi_{\alpha\beta}$ are the usual fluxes (7.10.1) from the Burnett equations and P, S, and ω are given in equations (7.11.17)–(7.11.20). If ω = const., power-like potentials and hard spheres, then a and b are numbers and $\Gamma_\alpha = 0$ in equation (7.11.21). Otherwise we obtain

$$\Gamma_\alpha = T \left[\frac{a'(T)}{a(T)} \nabla_\alpha \log \rho + \frac{b'(T)}{b(T)} \nabla_\alpha \log T \right].$$

The above formulas together with conditions of stability (7.11.16) yield the complete information for practical use of GBEs, though the important question of boundary conditions is not considered here.

The above described GBEs contain two free parameters, say, θ_1 and θ_2. The problem of reasonable choice of $\theta_{1,2}$ was considered in the paper [16] for Maxwell molecules. The purpose of that work was to study the half-space stationary problem for GBEs. The authors of [16] wanted to find such values of $\theta_{1,2}$ that the solutions of

GBEs are qualitatively similar to corresponding solutions of the Boltzmann equation. They found that the best result corresponds to the choice

$$\theta_1 = 1, \quad \theta_2 = 0. \tag{7.11.22}$$

Then we obtain from equations (7.11.1) that $\theta_3 = 0$. The same criterion can be applied to other molecular models. The standard approximation (7.10.11) leads to similar values (7.11.22) for the general case. Thus we obtain a uniquely defined set of GBEs. These equations were used for the shock structure problem in [35] and for the sound propagation in [55]. For both problems the results seem to be more close to the solution of the Boltzmann equation or experimental data than similar results obtained from the Navier–Stokes equations. Group properties and some invariant solutions of GBEs are studied in [51]. Surprisingly, GBEs with the choice of parameters (7.11.22) are even simpler than the original Burnett equations because of the reduced number of third derivatives. On the other hand, this choice of "optimal" replacement for ill-posed Burnett equations is not justified enough from a mathematical point of view. Therefore, this question remains open. We just have proposed one of the possible solutions.

7.12 Concluding remarks

The general topic "Boltzmann equation and hydrodynamics" is a very large research area. The materials of Chapter 7 cover only a small part of it. Below we give some very short comments on other parts of the topic.

1. The "conventional" scaling (7.1.9) of the Boltzmann equation formally leads at the limit $\varepsilon = 0$ to a solution in the form of a local Maxwellian with hydrodynamic parameters satisfying Euler equations (7.1.11) (some related mathematically rigorous results can be found in [62, 131]). There exists, however, another interesting scaling, which leads, as $\varepsilon \to 0$, to an absolute Maxwellian with a correction term of the order $O(\varepsilon)$. This term is asymptotically defined by a solution of incompressible Navier–Stokes equations. The corresponding hydrodynamic limit is considered in detail in many publications, beginning with papers [11, 12, 76]. More recent results in the area can be found, in particular, in [89, 131, 132]. The Hilbert and Chapman–Enskog expansions for corresponding solutions are discussed in [132]. In particular, it is shown there that problems with ill-posedness (see Section 7.5) do not appear in that kind of asymptotic solutions. Roughly speaking, the reason is that the corresponding Mach number has the order $O(\varepsilon)$ and this results in automatic regularization of the Chapman–Enskog expansion.
2. We did not discuss the hydrodynamic limit for stationary solutions of the Boltzmann equation in this book. It is another important area which should be considered separately (see [84] for a review). We mention just one point directly related to Burnett equations. There exists a class of problems where some Burnett terms

must be present in equations of hydrodynamics even in the continuum limit $\varepsilon = 0$. This class, the stationary non-isothermal flows with low Mach numbers $Ma = O(\varepsilon)$, was introduced in the 1960s and 1970s in a series of papers (see [104, 105] and references therein) by Kogan, Galkin, and Fridlender, where the corresponding modified Burnett equations were derived and studied. The equations were derived by both Hilbert and Chapman–Enskog methods, which are, roughly speaking, equivalent in this case [105]. The authors found and discussed in detail many interesting phenomena related to new types of gas convection of order $O(\varepsilon)$. It was proved in 1995 [29] that the modified Burnett equations used in this theory contradict the conventional heat equation in the continuous limit $\varepsilon = 0$. This result can be briefly described in the following way. Consider the stationary Boltzmann equation in its standard dimensionless form in a non-symmetric domain $\mathcal{D} \subset \mathbb{R}^3$ with diffusive (complete accommodation) boundary conditions on non-isothermal boundaries of $\mathcal{D}$. Assume that the gravity is absent and pass to the limit $\varepsilon = 0$. It is clear that the limiting solution is a Maxwellian with zero bulk velocity and with some temperature $T(x)$ and density $\rho(x)$ such that $\rho T = \text{const.}, x \in \mathcal{D}$. One can expect that $T(x)$ satisfies the non-linear heat equation which follows from Navier–Stokes equations in this case. The statement proved in [29] means that this cannot be true provided the Kogan–Galkin–Fridlender theory is correct. Then, in order to find $T(x)$, we need to solve the full system of their equations instead of the relatively simple heat equation.
The comprehensive investigation of this problem for the BGK model [70] of the Boltzmann equation was performed by the Kyoto group of Sone and Aoki [138]. The results have confirmed that the theory was right and the heat equation was wrong. The paper [138] became a starting point for many interesting publications by the Kyoto group on the similar "ghost effect" in quite different problems of rarefied gas dynamics (see book [137]). The only mathematically rigorous related result known to the author can be found in the paper [5].

3. It was already mentioned that there are several different approaches to regularization of the Chapman–Enskog expansion. We cite only a small part of related publications: [90, 91, 100, 130, 133–135, 139, 142]. In our view there are at present two most developed alternative approaches to the hydrodynamics beyond Navier–Stokes. One of the approaches was discussed in detail in Sections 7.5–7.11. The other approach, proposed by H. Struchtrup and M. Torrilhon (see [139, 141] and references therein), is based on generalization of the Grad moment method combined with direct expansion of moment equations in power series in Knudsen number. Both approaches have their advantages and disadvantages. It can depend on the specific problem which approach is more convenient for solving it.

Bibliography

[1] Ahlfors L. Complex analysis. McGraw-Hill, 1979.

[2] Alexandre R, Morimoto Y, Ukai S, Xu C-J, Tong Y. Smoothing effect of weak solutions for the spatially homogeneous Boltzmann equation without angular cutoff. Kyoto J Math 2012, 52, 433–463.

[3] Arkeryd L. On the Boltzmann equation. Arch Ration Mech Anal 1972, 34, 1–34.

[4] Arkeryd L. L^{∞}-Estimates for the space homogeneous Boltzmann equation. J Stat Phys 1982, 31, 347–361.

[5] Arkeryd L, Esposito R, Marra R, Nouri A. Ghost effect by curvature in planar Couette flow. Kinet Relat Models 2011, 4, 109–138.

[6] Arnold VI. Geometrical methods in the theory of ordinary differential equations. Springer-Verlag, 1988.

[7] Arsen'ev AA. The Cauchy problem for the linearized Boltzmann equation. Zh Vych Mat Mat Fiz 1965, 5, 864–882. English translation in: USSR Comput Math and Math Phys 1965, 5, 110–136.

[8] Arsen'ev AA. Lecture notes on kinetic equations. Nauka, Moscow, 1992. (in Russian).

[9] Baldassarri A, Marini Bettolo Marconi U, Puglisi A. Influence of correlations on the velocity statistics of scalar granular gases. Europhys Lett 2002, 58, 14–20.

[10] Balescu R. Statistical mechanics of charged particles. Interscience Publishers, Wiley and Sons, Ltd, London, 1963.

[11] Bardos C, Golse F, Levermore CD. On asymptotic limits of kinetic theory leading to incompressible fluid dynamics. C R Acad Sci Paris Sr I Math 1989, 309, 727–732.

[12] Bardos C, Golse F, Levermore CD. Fluid dynamic limits of kinetic equations, I. Formal derivation. J Stat Phys 1991, 63(1/2), 323–344.

[13] Barnsley M, Cornille H. On a class of solutions of the Krook–Wu model of the Boltzmann equation. J Math Phys 1980, 21, 1176–1193.

[14] Ben-Naim E, Krapivsky PL. Multiscaling in inelastic collisions. Phys Rev E 2000, 61, R5–R8.

[15] Bisi M, Carrillo JA, Toscani G. Decay rates towards self-similarity for the Ernst–Brito conjecture on large time asymptotics of the inelastic Maxwell model. J Stat Phys 2006, 124(2–4), 625–653.

[16] Bisi M, Cassinari MP, Groppi M. Qualitative analysis of Generalized Burnett Equations and applications to half-space problems. Kinet Relat Models 2008, 1, 295–312.

[17] Bobylev AV. Fourier transform method in the theory of the Boltzmann equation for Maxwell molecules. Dokl Akad Nauk SSSR 1975, 225, 1041–1044. English translation in Sov Phys Dokl 1976, 20, 820–822.

[18] Bobylev AV. On exact solutions of the Boltzmann equation. Dokl Akad Nauk SSSR 1975, 225, 1296–1299. English translation in Sov Phys Dokl 1976, 20, 822–824.

[19] Bobylev AV. A class of invariant solutions of the Boltzmann equation. Dokl Akad Nauk SSSR 1976, 231, 571–574. English translation in Sov Phys Dokl 1976, 21, 632–635.

[20] Bobylev AV. Green function of the linear Boltzmann equation. Dokl Akad Nauk SSSR 1979, 249, 340–343. English translation in Sov Phys Dokl 1980, 25, 257–259.

[21] Bobylev AV. On the structure of spatially homogeneous normal solutions of the nonlinear Boltzmann equation for a mixture of gases. Dokl Akad Nauk SSSR 1980, 250, 340–344. English translation in Sov Phys Dokl 1982, 25, 50–53.

[22] Bobylev AV. Poincaré theorem, the Boltzmann equation, and KdV-type equations. Dokl Akad Nauk SSSR 1981, 256, 1341–1346. English translation in Sov Phys Dokl 1981, 26, 166–169.

[23] Bobylev AV. Chapman–Enskog and Grad methods of solving of the Boltzmann equation. Dokl Akad Nauk SSSR 1982, 262, 71–75. English translation in Sov Phys Dokl 1982, 27, 29–31.

https://doi.org/10.1515/9783110550986-009

[24] Bobylev AV. Exact solutions of the Boltzmann equation and its models. In: Molecular gas dynamics. Proc. 5th Natl. RGD Conf., Dolgoprudnyi, 1978. Nauka, Moscow, 1982, 50–53. English translation in Fluid Mechanics – Soviet Research, 1984, 13, 105–110.

[25] Bobylev AV. Exact solutions of the Boltzmann equation and the theory of relaxation in Maxwell gas. Theor Math Phys 1984, 60, 280–310. English translation in: USSR Theor and Math Phys 1984, 60.

[26] Bobylev AV. Exact and approximate methods in the theory of nonlinear kinetic equations of Boltzmann and Landau. Keldysh Institute of Applied Mathematics, USSR Academy of Sciences, Moscow, 1987 (in Russian).

[27] Bobylev AV. The theory of the nonlinear spatially uniform Boltzmann equation for Maxwell molecules. Sov Sci Rev, C, Math Phys Rev, 1988, 7, 111–233.

[28] Bobylev AV. The Boltzmann equation and the group transformations. Math Models Methods Appl Sci 1993, 3, 443–476.

[29] Bobylev AV. Quasistationary hydrodynamics for the Boltzmann equation. J Stat Phys 1995, 80, 1063–1083.

[30] Bobylev AV. Moment inequalities for the Boltzmann equation and applications to spatially homogeneous problems. J Stat Phys 1997, 88, 1183–1214.

[31] Bobylev AV. Instabilities in the Chapman–Enskog expansion and hyperbolic Burnett equations. J Stat Phys 2006, 80, 1063–1085.

[32] Bobylev AV. Boltzmann equation and hydrodynamics beyond the Navier-Stokes level (Harold Grad Lecture). In: Rarefied gas dynamics. Publishing House of the Siberian Branch of the Russian Academy of Sciences, Novosibirsk, 2007, 3–12.

[33] Bobylev AV. Generalized Burnett hydrodynamics. J Stat Phys 2008, 132, 569–580.

[34] Bobylev AV. Boltzmann equation and hydrodynamics beyond Navier–Stokes. Philos Trans R Soc A, Math Phys Eng Sci 2018, 376. DOI 10.1098/rsta.2017.0227.

[35] Bobylev AV, Bisi M, Cassinari MP, Spiga G. Shock wave structure for generalized Burnett equations. Phys Fluids 2011, 23, 030607–030617.

[36] Bobylev AV, Carrillo JA, Gamba IM. On some properties of kinetic and hydrodynamic equations for inelastic interactions. J Stat Phys 2000, 98, 743–773.

[37] Bobylev AV, Cercignani C. On the rate of entropy production for the Boltzmann equation. J Stat Phys 1999, 94, 603–618.

[38] Bobylev AV, Cercignani C. Moment equations for a granular material in a thermal bath. J Stat Phys 2002, 106, 547–567.

[39] Bobylev AV, Cercignani C. Exact eternal solutions of the Boltzmann equation. J Stat Phys 2002, 106, 1019–1038.

[40] Bobylev AV, Cercignani C. Self-similar solutions of the Boltzmann equation and their applications. J Stat Phys 2002, 106, 1039–1071.

[41] Bobylev AV, Cercignani C. The inverse Laplace transform of some analytic functions with an application to the eternal solutions of the Boltzmann equation. Appl Math Lett 2002, 15(7), 807–813.

[42] Bobylev AV, Cercignani C. Self-similar asymptotics for the Boltzmann equation with inelastic and elastic interactions. J Stat Phys 2003, 110, 333–375.

[43] Bobylev AV, Cercignani C, Gamba IM. Generalized kinetic Maxwell type models of granular gases. In: Capriz G, Giovine P, Mariano PM (eds.) Mathematical models of granular matter. Lecture notes in mathematics, Vol. 1937. Springer, Berlin–Heidelberg–New York, 2008, 23–58.

[44] Bobylev AV, Cercignani C, Gamba IM. On the self-similar asymptotics for generalized non-linear kinetic Maxwell models. Commun Math Phys 2009, 291, 599–644.

[45] Bobylev AV, Cercignani C, Toscani G. Proof of an asymptotic property of self-similar solutions of the Boltzmann equation for granular materials. J Stat Phys 2003, 111, 403–417.

[46] Bobylev AV, Dorodnitsyn VA. Symmetries of evolution equations with non-local operators and applications to the Boltzmann equation. Discrete Contin Dyn Syst 2009, 24(1), 35–57.

[47] Bobylev AV, Gamba IM. Boltzmann equations for mixtures of Maxwell gases: exact solutions and power like tails. J Stat Phys 2006, 124, 497–516. https://doi.org/10.1007/s10955-006-9044-8.

[48] Bobylev AV, Gamba IM. Solutions of the linear Boltzmann equation and some Dirichlet series. Forum Math 2012, 24, 239–252.

[49] Bobylev AV, Gamba IM. Upper Maxwellian bounds for the Boltzmann equation with pseudo-Maxwell molecules. Kinet Relat Models 2017, 10, 573–585.

[50] Bobylev AV, Gamba IM, Panferov VA. Moment inequalities and high-energy tails for Boltzmann equations with inelastic interactions. J Stat Phys 2004, 116(5.6), 1651–1682.

[51] Bobylev AV, Meleshko SV. Group analysis of the generalized Burnett equations. J Nonlinear Math Phys 2020, 27(3), 1–15.

[52] Bobylev AV, Nota A, Velazquez JL. Self-similar asymptotics for a modified Maxwell–Boltzmann equation in systems subject to deformations. Commun Math Phys, 2020 (accepted for publication in July 2020).

[53] Bobylev AV, Toscani G. On the generalization of the Boltzmann H-theorem for spatially homogeneous Maxwell gas. J Math Phys 1992, 33(7), 2578–2586.

[54] Bobylev AV, Windfall A. Kinetic modeling of economic games with large number of participants. Kinet Relat Models 2011, 4, 169–185.

[55] Bobylev AV, Windfall A. Boltzmann equation and hydrodynamics at the Burnett level. Kinet Relat Models 2012, 5, 237–260.

[56] Bogolyubov NN. Problems of a dynamical theory in statistical physics (Moscow: State Technical Press, 1946, in Russian); English translation in 'Studies of Statistical Mechanics I', ed de Boer J, Uhlenbeck GE. Part a, Amsterdam, North-Holland, 1962.

[57] Boltzmann L. Weiter Studien uber das Warmegleichgewicht unte Gasmolekulen. Wien Akad Sitzungsber, 1872, 66, 275–370.

[58] Burnett D. The distribution of velocities in a slightly non-uniform gas. Proc Lond Math Soc 1935, 39, 385–430.

[59] Burnett D. The distribution of molecular velocities and the mean motion in a non-uniform gas. Proc Lond Math Soc 1935, 40, 382–435.

[60] Cabannes H. Proof of the conjecture on "eternal" positive solutions for a semi-continuous model of the Boltzmann equation. C R Acad Sci Paris Ser I Math 1988, 327, 21–222.

[61] Cabannes H. "Eternal" solutions for a model of the Boltzmann equation. Math Models Methods Appl Sci 1999, 9, 12–137.

[62] Caflisch RE. The fluid dynamic limit of the nonlinear Boltzmann equation. Commun Pure Appl Math 1980, 33, 651–666.

[63] Cannone M, Karch G. Infinite energy solutions to the homogeneous Boltzmann equation. Commun Pure Appl Math 2010, 63 (6), 747–778.

[64] Carleman T. Problemes mathematiques dans la theorie cinetique des gaz. Almqvist and Wiksell, Uppsala, 1957.

[65] Carlen EA, Carvalho MC. Entropy production estimates for Boltzmann equations with physically realistic collision kernels. J Stat Phys 1994, 74, 743–782.

[66] Carlen EA, Carvalho MC, Gabetta E. Central limit theorem for Maxwellian molecules and truncation of the Wild expansion. Commun Pure Appl Math 2000, LIII, 370–397.

[67] Carlen EA, Carvalho MC, Gabetta E. On the relation between rates of relaxation and convergence of Wild sums for solutions of the Kac equation. J Funct Anal 2005, 220(2), 362–387.
[68] Cercignani C. Mathematical methods in kinetic theory. Plenum Press, 1969.
[69] Cercignani C. H-theorem and trend to equilibrium in the kinetic theory of gases. Arch Mech 1982, 34, 231–241.
[70] Cercignani C. The Boltzmann equation and its applications. Springer-Verlag, 1988.
[71] Cercignani C, Frezzotti A, Grosfils P. The structure of an infinitely strong shock wave. Phys Fluids 1999, 11, 275–2764.
[72] Cercignani C, Illner R, Pulvirenti M. The mathematical theory of dilute gases. Springer-Verlag, 1994.
[73] Chapman S. On the law of distribution of molecular velocities, and on the theory of viscosity and thermal conduction, in a non-uniform simple monoatomic gas. Phil Transactions Royal Society, London 1916, A216, 279–348.
[74] Chapman S, Cowling TG. The mathematical theory of non-uniform gases. Cambridge University Press, 1990.
[75] Cordier S, Pareschi L, Toscani G. On a kinetic model for a simple market economy. J Stat Phys 2005, 120, 253–277.
[76] De Masi A, Esposito R, Lebowitz J. Incompressible Navier–Stokes and Euler limits of the Boltzmann equation. Commun Pure Appl Math 42, 1989, 1189–1214.
[77] Desvillettes L. Some applications of the method of moments for the homogeneous Boltzmann and Kac equations. Arch Ration Mech Anal 1993, 123, 387.
[78] Ellis R, Pinsky M. The first and second approximation to the linearized Boltzmann equation. J Math Pures Appl 1975, 54, 125–156.
[79] Enskog D. Kinetische Theorie der Vorgangein massig verdunnten Gasen. Almqvist and Wiksell, Uppsala, 1917.
[80] Ernst MH. Nonlinear model Boltzmann equations and exact solutions. Phys Rep 1981, 78, 171.
[81] Ernst MH. Exact solutions of the nonlinear Boltzmann equation, J Stat Phys 1984, 34, 1001–1017.
[82] Ernst MH, Brito R. Scaling solutions of inelastic Boltzmann equations with overpopulated high energy tails. J Stat Phys 2002, 109, 407–432.
[83] Ernst MH, Hendriks EM. An exactly solvable nonlinear Boltzmann equation. Phys Lett 1979, 70 A, 183–185.
[84] Esposito R, Guo Y, Kim C, Marra R. Stationary solutions to the Boltzmann equation in the hydrodynamic limit. Ann PDE 2018, 4(1), 1–39.
[85] Feller W. An introduction to probability theory and applications, Vol 2. Wiley, NY, 1971.
[86] Ferziger JH, Kaper HG. Mathematical theory of transport processes in gases. North-Holland, 1972.
[87] Gabetta E, Toscani G, Wennberg B. Metrics for probability distributions and the trend to equilibrium for solutions of the Boltzmann equations. J Stat Phys 1995, 81(5.6), 901–934.
[88] Gamba IM, Panferov V, Villani C. Upper Maxwellian bounds for the spatially homogeneous Boltzmann equation. Arch Ration Mech Anal 2009, 194, 253–282.
[89] Golse F, Saint-Raymond L. The incompressible Navier–Stokes limit of the Boltzmann equation for hard cutoff potentials. J Math Pures Appl 2009, 91, 508–552.
[90] Gorban AN, Karlin I. Method of invariant manifolds and regularization of acoustic spectra. Transp Theory Stat Phys 1994, 23, 559–632.
[91] Gorban AN, Karlin I. Hilbert's Sixth Problem: exact and approximate hydrodynamic manifolds for kinetic equations. Bull Am Math Soc 2014, 51, 186–246.

[92] Grad H. Principles of the kinetic theory of gases. In: Handbuch der Physik, Vol. 12. Springer, 1958, 205–251.
[93] Grad H. Singular and nonuniform limits of solutions of the Boltzmann equation in transport theory. In: Proc. Sympos. Appl. Math., New York, 1967. SIAM-AMS proc Vol. I. Amer Math Soc, Providence, RI, 1969, 26–308.
[94] Gradshtein IS, Ryzhik IM. Table of integrals, series, and products. Academic Press, New York, 1965.
[95] Hendriks EM, Nieuwenhuizen TM. Solution to the nonlinear Boltzmann equation fro Maxwell models for nonisotropic initial conditions. J Stat Phys 1982, 29, 591–615.
[96] Hilbert D. Begrundung der kinetische Gastheorie. Math Anal 1912, 72, 562–577.
[97] Hille E, Fillips R. Functional analysis and semigroups. Am Math Soc, 1957.
[98] Ikenberry E, Truesdell C. On the pressure and the flux of energy according to Maxwell's kinetic energy. J Ration Mech Anal 1956, 5, 1–54.
[99] James RD, Nota A, Velazquez JL. Self-similar profiles for homoenergetic solutions of the Boltzmann equation: particle velocity distribution and entropy. Arch Ration Mech Anal 2019, 231 (2), 787–841.
[100] Jin S, Slemrod M. Regularization of the Burnett equations via relaxation. J Stat Phys 2001, 103, 1009–1033.
[101] Kac M. Probability and related topics in the physical sciences. Interscience, London–New York, 1959.
[102] Kato T. Perturbation theory for linear operators. Springer-Verlag, 1995.
[103] Kogan MN. Rarefied gas dynamics. Plenum Press, 1969.
[104] Kogan MN. Kinetic theory in aerodynamics. Prog Aerosp Sci 1992, 29, 271–354.
[105] Kogan MN, Galkin VS, Fridlender OG. Stresses produced in gases by temperature and concentration inhomogeneities. New types of free convection. Sov Phys Usp 1976, 19(5), 420–428.
[106] Kolmogorov AN, Fomin SV. Elements of theory of functions and functional analysis. Nauka, Moscow, 1981 (in Russian).
[107] Krook M, Wu TT. Formation of Maxwellian tails. Phys Rev Lett 1976, 36, 110–1189.
[108] Krook M, Wu TT. Exact solutions of the Boltzmann equation. Phys Fluids 1977, 20, 1589–1595.
[109] Krupp RS. A nonequilibrium solution of the Fourier transformed Boltzmann equation. Submitted to the Department of Aeronautics and Astronautics in partial fullfillment of the requirements for the degree of Master of Science. MIT, Cambridge, Mass, 1967.
[110] Landau LD, Lifshitz EM. Course of theoretical physics, 1. Mechanics. Pergamon Press, 1981.
[111] Lanford III O. The evolution of large classical systems. In: Moser J (ed.) Dynamical systems, theory and applications. LNP, Vol. 35. Springer, Berlin, 1975, 1–111.
[112] Lifshitz EM, Pitaevskii LP. Course of theoretical physics, 10. Physical kinetics. Pergamon Press, 1981.
[113] Liu T-P. Hyperbolic and viscous conservation laws. Society for Industrial and Applied Mathematics (SIAM), Philadelphia, PA, 2000.
[114] Lukacs E. Characteristic functions. Griffin, London, 1970.
[115] Maxwell JC. On the dynamical theory of gases. Phil Trans Royal Soc, London 1867, 157, 49–88.
[116] McKean HP, Jr. Speed of approach to equilibrium for Kac's caricature of a Maxwellian gas. Arch Ration Mech Anal 1966, 21, 343–367.
[117] Morgenstern D. General existence and uniqueness proof for spatially homogeneous solution of the Maxwell–Boltzmann equation in the case of Maxwellian molecules. Proc Natl Acad Sci USA 1954, 40, 719–721.
[118] Morgenstern D. Analytical studies related to the Maxwell–Boltzmann equation. Doctorate Thesis in Mathematics, Indiana University, Bloomington, Indiana, 1954.

[119] Morimoto Y, Tong Y, Zhao H. Convergence to the self-similar solutions to the homogeneous Boltzmann equation. J Eur Math Soc 2017, 19, 2041–2067.
[120] Muncaster RG. On generating exact solutions of the Maxwell–Boltzmann equation. Arch Ration Mech Anal 1979, 70, 79–90.
[121] Nikol'skii AA. On a motion of monoatomic dilute gas in uniformly expanding space. Dokl Akad Nauk SSSR 1963, 153, 543–546 (in Russian).
[122] Nikol'skii AA. The simplest exact solution of the Boltzmann equation for the motion of a rarefied gas. Sov Phys Dokl 1964, 8, 633–635 (in Russian).
[123] Nikol'skii AA. The three-dimensional expansion–contraction of a rarefied gas with power interaction functions. Sov Phys Dokl 8, 639–641 (in Russian).
[124] Ovsiannikov LV. Group analysis of differential equations. Academic Press, N. Y., 1982.
[125] Painlevé P. Sur les equations differentielles du second ordrea points critiques fixes. C R Acad Sci Paris, Ser. I, 1906, 143, 111–117.
[126] Pomeau Y. Shock at very large Mach number in simple gases: a physicist approach. Transp Theory Stat Phys 1987, 16, 727–734.
[127] Poschel T, Brilliantov N (eds.). Granular gas dynamics. Springer, Berlin, 2003.
[128] Postnikov MM. Stable polynomials. Nauka, Moscow, 1981 (in Russian).
[129] Pulvirenti A, Wennberg B. A Maxwellian lower bound for solutions to the Boltzmann equation. Commun Math Phys 1997, 183(1), 145–160.
[130] Rosenau R. Extending hydrodynamics via the regularization of the Chapman–Enskog expansion. Phys Rev A, 1989, 40, 7193–7196.
[131] Saint-Raymond L. Hydrodynamic limits of the Boltzmann equation. Lecture notes in mathematics, Vol. 1971. Springer-Verlag, Berlin 2009.
[132] Saint-Raymond L. A mathematical PDF perspective of the Chapman–Enskog expansion. Bull Am Math Soc 2014, 51, 247–275.
[133] Sakurai JJ. Modern quantum mechanics. Addison-Wesley Publishing Company, NY, 1994.
[134] Slemrod M. A normalization method for the Chapman–Enskog expansion. Physica D 1997, 109, 257–273.
[135] Slemrod M. Constitutive relations for monoatomic gases based on a generalized rational approximation to the sum of the Chapman–Enskog expansion. Arch Ration Mech Anal 1999, 150, 1–22.
[136] Slemrod M. In the Chapman–Enskog expansion the Burnett coefficients satisfy the universal relation $\omega_3 + \omega_4 + \theta_3 = 0$. Arch Ration Mech Anal 2002, 161, 339–344.
[137] Sone Y. Kinetic theory and fluid dynamics. Birkhauser, 2002.
[138] Sone Y, Aoki K, Takata S, Sugimoto H, Bobylev AV. Inappropriateness of the heat-conduction equation for description of a temperature field of stationary gas in the continuum limit. Phys Fluids 1996, 8, 628–653.
[139] Struchtrup H. Macroscopic gas models for rarefied gas flows: a brief review. IMA J Appl Math 2011, 76(5), 672–697.
[140] Takata S, Aoki K, Cercignani C. The velocity distribution function in an infinitely strong shock wave. Phys Fluids 2000, 12, 2111–2127.
[141] Torrilhon M. H-theorem for nonlinear regularized 13-moment equations in gas kinetic theory. Kinet Relat Models 2012, 5(1), 185–201.
[142] Toscani G, Villani C. Sharp entropy dissipation bounds and explicit rate of trend to equilibrium for the spatially homogeneous Boltzmann equation. Commun Math Phys 1999, 203, 667–706.
[143] Toscani G, Villani C. Probability metrics and uniqueness of the solution to the Boltzmann equation for a Maxwell gas. J Stat Phys 1999, 94, 619–637.
[144] Truesdell C, Muncaster RG. Fundamentals of Maxwell's kinetic theory of a simple monoatomic gas. Academic Press, 1980.

[145] Vedenyapin VV. Anisotropic solutions of the nonlinear Boltzmann equation for Maxwellian molecules. Dokl Akad Nauk SSSR 1981, 256, 338–342 (in Russian).
[146] Vedenyapin VV. Differential forms in spaces without a norm. A theorem of the uniqueness of Boltzmann's H-function. Russ Math Surv 1988, 43(1), 193–219.
[147] Villani C. A review of mathematical topics in collisional kinetic theory. In: Friedlander S, Serre D (eds.) Handbook of mathematical fluid dynamics, Vol. 1. Elsevier/Noth-Holland, Amsterdam, 2002, 71–307.
[148] Villani C. Cercignani's conjecture is sometimes true and almost always true. Commun Math Phys 2003, 234(3), 455–490.
[149] Vlasov AA. On vibrational properties of electron gas (in Russian). J Exp Theor Phys 1938, 8, 291–315.
[150] Wang Chang CS, Uhlenbeck GE. The kinetic theory of gases. Studies in statistical mechanics. North-Holland, Amsterdam, 1970.
[151] Wennberg B. Stability and exponential convergence for the Boltzmann equation. PhD thesis, Department of Mathematics, Gothenburg, 1993.
[152] Wennberg B. On moments and uniqueness for solutions to the space homogeneous Boltzmann equation. Transp Theory Stat Phys 1994, 23(4), 533–539.
[153] Wennberg B. Entropy dissipation and moment production for the Boltzmann equation. J Stat Phys 1997, 86, 1053–1066.
[154] Wigner EP. Group theory and its applications to the quantum mechanics of atomic spectra. Academic Press, NY, 1959.
[155] Wild E. On Boltzmann equation in the kinetic theory of gases, Math Proc Camb Philos Soc, 1951, 47, 602–609.

Index

De Gruyter Series in Applied and Numerical Mathematics

Volume 4
Claude Le Bris, Pierre-Louis Lions
Parabolic Equations with Irregular Data and Related Issues. Applications to Stochastic Differential Equations, 2019
ISBN 978-3-11-063313-9, e-ISBN (PDF) 978-3-11-063550-8,
e-ISBN (EPUB) 978-3-11-063314-6

Volume 3
Dominic Breit, Eduard Feireisl, Martina Hofmanová
Stochastically Forced Compressible Fluid Flows, 2018
ISBN 978-3-11-049050-3, e-ISBN (PDF) 978-3-11-049255-2,
e-ISBN (EPUB) 978-3-11-049076-3

Volume 2
Zahari Zlatev, Ivan Dimov, István Faragó, Ágnes Havasi
Richardson Extrapolation. Practical Aspects and Applications, 2017
ISBN 978-3-11-051649-4, e-ISBN (PDF) 978-3-11-053300-2,
e-ISBN (EPUB) 978-3-11-053198-5

Volume 1
Anvarbek Meirmanov, Oleg V. Galtsev, Reshat N. Zimin
Free Boundaries in Rock Mechanics, 2017
ISBN 978-3-11-054490-9, e-ISBN (PDF) 978-3-11-054616-3,
e-ISBN (EPUB) 978-3-11-054504-3

www.ingramcontent.com/pod-product-compliance
Ingram Content Group UK Ltd.
Pitfield, Milton Keynes, MK11 3LW, UK
UKHW050953290926
15470UKWH00006B/1800

9 783110 550122